Generating Difference

Generating Difference

Race and Reproduction in the British Empire, 1660–1840

ANDREW WELLS

Johns Hopkins University Press

Baltimore

2 4 6 8 9 7 5 3 1

Johns Hopkins University Press
2715 North Charles Street
Baltimore, Maryland 21218
www.press.jhu.edu

Library of Congress Cataloging-in-Publication Data is available.
A catalog record for this book is available from the British Library.

ISBN 978-1-4214-5360-6 (paperback)
ISBN 978-1-4214-5361-3 (ebook)

Special discounts are available for bulk purchases of this book. For more information, please contact Special Sales at specialsales@jh.edu.

EU GPSR Authorized Representative
LOGOS EUROPE, 9 rue Nicolas Poussin
17000, La Rochelle, France
email: contact@logoseurope.eu

For Jamie and Katrin, the reasons why

CONTENTS

This book has been many years in gestation and in that time I have amassed many debts, which it is my pleasure here to acknowledge. As befits a first monograph, I also wish to thank all those who have offered support, guidance, kindness, generosity, inspiration, and—not unimportantly—employment over the years. At Bristol, my alma mater, I owe much to the inspired teaching and warm support of Robert Bickers, Tim Cole, William Doyle, Ronald Hutton, Josie McClellan, Catherine Merridale, Richard Sheldon, James Thompson, and Ian Wei. At the University of British Columbia, my interest in eighteenth-century culture in Britain and its colonies was stoked by Nicholas Hudson (whose generosity during my time in Oxford is also fondly and gratefully remembered) and Peter Moogk. Joy Dixon and Tim Cheek provided cheerful intellectual and moral support to a young Englishman many miles from home.

Generating Difference started life as a doctoral project at the University of Oxford, supported by the Arts and Humanities Research Council. This volume differs substantially from the dissertation that resulted, but my gratitude to my supervisors, Brian Young and Faramerz Dabhoiwala, remains unaltered. I owe a tremendous amount to their professional example and careful guidance. I am also greatly indebted to my college, Merton, for intellectual and material support throughout my studies. In particular, the unfailing generosity of Steven Gunn, Jason McElligott, and Philip Waller deserves my profound thanks. So, too, do the members of the Graduate Seminar in History 1680–1840 for their engagement and insightful comments and questions, especially its moving spirits, Perry Gauci and Joanna Innes. For their assistance at various stages of my doctorate, I thank Pietro Corsi, Margaret Pelling, and Simon Skinner. I am also very grateful for the careful reading and thoughtful recommendations made by my examiners, Kathryn Gleadle and Peter Kitson.

As the book began to take shape, I embarked on an itinerant academic career, first in Scotland and later in Germany. For two fellowships that helped to mold this book into its current state, I am grateful to the Huntington Library and the Lewis Walpole Library at Yale. As the Newby Trust Postdoctoral Fellow at the Institute for Advanced Studies in the Humanities at the University of Edinburgh, I was fortunate indeed to have known the late Susan Manning, whose brilliance was clear from even the most fleeting of encounters. I owe much to her kind assistance and that of Pauline Phemister, Karina Williamson, Charlie Withers, and many of the other visiting fellows, several of whom became close friends. In the School of History, Classics, and Archaeology, where I spent three wonderful years, I was also lucky to have such dedicated, generous, supportive, and welcoming colleagues as Thomas Ahnert, Pertti Ahonen, Cordelia Beattie, Donald Bloxham, Stephen Bowd, Jill Burke, Ewen Cameron, Frank Cogliano, Gayle Davis, Tom Devine, Adam Fox, James Fraser, Julian Goodare, Trevor Griffiths, Fabian Hilfrich, David Kaufman, Louise Jackson, Paul Quigley, Tom Webster, and Nuala Zahedieh. Despite having already retired when I arrived, Harry Dickinson was unfailingly kind and selfless with his time and assistance, as was John Cairns, whose generosity went above and beyond.

In Germany, my thanks go to colleagues at the Universities of Dresden, Erfurt, Göttingen, Greifswald, and Kiel. At the Georg-August-Universität Göttingen, particular thanks for intellectual and career support are due to Marian Füssel, Nele Hoffmann, Heinz-Günther Nesselrath, and Barbara Schaff. I am grateful to the Fritz Thyssen Stiftung for the opportunity of a research fellowship in the beautiful surroundings of the Schloss Friedenstein, then home to the Forschungszentrum Gotha of the Universität Erfurt. At the Alfried Krupp Wissenschaftskolleg Greifswald, I had the good fortune to be part of a cohort of fellows whose warm friendship and acute insights constantly reminded me of what is best about an academic career. Thanks go to Bruno Bleckmann, Elhadji Ibrahima Diop, Andrea Esser, Brigitte Geißel, Matthias Lorenz, Joseph Rosen, Christian Suhm, Juana Vasella, and Benno Zabel. The untimely loss of Jan Plamper remains keenly felt by us all. I am also grateful to the Max Weber Stiftung and the German Historical Institute in London for the opportunity for a research stay in the United Kingdom just before the Covid-19 pandemic.

I owe a tremendous amount to my colleagues at the Christian-Albrechts-Universität zu Kiel for their unfailing friendship, support, and collegiality. Special thanks go to Theo Fischer, Kai Gräf, Max Linnemannstöns, Tobias Rusch, Regina Schneider, Kathrin Stocker, Erik Stoffer, Kai Wittmacher, Franka Zacharias, and above all Susan Richter.

Several scholars in the United Kingdom, Germany, and worldwide have been generous with suggestions and assistance throughout this project. In the United

Kingdom, particular thanks go to Jonathan Barry, Chris Bayly, Trevor Burnard, Linda Andersson Burnett, Sarah Cockram, Patricia Fara, Margot Finn, Karen Harvey, Tim Hitchcock, Jennie Jordan, Ben Marsh, Matthew McCormack, Steve Poole, Mike Rapport, Roey Sweet, Corin Throsby, Sarah Toulalan, James Walvin, and Robert Wokler. In Germany, I am grateful for the generous help offered by Susanne Lettow, Rebekka von Mallinckrodt, Mieke Roscher, Alan Ross, and Gerd Schwerhoff. For his particular generosity, I owe tremendous thanks to Renato Mazzolini. Beyond Europe, my gratitude extends to John Brewer, Bruce Buchan, Ted Burrows, Linda Colley, Emma Hart, Theresa Levitt, Simon Middleton, Deirdre Cooper Owens, Peter Hanns Reill, Robert Ritchie, and Londa Schiebinger for their sage advice and support, and for the opportunity to present earlier versions of my research.

I owe much to the diligence and expertise of library and archive professionals at a wide range of public and private institutions in Britain, Germany, and the United States. Special and particular thanks go to those colleagues who have read the manuscript in whole or in part and whose suggestions have immeasurably improved this book. I am especially grateful to Camillia Cowling, Andrea Esser, and Silvia Sebastiani. Thanks also to the anonymous reviewers for their careful reading and insightful comments. The remaining errors are, of course, all my own.

The journey from manuscript to printed book would not have been possible without the dedicated staff at Johns Hopkins, among whom Matt McAdam, Charles Dibble, Carol McGillivray, and Phoebe Oathout deserve particular thanks for their diligence and commitment to the project.

Friends and family have accompanied me along every step of this journey, and I owe them so much for their good cheer, encouragement, and indulgence. From Harborough—where it all began—thanks are due to Alex Brown, Alex Langford, Patrick Quinn, Michael Smith, and Michael White. They are also due to the Evingtonians Richard Cooper, Colin Massy, James O'Callaghan, Adam Weafer, and James Young (Committed? Commended.). I am especially grateful for the inspiration from Mike Newbold: those who knew him know why. From Bristol, my gratitude goes to Charlie Barton, Linda Browne, Hannah Christmas and Jon Buick, Andrea Dell, Christo Dicks, Liz England, Helen Kottenstette, Sarah Lusk, Nic and Mela Miles, Ray Stafford, and John Watts. In Vancouver, the friendship of Nic Clarke, Margaret Inoue, Jas Kambo, and Adam and Kate King deserves special mention. While at Oxford, I learned much and gained more from such friends as Andrew Beaumont, David Fallon and Felicia Gottmann, Ean and Rebecca Hernandez, Matt

Johnson (to whom I owe a lasting debt for introducing me to Harry Flashman), Miranda Kaufmann, George Newberry, Tiffany Shumaker, Jenny Skipp, Will Pettigrew, and Anna von der Goltz. The friendships forged in Scotland and Germany sustained me when I thought this book would never get done. My gratitude goes to Katherine Nicolai, Emma and Gregory Norminton, Chelsea Sambells, Miriam Schaub, Tom Turpie, Jens Elze, Gösta Gabriel, Julia Hauser, Erika Manders, Claudia Nickel, Daniele Panizza, Christiana Werner, Robert-Jan Wille, Matthias Bähr, Alexander Kästner, Peter and Katja Arndt, Jennifer Henke, Susi and Mark Leuchtmann, Constanze Späte and Loïc Paoli, and Karen Wendland. Other friendships, particularly those of Hazel Duke, Chris Eastburn, and Joe Stubenrauch, resist exact placement but are no less heartfelt.

To my family, to whom I owe everything, I express my love and gratitude. To my parents, Alan and Elaine; my brother, Stephen; sister-in-law, Yvonne; and their children, Harry and Amelia: your unfailing support, love, encouragement, and sheer patience with a history-obsessed son, brother, and uncle have made every difference; thank you so much. Meinen Schwiegereltern Annelie und Ehrhardt Berndt spreche ich meinen allerherzlichsten Dank aus: eure Herzlichkeit und Freundschaft haben mich und dieses Buch mehr unterstützt als ihr wissen könnt. To the wider Wells, Sherlock, Gardner, Berndt, and Ellermann families, I offer my grateful thanks for all your help, love, and support. To Sheila, thank you so much, for everything.

This book is dedicated to my son, Jamie, and my wife, Katrin, who have both lived with this project far longer than it was reasonable of me to expect. Jamie: you made the burden lighter in uncountable ways and brought sunshine with you at every turn. Katrin: ohne Dich wäre nichts möglich, nichts vorstellbar. Du schenkst mir so viel Freude und hast nie aufgehört, an mich oder dieses Buch zu glauben. I can never thank you enough; I'm glad I ordered the marshmallows.

As with any scholarly work that seeks to understand and explain patterns of thought in past societies, this book deals with concepts and terms that are archaic and may be considered offensive on that basis alone. The present work, however, addresses an especially sensitive theme—the development of racial thought and its relationship with sexual ideas—and its source material is replete with assumptions, prejudices, and terminology that would be shocking if endorsed today and in some cases appeared extreme even in the early modern era. Handling such material requires sensitivity both to the reader and to the historian's duty to seek to understand. As a result, the following account has been written with these principles in mind:

— In general, there has been no systematic attempt to translate early modern racial or sexual terms into a less offensive modern equivalent. To do otherwise would attract the justified charge of bowdlerization but, above all, would risk obscuring the history that this book seeks to explain. For example, it is significant for this history that the most common racial terms—above all, "Negro"—were of Iberian origin (as were ideas about hereditary Muslim- and Jewish-ness that could not be eradicated by conversion to Christianity), that the label "mulatto" carried significant reproductive implications, and that "Portuguese" was shorthand in India for mixed-race people.

— However, as Marc Bloch rightly cautioned, it is unsafe simply to copy the terminology of the past.* In the context of this history, to do so would appear gratuitous, even were contemporaneous labels to appear uniformly within "scare quotes," which furthermore reduce readability. Consequently, the terms "black" and "white" are used adjectivally (most frequently alongside the noun "people"),

---

* Bloch, *Historian's Craft*, 131.

in lower case and without quotation marks. Where appropriate, "sub-Saharan African" is used in the interests of clarity, but only very occasionally, since the nouns "Black" or "Negro" often referred to any and all dark-skinned peoples. These substantive nouns—along with "White"—have themselves been avoided, except where dictated by the source material, in which case they are placed in quotation marks. This is also the case with the specific categories of racial theories, for example, "Quadroon." The term "savage," as a generic label applying to people deemed uncivilized, appears without quotation marks only when paraphrasing sources.

— The choice to retain "black" in lowercase may surprise some readers. After all, there are powerful reasons for capitalizing the term "Black," not the least of which is the recognition of a shared history and identity, forged in the face of often terrible suffering. Hence the decision of many news outlets and publishers is to use "Black" adjectivally in uppercase. The choice to deploy the term consistently in lowercase is based principally on the historical nature of *Generating Difference*. The shared identity signified by "Black" did not yet exist (although it was clearly under development) during the period covered by this book, and as the story told within its pages includes the racialization of a number of the world's peoples—of which Africans were by far the most prominent but alongside whom were the native inhabitants of the Americas, Asia, and Oceania— the choice has been made not to single out one group, especially when the capitalized term used to do so was both ambiguous and endlessly repeated in the most offensive sources themselves.

— The terms "creole" (meaning born in the Caribbean, applicable to both white and black people), and "mulatto" are used both adjectivally and nominally, without capitalization or highlighting via quotation marks. Most often, the phrases "mixed-race people" or "people of color" (along with the adjective "colored") are used instead of "mulatto" except where paraphrasing sources or for reasons of precision (to denote specifically an individual with one black and one white parent). This choice has been dictated by the lack of uniformity in eighteenth-century sources, by which "mulatto" could mean the offspring of black and white parents, but often connoted the entire class of people of mixed parentage, including parents who were not black or were themselves of mixed origin.

— Similarly, the broad range of eighteenth-century applications of "hermaphrodite" (which most often referred to intersex, but could also denote transvestism or transsex/gender) means that "hermaphrodite" also appears when paraphrasing sources, without capitalization or quotation marks, but where it is clear that intersex is meant, this term is used.

— Where particular labels denoting ethnicity or nationality—for example "Coromanti," "Ibo," "British," "English," "Scottish," "French," "European," etc.—are relevant they are used. "Amerindian" appears in preference to "American Indian" or just "Indian," to avoid both confusion with South Asian peoples and the anachronism of applying the official label, "Native American," and in recognition of eighteenth-century usage (where the native peoples of the Americas were often labelled simply "Americans").

— This account remains agnostic on the question of preferring "enslaved person" to "slave." Where possible without undue circumlocution, "enslaved" has been used, as the intention behind this term—avoiding the reduction of enslaved people to their legal status and recognizing the illegitimacy of slavery—is unimpeachable. But it should be remembered that slaveowners, too, sought to avoid the use of the word "slave" in their efforts to euphemize chattel slavery, especially while fighting for its survival; this was a not insignificant force propelling the development of racial ideas, as discussed in chapter 6. Hence wherever "slave" appears in these pages, it is with the same intention as that underpinning "enslaved."

Perhaps not all readers will agree with these choices, but they have been made in good faith and in the hope of shedding further light on a difficult chapter of human history.

Generating Difference

Introduction
(Re)producing Bodies and Identities

This book explores the relationship between race and sex in eighteenth-century British culture, both in Great Britain and its empire, by investigating the reproductive common ground they shared. Few products of that culture can better encapsulate the themes of the following chapters than the fourth book of *Gulliver's Travels* (1726). On a hot day in Houyhnhnmland, Gulliver prevailed on his companion and protector, the "Sorrel Nag," to allow him to take a cooling dip in a nearby river. He was watched by a female Yahoo who, "enflamed by Desire," leaped into the water next to Gulliver and embraced him "after a most fulsome manner." Hearing Gulliver's screams, the nag galloped back from where he had been grazing and the Yahoo scrambled to the opposite bank, "where she stood gazing and howling all the time I was putting on my Cloaths." For the Houyhnhnms, this incident was a source of amusement; not so for Gulliver, "for now I could no longer deny, that I was a real *Yahoo*, in every Limb and Feature, since the Females had a natural Propensity to me as one of their own Species: Neither was the Hair of this Brute of a Red Colour, (which might have been some Excuse for an Appetite a little irregular) but Black as a Sloe, and her Countenance did not make an Appearance altogether so hideous as the rest of the Kind; For, I think, she could not be above eleven Years old."[1]

Of the many themes addressed by this tale, perhaps the clearest is the power of sexual desire. Gulliver's careful and sustained efforts to erect barriers between himself and the Yahoo population were swept away in an instant, thanks to his sex appeal for a young Yahoo female. As subsequent chapters will show, sex was a powerful motive force driving the development of categories of identity, serving both to construct and demolish boundaries between groups.

The most fundamental of these borders was that separating species. Gulliver suspected from the outset that he was an actual Yahoo, as his physical form was by and large identical to theirs. But he was uncertain, and the only means available to establish conclusively whether he was of the same species was to procreate; fertile offspring would prove their identity, infertile their dissimilarity. The mere production of offspring itself would prove nothing—despite belonging to different species, horses and donkeys can nevertheless reproduce together—and only the fertility of that offspring was decisive. Such an experiment was altogether out of the question: Gulliver's intense loathing of the Yahoos not only made an attempt unpalatable but also strongly suggested its likely failure. Similar experiments went largely untried by naturalists in the eighteenth century (perhaps due to the lengthy wait for results), so mutual sexual "repugnance" was taken as presumptive proof of species difference. Gulliver's world was shattered because his Yahoo molester manifestly lacked this repugnance for him; his own aesthetic judgment on his attacker ("not . . . altogether so hideous as the rest") was also somewhat equivocal, hinting at his own desire while urgently seeking to suppress it.

Sexual desire (or lack thereof) is only the most obvious of judgments made about physical bodies. Bodies carried in, on, and with them a wealth of culturally determined meanings. The human body, which was understood until well into the nineteenth century as a porous entity, embedded in its environment and subject to its influence, was a signifier of moral as well as physical wellbeing or corruption. The youth and hair color of the Yahoo girl were bodily signs that Swift's readership could easily interpret. Red hair was widely understood as a marker of sexual insatiability, one that had a powerful geographical association with Britain's semicivilized Celtic fringe, and the tender age of Gulliver's assailant pointed toward the tendency of hot climates to accelerate female sexual development.[2] Furthermore, it was Gulliver's own physical form that led the Houyhnhnm Grand Assembly to acknowledge that he was a Yahoo and to order his expulsion from their midst. His description—"I was an exact *Yahoo* in every Part, only of a whiter Colour, less hairy, and with shorter Claws"—carries an unmistakable racial overtone in the mention of Gulliver as a *whiter* Yahoo. And, finally, the status of Yahoos as an inferior form of life was clearly conveyed in the story of their creation: "Many Ages ago, two of these Brutes appeared together upon a Mountain, whether produced by the Heat of the Sun upon corrupted Mud and Slime, or from the Ooze and Froth of the Sea, was never known."[3] Such theories of spontaneous generation had been used to account for the reproduction of insects and vermin, and although increasingly discredited after the late seventeenth century, the use of these ideas by Swift clearly and concisely marked the Yahoos as a less worthy form of life.[4]

Fantastic though *Gulliver's Travels* may have been, in drawing attention to sex, reproduction, and the physical body as makers, breakers, and repositories of identity, it participated in broader cultural developments that were underway in the eighteenth century. The rooting of certain categories of identity in the fabric of the physical body was part of an ongoing process of basing profoundly *human* institutions and standards on *natural* foundations.[5] The conceit at the heart of this process was that such "somatic" (or bodily) identities were uniformly and unproblematically legible. The fallacy of this belief and the privileging of corporeal empirical markers over other sources of knowledge (such as scripture) are key themes of this book. Beginning in the seventeenth century, empiricist epistemologies gathered pace and became dominant in learned culture during the Enlightenment. But they did not carry all before them, and alternative forms of knowledge and ignorance about bodies continued to inform popular debates and concepts until beyond our period. Nowhere was this more apparent than in the ebb and flow of ideas about sexual reproduction. This was the crucial nexus for all forms of somatic identity: all bodies must be born, and the customs, theories, and practices of generation played a critical role in shaping ideas about sex and race.

Consequently, *Generating Difference* contends that the development of racial thought and practice in eighteenth-century Britain and its empire cannot be properly understood without recognizing the major role played by reproduction and the reproductive body. The argument proceeds in three stages. The first asserts that sexual reproduction was a matter of such supreme importance for imperial security and prosperity that a transatlantic culture of pronatalism emerged after the mid-seventeenth century. For reasons of geopolitics and political economy that centered on Britain's competition with France and other Catholic maritime powers, Britons sought to augment their country's small population. Yet due to the unpopularity or impracticability of alternatives, the only means available were reproductive. In Britain's Caribbean slave colonies, the abolition of the slave trade in 1807 meant that the only hope to maintain slavery lay in promoting reproduction, but the lackluster uptake of pronatalist measures was due both to commercial pressures and to the reluctance of planters to countenance the monogamous marriage of slaves, which in a Protestant empire entailed baptism and hence literacy.

Having established the ubiquity of reproduction as an imperial economic and political priority, the second stage of the argument proceeds to explore the place of the reproductive body in racial thought and the role played by race in embryology. The intellectual scaffolding of racial ideas was based on reproductive sex at every turn, whether in the notion of "species" as a reproductive community, in the enduring meaning of "race" as lineage and descent (see below), or in the prominent place given in racial theory to sexual characteristics, both physical and cultural. Embryology was so powerfully associated with matters of race that it set the parameters of racial thought: every author who discussed racial ideas thereby also took a position on embryological theory, and vice versa. There was substantial overlap in the data used by both bodies of thought, but the fundamental datum they shared concerned mixed-race people, who proved not only that both parents contribute to the fetus but also that they belong to the same species. This had far-reaching implications for the shape of ideas on heredity and hybridity in racial and embryological thought, and it powerfully influenced the discussion of associated phenomena, such as albinism, hermaphroditism, and the power of the mother's imagination to shape her fetus.

The final part of the book explores these pronatalist priorities and the imbrication of racial and reproductive thought in contexts beyond the British Atlantic world. The Pacific had long been portrayed as a uniquely sexualized zone, but the confusion produced by actual contact with island peoples and their sexual and social customs far exceeded anything expected by Europeans. The prominence of the South Seas in European racial thought affords an opportunity to explore the aesthetic and empirical judgments made by researchers pursuing explanations of human diversity, and the crucial feature used to sort peoples into distinct groups—language—highlights one of the fundamental limits of the abstract concept of race. Its attempt to inscribe identity unambiguously in/on the human body foundered because extracorporeal knowledge (such as genealogy, in this case indicated by linguistic evidence) is necessary to make a conclusive determination. The extent to which the body could not function as intended by racial models is especially clear from the processes of ethnogenesis that took place in the Caribbean and British India. In the first of these contexts, an elaborate typology of racial mixture was developed that was wholly absent from the latter. Consideration of the near-identical ways in which these mixed populations arose and were viewed makes it clear that the driving force behind the development of this taxonomy was the need to maintain chattel slavery. And the overwhelming power of this imperative explains why the contradictions in race that are exposed on close inspection of its reproductive elements failed to undermine the concept in any substantial manner: it was simply too useful to do without.

Concepts in Caricature: George Cruikshank's
The New Union Club (1819)

In making this argument, *Generating Difference* builds on a broad and diverse scholarship, using and further developing concepts that have been shaped by this literature. Specifically, this book is situated at the crossroads of two major bodies of historical writing—on racial and sexual difference—both of which intersect with the history of the body. Each of these literatures has identified one or more turning points in the long eighteenth century, so another product of that culture, one that encapsulates many of the themes of *Generating Difference*, will be used to guide the following discussion of its scholarly and conceptual background.

George Cruikshank's engraving, *The New Union Club* (1819, fig. 0.1), was a collaborative effort between Cruikshank and Frederick Marryat, a naval officer with an interest in graphic humor whose father, Joseph, was an MP and colonial agent for Trinidad and Grenada. It was inspired by a pamphlet written by Joseph that sought to discredit the work of the African Institution to enforce the ban on the slave trade and which bemoaned what he thought was misdirected charity that ignored "fellow-subjects, but . . . is reserved for foreigners and savages."[6] Joseph described a dinner held in March 1816 by "the members and friends of the African and Asiatic Society," attended by several leading abolitionists, such as William Wilberforce, Zachary Macaulay, and James Stephen. The company was segregated during the meal, but first (mixed-race) children and later "a medley of blacks and mulattoes appeared; many of them mendicants," and Joseph hinted at the resulting disorder by stating that the excitement of the children had drowned out one of the key speakers and that wine was given to the adults "to drink with their benefactors."[7]

Nothing in Joseph's account suggested the scene of debauched chaos rendered by Cruikshank and Frederick that borrowed heavily from James Gillray's *Union Club* (1801). But the mayhem offers one of several points of entry into the subject matter of *Generating Difference*. The sheer busyness and overpopulation of the scene reflects the change in perceptions of Britain's demographic

Fig. 0.1. George Cruikshank, *The New Union Club* (1819). BM 13249. Courtesy of the Lewis Walpole Library, Yale University.

UNION CLUB.
celebrated ——— society. ——— Vide Mr. M——'s Pamphlet entitled "More Thoughts &c &c."
Pubd July 19 1819 by G Humphrey 27 St James's Street London

situation associated with the writings of Thomas Malthus. As detailed in chapter 1, the concern that Britain was dangerously underpopulated transformed rapidly around 1800 into an anxiety that the population was growing fast beyond its capacity to feed itself and that misdirected charity was largely responsible. The ensuing social unrest, manifested in the Peterloo Massacre that took place less than a month after *The New Union Club* was published, endangered national security every bit as much as Britain's alleged underpopulation (in comparison to France and other competitors) was previously thought to do. For Malthus and those who based their legislation on his ideas, individuals who were disabled in the pursuit of that security deserved charity, while poor relief should be denied to the able-bodied poor. Yet an able-bodied black sailor is shown aggressively pushing a disabled comrade out of the bottom-right corner of the frame, symbolizing the profoundly misguided philanthropy that exercised Joseph Marryat.

The fact that the vast majority of figures in *The New Union Club* are black reflects more than errant humanitarianism. It testifies to the visibility of Britain's black population, especially as several figures in the image—from the aggressive sailor, Charles McGee, to popular beggars such as Billy Waters (shown playing his fiddle) and Joseph Johnson—would have been familiar to Cruikshank and Frederick's audience. Black people populated British cities and participated in British social, cultural, economic, and public life across the century, which makes it even more remarkable that despite pioneering work undertaken in the 1970s and 1980s by Peter Fryer, Folarin Shyllon, James Walvin, and others, scholarly attention to race and racism in Georgian Britain dates largely to the last thirty years.[8] The black grocer, musician, patron, servant, voter, and writer Ignatius Sancho (*c.* 1729–1780) is a case in point: born on a slave ship during the middle passage, he was taken under the wing of the Duke of Montagu and became extraordinarily well connected, encountering aristocracy and royalty during his time in service; writing to his literary hero, Sterne, to recruit his pen in support of enslaved Africans; publishing music; and participating in a network of Afro-Britons who fought against slavery overseas and prejudice and discrimination at home. Sancho provides an object lesson that eighteenth-century Britons were "at home with the empire," something that has become a topic of major interest, especially as historians have turned toward the global in recent years.[9]

It is, however, crucial to recognize that *The New Union Club* says nothing substantive about such lives as Sancho's, or even about the individuals it depicts: Johnson, McGee, and Waters. They are all ciphers for the satirical goals

of Cruikshank and the Marryats; as such, the engraving is an eloquent statement of metropolitan culture concerning race and slavery in the late 1810s but
a poor representation of black identities in Britain at this time.

Identity

The ubiquitous analytical category *identity* has come under sustained attack
since the millennium, most influentially by Rogers Brubaker and Frederick
Cooper. They put forward the compelling argument that the very ubiquity of
identity as a category has done untold damage to its usefulness for scholars. As
cultural, social, and historical investigation has become more sophisticated, a
growing range of qualifiers has been added to identity, such that it has come to
denote a flexible, contested, performed, intermittent, inchoate, and multiple
(among other adjectives) thing; in short, it is everything and nothing, and therefore is a concept that has outlived its utility.

As powerful as this argument undoubtedly is, it has not succeeded in eject-
ing identity from the scholarly toolkit. Brubaker and Cooper proposed several
alternatives that appear viable enough as like-for-like replacements of the various facets of identity they isolated—ultimately reducible to the three elements
of categorization, self-understanding, and groupness—but none successfully
captured the polyvalence that is at the heart of identity's continued attractive-
ness for scholars. This was, of course, the point: it was the very extent of this
polyvalence that led Brubaker and Cooper to argue that identity was no longer
fit for analytical purpose. But there is no obvious reason why a concept of identity that embraces a "constructed, fluid, and multiple" set of referents should
be ipso facto illegitimate.[10]

Nevertheless, the point that it is important to use identity clearly and consistently is well taken. In helping to define the concept more closely, Nira Yuval-
Davis has proffered the valuable insight that "identities should be understood
as specific forms of narratives regarding the self and its boundaries." These
narratives are themselves performative and dialogic and, as Stuart Hall has
stated, are perpetually in development. These are the hallmarks of the understanding of identity that will reappear in subsequent chapters.[11]

What will be less prominent, however, is discussion of *the self.* This may
appear problematic, as identity arguably "manifest[s] itself in the stories that
people tell about themselves as collectivities."[12] Indeed, one important register
for the scholarly discussion of identity is the philosophical, which has tended
to focus on the development of the modern self as a configuration of identity

characterized by interiority and uniqueness. Such approaches have been aug-mented by cultural historians, who have moved beyond elite and learned sources to incorporate the quotidian in their studies of ideas about the self.[13]

This book is, however, less about the making of the *modern self* than of the *modern other*. Most of the sources on which the analysis in subsequent chapters is based focus not on the identities of their largely white, largely male authors but on those of their non-white or nonmale subjects. Of course, discussion of the other always entails discussion of the self, however obscure or indirect this may be. In our case, white male authors, by proceeding from their own subjec-tive location, often established the white male body as a benchmark by which all others were judged; female or non-white bodies were implicitly and sometimes explicitly perceived as deviations from this norm. The fact remains, however, that such authors concentrated not on themselves but on others, and it is one of the contentions of this book that, by developing modern concepts of racial and sexual difference through discussions of non-white, nonmale people, authors were making a case less for the *superiority* of white males than for the *inferiority* of everyone else. The overall result (white male hegemony) was the same, but the focus was overwhelmingly on demonstrating deviation and difference from an imaginary ideal rather than describing a group's fulfillment of it.[14]

But does this grant too much power to these writers and thinkers—often themselves armchair philosophers ensconced within Europe—to dictate the terms of others' identities? After all, if identity is in part performative, then the agency of individuals to assume or divest all or part of an identity/identities, or at the very least to resist efforts to pigeonhole them into roles they do not wish to perform, must be acknowledged. Indeed, as myriad postcolonial scholars have shown, one of the most enduring of colonial myths is the irresistible and in-controvertible power of Europeans to assign identities to the colonized.[15]

Were this book primarily interested in a historical reconstruction of identi-ties as they were lived and experienced by individuals or groups themselves, then this objection would be formidable. But this is not a social history of race or sex, and important as the experiences of individuals and groups were, the his-tory offered in these pages is one of the fabrication of cultural and intellectual categories. This owed a great deal to a myopic perspective on lived reality and an often-deliberate ignorance of the experiential dimension of identity: to this extent, it is partly a history of white male blindness to the lives and worlds of others, exemplified par excellence by the representation of Britain's familiar black personalities in *The New Union Club*.

So how important could their shortsighted and self-important musings be? Sadly, very. The ideas concerning race and sex that Europeans elaborated during these years unquestionably had an often-devastating impact on people's lives, from defending slavery to naturalizing sexual inequality. These ideas were often courageously resisted and were (therefore) nowhere imposed without some modification, but they helped to frame the outlook of Europeans at home and abroad, and served to shape sexual and racial orders across the globe, even as their inadequacy and inaccuracy became increasingly apparent.

Sex and the Body

While *The New Union Club* turned a blind eye to experiential dimensions of identity, it nevertheless reflects the project to embed categories of identity in the fabric of the human body. The most obvious registers of this process are racial and sexual, but the engraving also uses the human body to convey differences and meanings along axes of age and disability. As such, it reflects the diversity of historical writing on the history of the body that has bloomed remarkably since the 1980s. Studies exist that chart the historical human body from birth to death, through childhood, puberty, adulthood, and senescence, and which examine the full range of the body's activities and states: dead and alive, diseased and healthy, hungry and sated, naked and clothed, idle and active, asleep and awake. To these may be added other aspects tied to European expansion and colonialism: enslaved and free, foreign and domestic, civilized and savage.[16]

The cultural and scientific understanding of the body's relationship with its environment has also been extensively scrutinized, especially concerning the influence of climate on the shape and functions of the body, whether mediated through "airs, waters, places," food, or other factors. The porosity of the human body was a fundamental tenet of the humoral model of medicine that remained powerful until the nineteenth century. According to this, four fluids (or humors: blood, phlegm, yellow bile, and black bile) were believed to determine an individual's complexion and health; an individual balance of humors was essential for maintaining physical and mental well-being. Each fluid had a unique mix of the properties of heat, cold, wetness, and dryness, all based on the four basic elements of earth, wind, fire, and water. Humors and their properties were not merely material explanations for individual personalities (phlegmatic, sanguine, melancholy, choleric) or states of health but were also the lens through which early modern concepts of sex and race were viewed. Women were possessed of the lesser qualities of coldness and wetness, and although heat and dryness were superior, their untrammeled preponderance—in places such as Africa—led to an imbalance that was also viewed negatively.[17]

This range of variables made reading the premodern body no simple matter and left open the possibility of radical change. This remained at the heart of orthodox discussions of human phenotypical difference until the nineteenth century but was closed off in the case of sex by the change described in Thomas Laqueur's *Making Sex* (1990), which argues for nothing less than that "sometime in the eighteenth century, sex as we know it was invented."[18] This invention consisted of the replacement of a *one-sex* by our *two-sex* model of sexual difference. Supported by Galenic medicine, the one-sex model imagined "male" and "female" as points on a spectrum rather than opposites, and it inverts the conventional ordering of sex and gender by placing sex as the cultural category and gender as the corporeal entity. This model, under which sex "was still a sociological and not an ontological category," was supplanted by the prevailing two-sex model in the late eighteenth century because the political need to restrict women to the "private sphere" made it necessary to invent, consciously or not, a scientific justification for their confinement.[19]

Though undeniably influential, Laqueur's account has been subject to stiff criticism. Several commentators have argued that his models are reductive and totalizing, ignoring clear differences between the sexes in Aristotelian thought, medieval medicine, and Renaissance skeletal anatomy.[20] Others have pointed out that concentration on the genitals is misplaced, as seventeenth-century anatomists identified sex differences particularly in the brain and nervous systems, and that Laqueur entirely neglects the experiential dimension of sexual difference.[21] Other gender historians have argued for a more uneven and lengthier transition from one-sex to two-sex, emphasizing continuity rather than rapid change.[22] Indeed, one-sex was itself apparently a humanist rediscovery that had lain dormant for a millennium after Galen's discussion of the idea.[23] Yet most criticism of *Making Sex* has taken the form of modifying rather than destroying its argument, and even Laqueur's harshest critics admit that there is something to his claims, arguing that the period 1600–1900 saw "a dramatic, if gradual, transformation in thinking about sex and sex difference, *rooted in a biologistic and medicalizing terminology* that slowly supplanted the earlier metaphysical language."[24]

Instances in *The New Union Club* of the embodiment of feminine and masculine ideals are not hard to spot, especially because—in keeping with its satirical vision—they are the inverse of European values. Drunken and (sexually) aggressive black women appear throughout, and in the center foreground two women are fighting while a nursing infant suckles from the enormous suspended breasts of its mother's antagonist. The myth of gigantic breasts that facilitate suckling over a mother's shoulder, discussed in chapter 3, is also referenced elsewhere in the engraving. The central figure standing with one arm raised (one Mr Paul, an African American preacher) hints at black male hypersexuality, given the risqué placement of his left hand behind a suggestive speech bubble and his ecstatic expression. Cruikshank and Frederick Marryat emphasize simultaneously that real sexual differences inhere to the human body, and yet that the black body is an unstable marker of sex.

This, too, presages modern scholarly developments. The role of nature as the foundation of sex poses conceptual problems that have not only persisted but seem to have worsened in the past decade. The concept of sex is complicated by at least two factors: its relation to gender and the supersession of binary perspectives. Since the 1970s, feminist critics have commented on the undesirability of the sex/gender binary, as the notion of *sex* as a purely biological substrate on which the purely cultural *gender* is overlaid misrepresents both concepts.[25] In the face of atavistic assertions that cultural differences were really biological, some feminists attempted to evacuate sex of its significatory power by making all apparently biological differences fundamentally cultural.[26] By the 1990s, scholars such as Joan Scott and Judith Butler acknowledged the real differences that subsist in the human body but argued that the human abstraction *sex* was discursively constituted even if it was more anchored to a corporeal reality than *gender*.[27] What has emerged from this debate is a concept of sex that remains fundamentally based on the physical body but that has far less prescriptive power than it had hitherto.[28]

This inability to extricate sex from the human body has meant that the concept has become more complicated as we have been made more aware of the messiness of corporeal reality. The male/female binary, for example, can no longer dominate concepts of sex in the way that it could before twentieth-century advances in endocrinology. Where before sexual difference could be based on anatomical observation, now it is founded on chromosomal biochemistry; sexual dimorphism is still the rule, but natural exceptions to that rule (i.e., intersex) are more understood, and cultural exceptions (i.e., transsex/transgender) have become more common of late. More frequent today than in the eighteenth century are known cases of transsexual people, although the past absence of surgical possibilities did not prevent individuals such as the Chevalier d'Eon from playing their part in problematizing clear-cut sexual distinctions. In the last analysis, sex certainly has *some* discursive limits, and it is important for any study of the concept to recognize this at the outset.[29]

These concepts have emerged from a rich and flourishing scholarship on sex in eighteenth-century Britain. Apart from pioneering studies of sexuality and relations between the sexes dating from the interwar period, serious historical interest in eighteenth-century sex only emerged with women's history in the 1960s and 1970s.[30] Following the turn to gender in the 1980s and the concurrent development of the history of sexuality, accounts of sex in the eighteenth century appeared with increasing frequency.[31] Nonetheless, older myths of the century as a period of comparative sexual abandon, sandwiched between early modern rambunctiousness and Victorian austerity, have proven difficult to eradicate.[32] The focus of most histories of sexuality on aberrant and illegitimate sexualities has greatly contributed to a notion of eighteenth-century exceptionalism.[33] Demographic change is one basis for this myth: the last half of the eighteenth century witnessed a rapid rise in population size without a commensurate drop in mortality, suggesting an increase in conceptions. The implications of this development were profound and are discussed at greater length in chapter 1.[34] At the same time, ideals of maternity and domesticity were increasingly at work to confine women within the so-called domestic sphere, as industrialization gathered momentum and the sexual division of labor became more pronounced.[35] By the early nineteenth century, home keeping and childbearing had become the only legitimate form of women's work, at least for the middle classes.[36] The growing dominance of these ideals is no indication that they were successful in practice—as with other prescriptive sources, such as conduct books or sex manuals, the repetitive insistence on behaving in a certain manner strongly suggests that this advice was not heeded—but they were held as a model to which to aspire. Unlike histories of race, those of sex and gender have been sensitive to issues of class differentiation, although there remains little work on the sexual lives of women from more humble ranks in the eighteenth century (partly, it must be said, due to the paucity of source material).[37]

Race

In contrast to the continuing debate on the natural foundations of sex, the notion that race might have some biological reality has been justly banished from the realm of scientific respectability. Treated as a relatively straightforward biological entity until the 1980s, it has remarkably quickly—within the space of a single generation—come to be viewed as a purely cultural construct.[38] The production of histories of race over the course of the twentieth century highlighted its historical roots at the same time as it became politically imperative to reject race as a biological fact.[39] Scholars since the 1990s have searched for it in locations beyond its traditional temporal habitat—the nineteenth and twentieth centuries—finding it in such diverse places and times as China, India, classical antiquity, and medieval and early modern Europe.[40]

This collective enterprise has, as Ann Stoler recognized in 1997, two principal aspects. First, at its heart is the project to locate the "origins" or "invention" of race or racism, a project that is now undeniably producing diminishing returns. This is partly because, second, the search for race has resulted in a proliferation of definitions of the concept that are suited to the particular periods and locations under investigation, with limited applicability beyond these contexts.[41] As the principal trajectory in the historiography of race and racism has been to identify these phenomena in ever-earlier periods, the most obvious change in definitions of these concepts has been to deemphasize their biological factors and to highlight the cultural. A key example relating to eighteenth-century Britain is Roxann Wheeler's catch-all term "multiplicity," which incorporates both biological and cultural components of race, leading her to treat clothing and religion as racial attributes.[42] There is certainly something to be said for this approach, given that contemporary racism is primarily cultural rather than biological—meaning that clothing and costume are potent signifiers—and that premodern forms of (proto-)racism often posited hereditary Jewishness, Muslimness, or heathenism as quasi-biological characteristics.[43]

But there are serious drawbacks to such capacious definitions of race. It is difficult to see, for instance, why *race* should be the analytical category of choice when examining cultural prejudices based on dress. More fundamentally, a focus on "culture" rather than "biology" (used here to mean corporeal features rather than the scientific discipline) begs the tricky question of the shifting relationship between these two elements of racial thought. It seems clear that for something to be described as "race," "racism," or "racial," the biological must be present to a certain extent, even when cultural aspects are thought to predominate. This is the case for both contemporary and remote periods. In the words of George Fredrickson, contemporary cultural racism "reifies and essentializes culture rather than genetic endowment" and in so doing recycles many of the tropes of earlier scientific "racialism" (Kwame Appiah's term for the belief in distinct races, based on shared and heritable characteristics).[44] In like manner, several studies of race and racism in antiquity, the Middle Ages, and the early modern era have argued that cultural features, particularly those associated with religion, were just as permanent, heritable, inescapable, and prescriptive in former societies as biological characteristics are today.[45] Modern science thus continues to be invoked, even if only as a metaphor to communicate to the contemporary reader the permanence and innateness of cultural characteristics.

As a product of the period in which science was acquiring this authority, the references to race in *The New Union Club* point to precisely this complex of biological and cultural issues. On the one hand, biology was paramount and the permanence of blackness asserted: one of the paintings in the background (top right of fig. 0.1) shows women attempting to bathe a black woman in the vain attempt to "wash an Ethiop white"; another (immediately behind Waters) shows the "Hottentot Venus," Saartje Baartman, a key figure in the racial science of the 1810s who appears in chapter 3. On the other, this permanence was paradoxically reinforced by its subversion for satirical ends. According to Joseph Marryat, the abolitionist James Stephen had stated at the dinner "that he actually felt" "ashamed of his own complexion," and for that reason a black woman (to the left of Mr Paul) is shown blackening his face with soot from a burnt cork.[46] Others had different motives for racial counterfeit, a topic revisited in the conclusion.

Efforts to move entirely beyond the human body in defining race have had at best mixed success. Geraldine Heng, for example, has claimed that "race is a structural relationship for the articulation and management of human differences, rather than a substantive content."[47] Such definitions facilitate comparison across the *longue durée* and deal with the culture-biology relationship (problematically) by excluding both elements, but the abstraction of race into what is essentially a metaphysical category, a means of structuring knowledge, greatly weakens its analytical value. The principal and most obvious drawback with such redefinitions is that, in seeking to produce a concept sufficiently broad to capture everything the researcher believes should be classified as racial, it ultimately loses sight of the human body. There is, therefore, the need for a notion of race sufficiently broad to include cultural elements, but which must remain based on the flesh-and-blood reality of the human form for the sake of clarity, comparability, and accuracy.[48]

The most valuable recent definitions both of racism and race have taken this form. Benjamin Isaac's characterization of "racism" (2003) describes it as "an attitude towards individuals and groups of peoples which posits a direct and linear connection between physical and mental qualities. It therefore attributes to those individuals and groups of peoples collective traits, physical, mental, and moral, which are constant and unalterable by human will, because they are caused by hereditary factors or external influences, such as climate or geography."[49] Even more starkly, the art historian David Bindman, declining in

2023 to offer a definition of race, nonetheless stated that "it always contains one or more of the following: appeal to a common ancestry, the inheritance of 'blood,' a certain biological permanence, and an imagined community based on one or more of the former."[50] Important for what follows is the common ground shared by these definitions: hereditary factors, common ancestry, inheritance, and "constant" or "permanent" characteristics.

Correspondingly, this book applies the terms "race" and "racial" somewhat loosely, using them to refer to the set of discourses on the world's peoples, their origins, diversity, and characteristics, that concentrated primarily but not exclusively on physical features. Not only must cultural aspects be included in this loose definition, but even some bodily phenomena, such as "complexion" (which could mean skin pigmentation, an individual's baseline healthy state or a quasi-biological notion of personality type), cannot be simply incorporated without comment. A lengthier discussion of the terminology and content of *race* in the period under investigation is offered in chapter 3.[51]

These concepts are also based in a rich historiography exploring race in the eighteenth century, in which most are agreed that the modern concept of race emerged probably in its final decades.[52] Among the most common explanations offered by historians are the campaigns against slavery, the development of modern biological sciences, the emergence of modern aesthetics, and the role of political and imperial developments, including the upheavals of the "age of revolutions" and the associated rise of nationalism.[53] Others focus on the importance of particular philosophers for ideas of race, especially David Hume and Immanuel Kant, contributing to the revisionist picture of the Enlightenment as productive of discrimination and cruelty.[54] But while there exists broad agreement that the turn to modern scientific racialism took place during these years, there is less consensus on what it replaced and whether this replacement was total.

One important reason why historians of eighteenth-century Britain have only lately attended to race is because it was neither the only nor the most important means of describing human populations. Religion had long been a crucial register of difference and remained so until well into the nineteenth century.[55] As the following chapters will show, its influence was also broad; it contributed, for example, to the more secular categories of civilization and savagery. This latter category came in "noble" and "ignoble" forms, pointing to an ambivalence at the heart of the notion of savagery that was neatly encapsulated by the problematic relationship of stadial theory to race.[56] Stadial theory—the belief that all human societies passed through a set number of stages, usually characterized by the mode of subsistence, on their progress from savagery to civilization—was an innovation of French and Scottish political economists after 1750.[57] This totalizing interpretation of human history appeared to leave little scope for modes of thought that posited absolute and unbridgeable difference, such as race. As Wheeler and especially Silvia Sebastiani have shown, however, racist assumptions permeated the cultural anthropology of stadial theorists.[58] Race appears to lurk in the shadows of such seemingly universal models after all.

Yet there were also substantial continuities in the idea of race between the seventeenth and nineteenth (or even twentieth) centuries that are made especially visible by viewing the development of the concept through a reproductive lens. Exploring and highlighting these continuities is one of the key contributions made by *Generating Difference*. Chief among these persistent features of race is its original meaning of "lineage" or "descent." Far from being an etymological curiosity of its early history that was discarded once *race* assumed the full panoply of its modern signification, the retention of these meanings into late modernity powerfully shaped the concept. Sixteenth-century Iberian ideas concerning hereditary Jewishness and Muslimness (mentioned earlier) had a wide impact, especially in the extra-European world, that arguably led to the elaborate system of racial gradations in Britain's Caribbean colonies, discussed in chapter 6.[59] Even biblical ethnology, which explained human diversity through a genealogical model traceable to Noah's three sons, Ham, Shem, and Japhet, placed emphasis on the importance of a notion of heredity.[60] This was especially the case with Ham, whose curse was, as unconvincingly argued by enthusiastic exegetes and apologists of slavery, the cause at once of blackness and of its inferiority.[61] This genealogical schema was also used to support nascent English or British nationalism and fed into movements such as Gothicism and, later, Anglo-Saxon racialism, as the mixture of concepts that composed eighteenth-century racial thought were applied to intra-European diversity.[62]

Several studies of race in the early modern era, particularly those emerging from French universities, have highlighted the persistence of the concept's genealogical connotations. Several works have explored the close associations of race, (good) breeding, and aristocracy in what has been termed a "nobiliary paradigm" in racial thought. Underscoring the notion of hereditary excellence, this paradigm not only participates in longer patterns of French historical interpretation (it echoed the *thèse nobiliare*, which held that the ancient constitution descended from the Frankish conquest of the fifth century) but also helped to populate a wider semantic field for the term "breeding" by linking nobles and their animals. Mackenzie Cooley's work on renaissance texts about animal breeding, for example, points to the influence of the attempt to secure an improved *razza* for high-value animals (particularly horses) on racial thought in Italy, Iberia, and the Spanish and Portuguese colonies in America.[63]

Connecting Race and Sex/Gender

On the far left of *The New Union Club*, next to the precariously teetering William Wilberforce, stands the most powerful encapsulation of the relationship between race and sex that is the subject of *Generating Difference*. This three-person group consists of a black man, his white wife, and their mixed-race child, a "piebald pledge of Love." The baby is split vertically into black and white halves, unlike the actual "piebald" (i.e., leucodermic) child immediately behind this group, and bears an expression far more alarmed than the smiles of his proud parents. This expression betrays the aims of Cruikshank and Frederick Marryat, as the child is apparently aware that he is a monstrosity resulting from misguided abolitionist humanitarianism. Yet their rehashed miscegenationist fearmongering backfired in more than one way.

For one thing, they had overplayed their satirical hand. Dividing the child vertically in two undermined the largely successful effort, evidenced elsewhere in the composition, to emphasize the permanence and reliability of embodying sexual and racial identity. Giving racial mixture so obvious a manifestation only called attention to the fact that in real life it was often hardly obvious at all. Furthermore, they portrayed a couple that had done something remarkable. By producing a mixed-race child, these parents had proven both that they belonged to the same biological species and that they had contributed equally to the shape and form of their child. This "mixed-race datum" was at the heart of the common ground shared by discourses of race and sex, and is discussed at length in chapter 4.

This is one of the principal contributions made by *Generating Difference*, enabled by bringing together the scholarship on race and sex/gender and concentrating on reproduction. Others have explored some of the connections between race and sex/gender. A lively scholarship has emerged that explores their links in slavery, in the representation of non-European women in travel literature, science and medicine, art, fiction, and drama, and in the histories of empire and nationalism.[64] But with a few exceptions, the focus of this work has been *either* on race *or* on sex/gender. From the outset, the history of sexuality has acknowledged a close affinity between sex and race, yet works in this field have been content largely to hint at this linkage without developing a sustained account of it.[65] Laqueur stated that scientific race "developed at the same time and in response to the same sorts of pressures as biological sex" but refrained from discussing this in any greater depth.[66] This is typical of the treatment of race in much gender history, which has incorporated race as a supplement to its primary focus, the analysis of sex and gender.[67]

There exist three principal exceptions to the one-sided nature in which race and sex/gender are integrated in historical work. The first is the history of science, particularly in the scholarship of Londa Schiebinger and Nancy Stepan. Both moved from a perspective that tied race and sex together through "metaphor" or "analogy"—an influential approach, if one that obstructed a less mediated connection—to another that asserted a more integrated relationship.[68] Starting in the early 1990s, Schiebinger showed how gender influenced the eighteenth-century study of race but declined to explore how race might influence the study of sex.[69] Stepan's essay "Race, Gender, Science and Citizenship" (1998) called for the abandonment of appeals to nature and the body in framing political rights and subjectivities, pointing to the construction of racial and sexual difference through a process of "ontologising via embodiment."[70] The emphasis of this formulation, on the corporeal basis for racial and sexual identity, is at the heart of the argument put forward in the following chapters.

Also arguably situated within the history of science is the body of work that explores biopower, most pertinently in William Max Nelson's *Enlightenment Biopolitics* (2024). Much of this work focuses on continental European powers—especially France or cameralist Germany—or later colonial settings where the political will and means were available to promote reproductive schemes. But a detailed consideration of their conceptual underpinnings is often marginalized in favor of a discussion of the social and political ramifications and sources of these schemes.[71]

The second exception appears in the history of slavery. Since the work of pioneers like Elsa Goveia, Elizabeth Fox-Genovese, Hilary Beckles, and Barbara Bush, several historians have examined gender and New World slavery.[72] The majority of these have concentrated principally on women under chattel slavery, including how they were represented in abolitionist and emancipationist discourses, and on gender-specific aspects of the female experience, from work patterns to sexual abuse, reproduction, and child-rearing.[73] Reproduction is the specific focus of studies by Katherine Paugh and Sasha Turner, who have explored its social and cultural history in the Caribbean and its role in abolitionist debate.[74] Their work follows the trail blazed by Jennifer Morgan, one of the few historians to discuss the influence of reproduction under slavery on emergent theories of race.[75]

A final group of scholars whose work has examined the ties between race and sex/gender have focused on the slave colonies of mainland North America. Kathleen Brown, Kirsten Fischer, and Jennifer Spear have all explored the social history of (sexual) relations between white, enslaved, and indigenous communities and the influence that these relations had on the construction and development of racial thought. A crucial instrument for fixing racial and sexual categories was the law, and these three scholars have made extensive use of lawsuits, statutes, and legal sources to trace the development of these identities. So, too, in the case of Jamaica, has Brooke Newman, whose *Dark Inheritance* (2018) also emphasizes the importance of genealogy, but in the allocation of political rights rather than within a detailed analysis of racial thought.[76]

Plan of the Book

These are the only exceptions to the lopsided consideration of sex/gender and race in works that connect the two but focus primarily on one or the other. The extensive literature that engages with Laqueur, pro and con, manifests clearly the vital role played by the reproductive body in concepts of sexual difference; very few works consider this body in the development of racial ideas. This is the gap that *Generating Difference* fills. It offers an account of racial thought in the long eighteenth century that is global in scope—incorporating not only metropolitan Britain (and the significant contributions from mainland Europe, principally France and Germany) but also the Caribbean, British India, and the Pacific—yet remains squarely focused on the reproductive body. It uses a wide range of source material to build and sustain its case. The traditional stuff of intellectual history—detailed analysis of learned texts and fully developed argument—is combined with the social breadth of cultural history. Printed materials form the largest body of sources, not least because it helps to show that race was not something opaque or hidden in eighteenth-century Britain but rather was widely visible and accessible to all, as Catherine Molineux and others have shown. Cheaper printed materials, in particular, serve to demonstrate the social breadth of ideas concerning sex and race. For example (and as several scholars have shown), erotic or cheap medical texts play an important role in demonstrating that several models of identity could be available at one time and even in the same text. Manuscript sources are also used at various points to supply evidence of the development of these ideas before and after publication and of their social (mis)application.[77]

These sources provide an account of the common ground shared by race and sex, a common ground that was reproductive in origin and that was the crucible in which modern forms of somatic identity were forged. One of the most basic long-term causes of this development was overall demographic change, which was viewed in accordance with the lengthy tradition, stretching back to the ancient world, that a large population is an unqualified good. Chapter 1 explores the cultural and intellectual background of this idea before examining how it was applied in Great Britain across the long eighteenth century. Britain's population growth was stagnant for a century from *c.* 1640, and the concurrent need for economic and military manpower made overcoming this a matter of some urgency. The limitations of efforts to cut the death rate by enhancing public health or to promote immigration and restrict emigration were such that it soon emerged that the only viable means of boosting population growth was the encouragement of births. This had obvious moral implications as it was believed by all (with a few spectacular exceptions) that the only acceptable mode of reproduction was within monogamous matrimony. Consequently, sustained efforts were made to make marriage more respectable (Hardwicke's Marriage Act is a case in point) and to safeguard legitimate productive sex, not least by suppressing illegitimate or unproductive practices, such as contraception, abortion, infanticide, and polygamy.

What was good for Great Britain was also good for its colonies, and chapter 2 carries this account of populationism and pronatalism into Britain's Caribbean possessions and the debates about slavery and abolition at the turn of the nineteenth century. The so-called principle of population (that population expansion was both an intrinsic good and a sign of the health of a society) was mobilized as a key argument in the fight for and against slavery. Abolitionists had the best of this argument, as they were soon able to point to the disastrous demographic consequences of slavery that continued importation of slaves from Africa had obscured. With the abolition of the slave trade from 1808, planters ought to have encouraged—even if only from simple self-interest—the reproduction of their slaves. But they either did not or were so ambivalent toward this goal that little or no success attended their efforts. This was often due to the simple nature of sugar monoculture, in which slaves were worked to exhaustion at a time when workload increased due both to the attrition of irreplaceable slaves and the pressures of a world drop in sugar prices during the 1820s. But it was also due to planter prejudices that both suggested utterly ineffectual pronatalist measures, designed to appeal to a stereotypical simpleminded slave, as well as expressed unremitting hostility to slave marriage. Always chary of encouraging Christianity among their slaves, who might feel themselves on eye level with their masters, planters were also anxious to prevent the literacy that was attendant on meaningful Protestantism.

The second part of *Generating Difference*, "Theories of Race and Reproduction, 1600–1850," begins with a chapter that provides a detailed examination of racial theory, in which it is made clear that sex was absolutely central to definitions of race and species from the very outset. Notwithstanding the survival into the nineteenth century of the older, nonzoological concept of species as a logical category—something important to remember given the frequency with which racial authors discuss "species" of men—sex was central to speciation. Indeed, it somewhat scandalously formed the basis of Linnaean taxonomy and inspired a whole subgenre of botanic pornography. But the influence of sex on race was more direct than acting as a simple determinant of species: racial theory was awash with discussion of the sexual characteristics and behaviors of different communities of humans. These acted as means by which racial groups could be distinguished. The oft-quoted closing sentence of Charles White's *Account of the Regular Gradation in Man* (1799) is a case in point: "Where, except on the bosom of the European women, [are we to find] two such plump and snowy white hemispheres, tipt with vermillion?"[78] The sheer extent to which sexual characteristics featured in descriptions of different racial groups demonstrates the intellectually vital role played by empirically sensible features in new models of identity. That these features were often the same for racial and sexual difference—breasts, genitals, hair, and the skeleton were among the most prominent—goes to highlight how much these identities shared.

Chapter 4 shifts the focus to reproductive (particularly embryological) thought, which shared numerous points of contact with racial thinking. It is formed of three sections. The first sketches the principal embryological theories that were developed and debated between 1650 and 1850, and explores their broader political and cultural ramifications. The oscillation from epigenesis (where both parents contribute to the fetus that progressively develops during pregnancy) to preformation (one parent is the primary contributor to the fetus that is preformed at conception and simply increases in size during pregnancy) and back again was a product of several factors. One was the gradual shift from mechanism to vitalism as the putative organizing principle of life, which was related to wider philosophical and political commitments: patriarchalist preformation became the chosen theory for royalist and Jacobite investigators, while the debate between epigenesis and preformation became a proxy for Newtonian and rationalist approaches to science. An overlooked factor is the part played by colonialism and race, and this section highlights the use of non-European populations in embryology and the reproductive speculations of major racial theorists, such as Buffon, Blumenbach, and Kant. The second section explores "racial embryology": the principal terrain shared by discourses of race and reproduction. Here, the focus is on the mixed-race datum—the fertility of mixed-race couples that proves the unity of the human species and that both parents contribute to the fetus—and the light this shed on many aspects of reproductive debate, from hybridity to heredity. The final section uses the work of two influential authors on racial and reproductive topics of the mid-eighteenth century (Pierre Louis Moreau de Maupertuis and James Parsons) to explore other topics of racial embryology, including albinism, the power of the maternal imagination, and hermaphroditism.

The final two chapters of the book explore the issues already raised in a wider, global context. Chapter 5 begins with a vignette that highlights the uniquely sexualized view of the South Seas and its epistemological importance. The apparent discovery in 1768 that Tahitians might perceive sexual difference via olfaction not only emphasized the role of sex in European perspectives on the region but also highlighted the bodily alterity of Tahitians and their "aesthetic" (in the eighteenth-century sense, relating to perception by the senses) advantages over Europeans. The chapter's three sections explore these issues by examining, first, European discourses on the "South Seas" both prior to and following contact in the 1760s and the reproductive focus of much of the discussion of sexual features and customs that appeared in utopian literature and official travel reports. The chapter's second section uses the example of the impact of voyages of Pacific discovery on racial thought to explore the role of aesthetics and an empiricist epistemology in the construction of models of identity based on the physical body. The final section, however, demonstrates the severe limits to this process, as an extracorporeal marker—language—was necessary to produce the division of the Pacific's peoples into Polynesian and Melanesian, a development complete by 1832. Language offered something the physical body could not: a diachronic account of the Pacific's peoples. That this non-corporeal source of knowledge was essential to the process of identifying a racial division of the region was doubly ironic, as the vitalization of nature that enabled "scientific" concepts of race was predicated on a prioritization of natural history (*Naturgeschichte*) over the description of nature (*Naturbeschreibung*).

Chapter 6 examines the creation of mixed-race people ("ethnogenesis") in Britain's possessions in India and the Caribbean, its contexts and cultural footprint. It begins with an introductory section in which the chief difference in this phenomenon between these two locations is highlighted: in the Caribbean, a number of elaborate taxonomies of mixed-race, running from "white" and "negro" all the way to "quinteron" or "musteephino" (and, in one remarkable case, a mixture of 1 part "negro" to 8,191 parts "white") were developed; in India not only were there no such classifications but there was hardly agreement on what term should designate mixed-race people. The chapter argues that this difference is because there was scant need for such a system in India, where a pigmentocratic social order in support of chattel slavery did not exist. It makes this argument in two large sections at the heart of the chapter that respectively examine the Caribbean (principally Jamaica) and India. These sections are identically structured, first exploring attitudes to the non-white parent of mixed offspring before turning to deal with the emergence of the mixed-race population, attitudes toward them, and their struggles for political and social rights. The extensive parallels between these two locations make their differing approaches to racial taxonomy even more striking. The final section explores the place of slavery in both societies and in the making of race. It emphasizes that, even where taxonomies of mixed race did exist, such as the Caribbean, there was so little agreement about the categories and so much ambiguity inherent to the human form that knowledge derived from genealogy and even individual behavior had to be mobilized to situate an individual within such classifications. Race was self-evidently so slipshod a concept that only the enormous power of slavery and its supporters could sustain it until, through sheer longevity, utility, and popularity (especially among those who exploited it as a technology of rule), it insinuated its way into the cultural and scientific mainstream.

PRONATALISM IN THE BRITISH ATLANTIC WORLD

"The King's Honor"

Population and Pronatalism in Greater Britain

This chapter explores a cultural commonplace of the eighteenth century—the belief that a large and growing populace was a good thing—and relates it directly to the human body by showing that reproduction was the only widely acceptable means to this end. The argument proceeds in four stages. In the first section, the central pillars both of the belief in the desirability of a large population and of the drive for its growth are identified. These were spread widely throughout early modern culture: religious teaching, ancient writings, legal traditions, modern philosophical inquiry, and economic projects and policies all endorsed the idea that a large population was a desideratum. From the mid-seventeenth century this was combined with the belief that the population of England (later Britain) should be significantly enlarged. The magnitude of this consensus helps to explain the depth of the anguish felt by the few who, after 1750, came to think that a large population was not always a good thing. It also accounts for the violence of the reaction to Malthus's *Essay on the Principle of Population* (1798), which invoked the dreadful specter of population growth vastly outpacing any increase in food supply.

Before then, however, people concentrated on promoting population growth, which could be achieved by increasing the birth rate, cutting the death rate, or attracting migrants. These alternatives to reproduction are explored in the second section, which demonstrates that such projects were limited in scope and success, if not always in popularity. Attempts to improve public health were of extremely limited effect given the state of medical knowledge, and philanthropic endeavors to make individuals useful to the state (such as the Marine Society or the Foundling, Lock, and Magdalen Hospitals) had equally strong limitations. Likewise, efforts to discourage emigration were ineffective, while those to encourage immigration ran into a solid and broadly based wall of opposition.

All that remained was to improve the birth rate, but this could occur only under the strictest of conditions. Sexual activity in general, and procreation in particular, were only permissible within a marital framework. The third section of the chapter examines the strength of the connection between matrimony and reproduction. The successful attempt to reform English marriage law in 1753 was attacked and defended on the basis of its projected or actual effect on reproduction and population. Marriage sermons, medical works, popular literature, and a wide range of other texts all identified procreation with marriage. The strength of this connection is apparent from its longevity—only in 1947 did the House of Lords rule that procreation was not the essential purpose of matrimony—and from the protracted early modern debate on polygamy. The tendency of polygamous unions to promote or impede reproduction featured in discussions of biblical history and natural law throughout this controversy, in which the importance of a marital basis for reproduction was accepted by all participants, who differed merely on whether it should be monogamous or polygamous in nature.[1]

As marital procreation was the only widely accepted means to obtain population growth, substantial efforts were made to safeguard productive and combat unproductive sexual activity. The fourth and final section analyzes the discussion of such activities in eighteenth-century Britain. Procreation was encouraged almost everywhere sex was addressed, from the provision of advice on conception in popular sex manuals to developments in obstetric knowledge and practice that grew in scope and importance over the century. Conversely, contraceptive practices, abortion, and infanticide were widely attacked on populationist and other grounds. Arguments in favor of limiting fertility and family size were few and far between, confined to a radical fringe even once the influence of Malthus's *Essay* had led to a reassessment of population goals in the early nineteenth century. Fornication, although unquestionably productive of offspring, was nonetheless attacked for the same reason that polygamy was so widely discussed: children required married parents if they were to be useful for the nation or if the system of the hereditary transmission of property was to be maintained.

By demonstrating, on the one hand, that the increase of Britain's population was the summum bonum of social and economic policy and, on the other, that procreation was the only widely accepted means to this end, it becomes clear that the reproductive body was at the heart of British culture, politics, and society. This was the case not merely within Britain but also, as subsequent chapters will show, throughout its global empire. This argument builds on the scholarship of Leslie Tuttle (for France) and Lisa Cody and Mary Fissell (for Britain), but whereas Cody and Fissell have concentrated respectively on the emergence of man-midwifery and on popular ("vernacular") texts of the early modern era to the end of the seventeenth century, the analysis offered here concentrates principally on enlightened debate and policy on "population." Remaining chapters will explore the breadth and depth of the relationship between reproduction and new forms of bodily identity (race and sex), but the discussion in the following pages provides a foundation for one of the overall contentions of this book: that through this relationship and the importance of procreation in eighteenth-century British culture, such forms of identity—race as well as sex—mattered in Greater Britain.[2]

The Foundations of Populationism

Questions of population were deeply political in early modern Britain. The family was the fundamental unit and prototype of social and political order, and a well-governed realm had its analogue in the orderly household. So when population or politics came to be discussed, the family was unsurprisingly prominent in the discussion. Political questions, big or small, were all debated in relation to the common ancestor of every polity—the family—and how it was (or ought to be) governed.[3]

Such debates were especially rife in the turmoil of the sixteenth and seventeenth centuries. Yet remarkably for such a period of intellectual, social, and political ferment, the belief that population ought to be large found support in practically every area of early modern culture. This was perhaps due to the deep cultural roots of populationism, whose first and most fundamental source was God's word, transmitted in the Bible. The command to "be fruitful, and multiply" (Genesis 1:28, 9:1) was taken seriously and applied broadly; given when humankind consisted only of a single family, it related both to humanity as a whole *and* to the married couple. The Bible also advocated populationism at a third level of political organization between these two extremes, that of the kingdom: "In the multitude of the people is the king's honor: but in the want of people is the destruction of the prince" (Proverbs 14:28). Early modern glosses on this verse repeatedly made explicit the connection between population, increase, and good government.[4] But Scripture also supported the view that there were, or ought to be, limits to growth, as when Abram and Lot were forced to live apart because their combined families outstripped the capacity of the land to support them.[5] The Bible thus encapsulates three elements of the discourse on population that were to recur throughout the early modern period: a large population was a desideratum, limits to its growth may be necessary or desirable, and such constraints depended on economic and political circumstances.

These features were also visible in the second source of early modern populationism, ancient writings, especially political theory and law. Plato, Aristotle, and Cicero all linked the reproductive urge to the formation of society and increasingly complex forms of political community.[6] Aristotle's *Politics* (c. 335–323 BC), which was enormously influential for medieval population thought following its rediscovery and translation in the late thirteenth century, stated that an ideal *polis* required a well-sized but not invariably large population, for "experience shows that a very populous city can rarely, if ever, be well governed."[7] Aristotle did acknowledge that a larger populace might well produce a greater state, but he and Plato were also early advocates for limits to population growth.[8]

Ancient law more unequivocally favored increase. Plutarch's *Life* of Lycurgus (early second century AD) described populationist measures introduced by the Spartan legislator, including shaming bachelors, strengthening women to better withstand the rigors of childbirth, and deliberately raising and controlling sexual desire in regulations concerning female clothing and marital cohabitation.[9] The sexual communalism of the Lycurgan code (and of Plato's *Republic*) made it unpalatable for early modern commentators, who were more comfortable with the marital and adultery laws passed in Augustan Rome in AD 9 and 18. These also had populationist goals—including penalties for unmarried men and women, and benefits for those who produced children—but were firmly rooted within marriage.[10]

A third source of populationism was early modern political writing, which exemplifies the consensus behind the benefits of a large population. Contractarians and patriarchalists, despite differing fundamentally on the source of political authority,[11] both supported the ideal of a large population.[12] Among patriarchalists, Jean Bodin made the political stability of cities proportional to their size (the larger the better) and Sir Robert Filmer went further, arguing on the basis of Proverbs 14:28 that even a tyrant would try to preserve the lives and protect the goods of his subjects.[13] On this point, even Thomas Hobbes could agree with Filmer.[14] John Locke demonstrated persistent interest in these questions, using populationism to argue against some of Filmer's other claims.[15] Not all were agreed that growth should be unlimited—Hobbes in particular painted a vivid picture of war resulting from overpopulation—but theorists of every stripe recommended a large populace.[16]

Two developments after 1650 transformed the discourse on population. First, the belief took hold that Britain had too few inhabitants and that increase was urgent. The importance of this idea is difficult to overstate, as it permeated almost everything that was said and done regarding population for more than a century. Second, early forms of social science began to appear, introducing statistical methods that produced a more recognizably modern definition of "population" as a subject of study.

The Urgency of Increase

Several factors explain the growing conviction that Britain's population needed to increase fast. One was the prominence of the "ancients versus moderns" dispute, part of a wider current of cultural debate that characterized the early Enlightenment, and which had a significant demographic aspect.[17] Figures such as Montesquieu, the philologist Isaac Vossius, and the clergyman Robert Wallace argued that the ancient world was more populated than the modern, against the contrary position held by Hume, Buffon, and Voltaire.[18] While both sides argued that increase ought to be encouraged, such a conclusion stemmed more logically from the belief that there were now fewer people than in the past, which became the most common viewpoint on population in eighteenth-century Britain.[19]

Demography was another factor. Following a period of sustained growth, England's population hovered at around 5 million between 1640 and 1730, before increasing rapidly to 8.3 million by 1801.[20] During the period of growth prior to 1640, overpopulation was the predominant concern, voiced by such figures as Francis Bacon, Walter Ralegh, and Richard Hakluyt.[21] With stagnation came an anxious fixation on depopulation. Typical was the judgment of Roger Coke, who in 1671 complained that "the Countrey [has become] much more thin and un-inhabited."[22] Some dissented from this grim assessment, most notably Sir Matthew Hale, but we must be careful not to overstate the influence of demographic change.[23] Due primarily to a lack of reliable information on population trends, such changes had a limited and indirect effect on population discourse and were possible to overlook entirely. Indeed, the Scottish jurist Lord Monboddo was not alone in writing at length on the decline of the British population at precisely the point (the 1780s–1790s) that England achieved its highest growth rate between the sixteenth and nineteenth centuries.[24]

International rivalry also played a crucial role in keeping the state of Britain's population in the spotlight. From the late seventeenth century Britain became increasingly involved in lengthy and expensive foreign wars, stimulating an urgent need for manpower and matériel.[25] Population was most widely discussed when war was imminent, underway, or had recently concluded, and the twenty years from the early 1740s were a notable high point.[26] France's population dwarfed Britain's throughout the period, and the urgent need for men (particularly sailors) to combat this more populous foe was a powerful force for the encouragement of population growth.[27]

Such incentives existed also in peacetime. Britain's standing alongside its economic competitors was the central concern of the body of economic writing that has been traditionally labeled "mercantilist."[28] These texts often considered population, almost always recommending its increase, but it was less often the primary focus of discussion than a diagnostic indicator of economic and political power. This literature nevertheless made two substantial contributions to the discourse on population.

For one thing, it gave voice to novel forms of political economy that could not but enhance the status of populationism. These emphasized the range, quality, and quantity of manufactures as the primary sources of Britain's wealth, over and above the possession and cultivation of land. Where land is a limited commodity whose productive capacity can be exhausted, manufactures could potentially expand without any limit save one: the number of people to produce and consume manufactured goods. Such a doctrine of "hands not lands," which clearly prioritized the expansion of population as *the* key factor behind national wealth, has been associated with a "Whig" political economy that came to dominate after 1688. But it was also an important perspective in the years following the Restoration, as evidenced by innumerable discussions that juxtaposed Spain and the Netherlands: the territorially vast but impecunious imperial power compared unfavorably with the resource- and space-poor but population-rich economic superpower of the seventeenth century. Those unsympathetic or even hostile to this account of national wealth, such as Sir Josiah Child, nonetheless emphasized the importance of population growth for the increase of riches. Here as elsewhere, a populationist consensus ruled.[29]

Defining "Population"

"Mercantilist" literature also promoted new methods for dealing with population questions, chief among which was political arithmetic. Described by one of its pioneers as "the Art of Reasoning, by Figures, upon Things relating to

Government," political arithmetic was largely responsible for the rise of "population" as a fully-fledged concept and self-contained unit of study in its own right.[30] Compared with the earlier and more qualitative concept of "populousness," "population" (although pioneered in the sixteenth century by Bacon and Giovanni Botero) reflected a growing emphasis on the quantitative study of social phenomena that exploded onto the scene in the later seventeenth century, more quietly gathered pace throughout the eighteenth, and came into its own after 1800.[31]

The arguments supported by political arithmetic were often familiar. Almost all its major exponents toed the traditional line based on Proverbs 14:28, and not a few were zealously pronatalist in their recommendations for promoting population increase.[32] What was new about political arithmetic was its method: the use of numbers, often tabulated, to provide analyses of social and economic questions was a major innovation. But early modern population statistics were often wildly inaccurate, prompting authors to suggest a range of alternatives, such as extrapolating from the London Bills of Mortality, counting houses or burials, deriving a ratio of plague victims to survivors, making estimates on the basis of various household taxes, and sampling from parish records.[33] Only in Scotland, thanks to the efforts of Alexander Webster and James Steuart, was there anything approaching a sufficient empirical base for discussions of population. But even here the data were insufficient, and the flaws of other methods stoked the population debate until the eve of the first nationwide census in 1801.

Earlier attempts to introduce a census ran aground for a number of different reasons, ranging from the biblical to the strategic.[34] That which came closest to success (in 1753) failed because it was poorly managed in Parliament, was thought to be expensive and pointless, had the potential to incur God's wrath (as with David's census of Israel),[35] and because opponents remained unconvinced that there was no sinister purpose behind collecting such data. One hostile pamphlet claimed that, besides apportioning taxation, the census might be used to determine whether to naturalize all foreigners, how many men in a parish might be pressed into the army or navy, and "how many may be taken away for the Plantations [colonies]."[36] However paranoid this may seem, such suspicions were plausible because of political arithmetic's immanent pragmatism: from its inception, it sought directly to effect political, economic, or social change (especially in the unpublished but widely circulated manuscripts of William Petty).[37] As a source of ideas for change and an analytical technique that pioneered numerical argument, political arithmetic helped to create "population" as it measured it.

Yet the picture that emerged from this new data fueled a growing pessimism after the mid-eighteenth century. Some believed that the population was too large, its excessive size based on flimsy economic or political foundations. Several commentators rejected modern commercial society as deleterious to population, blaming luxury, manufacturing, government, or all three.[38] The enduring controversy on luxury generated a sizable literature that argued for a rejection of commercial society and a return to a simpler, more wholesome agrarian existence.[39] For writers such as William Bell, "the arts of refinement and luxury" amounted to nothing less than a subversion of "the strong propensities of human nature to the preservation and increase of the species."[40] What made the spread of luxury so alarming to contemporaries was that it ranged vertically as well as horizontally: not merely were more people developing a taste for exotic and refined goods, but the lower ranks of society were doing so. One telltale sign of "luxury" in the lower orders was their drinking habits, as beer was apparently abandoned in favor of tea or, worse, gin. Several authors blamed these drinks for supposed population decline: gin for its immediate health risks, tea for the loss of sailors required to transport it from Asia, and milk for the damage to arable farming caused by satisfying greater demand.[41]

The increase in pessimism after 1750 coincided with a more theoretical turn in the discourse on population and a growing tendency to compare British population trends unfavorably with American. The buoyant optimism of Ezra Stiles and Benjamin Franklin contrasted starkly with metropolitan pessimism. Franklin was deeply interested in questions of population, writing several related pieces on the topic and communicating relevant research to the Royal Society.[42] But despite his different outlook for population, he operated within the same discursive parameters as the pessimists: they both shared a desire for a large *but sustainable* population, and for both the chief limiting condition to its growth was the availability of food. As Franklin put it in 1755, "There is in short no Bound to the prolific Nature of Plants or Animals, but what is made by their crowding and interfering with each others Means of Subsistence."[43] Although the ultimate desirability of a large population was never in doubt, populationist discourse after 1750 did witness a sophisticated restatement of the need for limits to expansion. As we saw earlier, these were present in the Book of Genesis but were now based on a range of newly quantifiable political, social, and, above all, economic criteria.

The ne plus ultra of this tendency was the work of Malthus. Yet even the *Essay* was not the anti-populationist tract it was widely perceived to be.[44] Malthus's target was not a large populace per se, but rather the desire for growth irrespective of a population's capacity to sustain itself. As he explained in 1806, "In the desirableness of a great and efficient population, I do not differ from the warmest advocates of increase. I am perfectly ready to acknowledge with the writers of old, that it is not extent of territory but extent of population that measures the power of states. It is only as to the mode of obtaining a vigorous and efficient population that I differ from them."[45] In talking of a "vigorous and efficient" population, Malthus sought to combine the highest possible quantity of people—the focus of most discussions of population before 1750—with the best possible quality. But most commentators failed to get the message. As one complained, "Marriage is to be held in comparative disgrace, and offspring to be deprecated, whilst the unmarried and the barren are to receive the thanks of the state, and the applause of their own consciences."[46]

Admittedly, it was easy to draw the conclusion that Malthus was at best anti-populationist and at worst indifferent to the plight of the poor. He not only advocated the abolition of the poor laws but in one remarkable passage (that drew such criticism it only appeared in the second edition of 1803) he claimed that a child who could not obtain sustenance from its parents or its labor "has no claim of *right* to the smallest portion of food. . . . At nature's mighty feast there is no vacant cover for him."[47] And the (mis)use of the *Essay* by others helped matters none: William Cobbett unfairly accused him of proslavery leanings by correlating the disastrous demographic record of slavery to Malthus's arguments favoring checks to population, and William Pitt apparently withdrew a Poor Relief Bill on the strength of the *Essay* alone.[48]

Worse than his callousness toward poverty was the havoc Malthus was perceived to wreak on the institution of marriage. Believed to be simultaneously a betrayal of his vocation as a clergyman and an attack on received wisdom, Malthus's ideas concerning marriage explain, in large part, the ferocity of his critics. Take, for example, the response of Samuel Taylor Coleridge to Malthus's suggestion that the celebrant at every wedding should read a statement in which "the strong obligation on every man to support his own children" was to be emphasized alongside "the impropriety, and even immorality, of marrying without fair prospect of being able to do this." "The clergyman!" Coleridge exclaimed in the margins of his copy of the *Essay*, "No Mr. M. you must call in the Sow-gelder."[49] Social legislation, which was based on the *Essay* despite mounting evidence against its claim that the poor laws were responsible for the number of feckless marriages, continued to invoke the connection between marriage and reproduction. So, too, did the opponents of such measures. In 1821, for example, a bill to withhold poor relief from able-bodied single men was introduced and was immediately attacked as an "anti-matrimonial and anti-population scheme." Although this bill was withdrawn, the Malthusian legislation of the 1830s drew even more explicitly on the connection between marriage, population, and poverty. Why it was that alternatives to monogamous reproduction were unable to promote population growth is the subject of the following section.[50]

Alternatives to Reproduction

Population growth could be secured in three ways: increasing the birth rate, reducing the death rate, and managing migration. In eighteenth-century Britain, as this section will show, the latter two were ultimately ineffective. Among the many possible reasons for this, want of effort does not feature. The ethos of "improvement" often united with populationism to produce an impressive range of initiatives designed to reduce mortality. Sometimes beneficial effects were a happy coincidence. Urban improvement, for example, was frequently directed at removing nuisances and clearing thoroughfares but led to better sanitation as often by luck as by design.[51] Good fortune also played a part, as with the absence of famine and plague in England after the mid-seventeenth century.[52] But conscious effort too played its part in the struggle to cut the death rate, and foremost among such endeavors were the spread of voluntary hospitals, the growing adoption of smallpox inoculation, and the development of new forms of philanthropy.

Reducing Deaths I: The Hospital Movement

The "Hospital Movement" encapsulated a widespread drive to improve medical care for the laboring poor. From the first voluntary hospitals established in London after 1718 (Westminster and Guy's), provincial infirmaries were founded across Britain, from Edinburgh (1729) to Brighton (1828). As with many eighteenth-century charities, they were modeled on joint-stock enterprises, funded by subscription and governed by committee.[53] Regulations tended to be strict, but patients received free medical care and a comparatively balanced and wholesome diet. These hospitals were overwhelmingly aimed at the laboring ("deserving") poor, whose care was a national priority: they tended to fall through the cracks of existing health provision (the well-to-do could afford their own care, and those in receipt of poor relief enjoyed at least nominal health care courtesy of the parish) yet were viewed as the economic backbone of the country, on whom its prosperity depended.[54] The earliest promotional material made explicit that such hospitals are an "Advantage to a

Nation, for as many as are recovered in an Infirmary, are so many working Hands gained to the Country."[55] Besides satisfying individual and national self-interest, there is some evidence that voluntary hospitals may also have contributed modestly to the overall decline of the death rate from around 1730, particularly in their treatment of respiratory diseases.[56]

But there were serious shortcomings that limited the effect of hospitals in reducing mortality. Ironically, there is little evidence to support the claim that hospitals were "gateways to death" as the number of deaths in voluntary hospitals remained small.[57] Many—perhaps most—people were however unable to benefit from their care, as only the "laboring" poor (which excluded children younger than working age) were both eligible and willing to use these institutions. Even these had to overcome substantial obstacles before they could be treated.[58] Distance was one: at least before mid-century, rural dwellers were often forced to travel far to reach their nearest hospital, which perhaps explains why they constituted a high proportion of in-patients (locals could more conveniently receive out-patient care).[59] Connections was another: most hospitals insisted that all nonemergency cases were only admitted on recommendation by one of the governors.[60] Patients also had to be suffering from the right kinds of ailments. Venereal and infectious diseases were not handled, and pregnant and lying-in women tended not to be admitted.

The number treated was therefore very limited. Although the larger London hospitals were admitting up to 7,000 annually by the mid-1780s, demand far outstripped supply. The scale of many provincial hospitals was much smaller. For example, despite successive expansions that saw capacity increase from thirty beds to over ninety by the 1790s, the County Hospital in Northampton admitted barely more than 400 in-patients (and 900 out-patients) in 1781. Its report of that year listed a total of 14,680 patients cured since its foundation in 1744 (or little more than twice the *annual* admission rate of the London Hospital). When set against the backdrop of a total population in 1801 of 131,430 for Northamptonshire, the limited capacity for significant impact on mortality becomes clear.[61]

Reducing Deaths II: Inoculation and Variolation

Advances in medical techniques also promised to reduce mortality. The runaway success story of eighteenth-century medicine was undoubtedly smallpox inoculation.[62] Folk practices, in Wales and elsewhere, had long encouraged the early exposure of children to smallpox in the hope that they would receive only mild infection at a propitious time.[63] But interest in variolation only emerged in Britain during the first decades of the eighteenth century as a result of encountering the practice in Africa, China, and the Levant.[64] Its most celebrated popularizer was Lady Mary Wortley Montagu, wife of the British ambassador to the Ottoman Court, but (*pace* Voltaire) figures from the Royal Society such as Hans Sloane and James Jurin contributed more fundamentally to its medical and statistical respectability.[65] Concurrently with the first trial inoculations in London—conducted on Newgate prisoners promised a pardon in return for volunteering—the practice emerged independently in Boston. Knowledge about inoculation was provided by a slave named Onesimus and spread by the cleric Cotton Mather and medic Zabdiel Boylston, whose audience was particularly receptive thanks to an epidemic then raging in the city.[66]

Even once introduced, the uptake of inoculation tended to follow the peaks and troughs of epidemic smallpox. A first wave of inoculations coincided with outbreaks during the 1720s, but the practice apparently waned as the disease abated in the following decades. The severe epidemic of 1751–1753 resulted in the practice spreading more widely beyond the elite.[67] This rather reactive pattern of inoculation was probably due as much to cultural concerns as to anxiety about the procedure. Hostile clerics claimed it was a blasphemous attempt to limit God's power, some even going so far as to claim Satan was the first inoculator when he smote Job with boils.[68] Others fixated on the potential violation of the Hippocratic oath to do no harm, on the possible ineffectiveness of inoculation (by highlighting apparent cases of inoculees subsequently catching smallpox), and on the questionable sources of knowledge about the practice, such as old and poor women or slaves.[69] It probably took the imminent threat of death or disfigurement to overcome these cultural scruples.

Inoculation was also initially very expensive, as physicians sought to integrate it within established canons of knowledge and practice.[70] Lengthy periods of preparation, including a strict diet and regular bleeding, were deemed necessary to ensure optimum conditions for successful variolation. This expensive treatment remained beyond the reach of all but the wealthiest before cheaper methods were pioneered at mid-century. Robert and Daniel Sutton, a father-and-son team, were the most famous and successful of the innovators who widened access to immunization. They managed to reduce variolation fatalities and shorten preparation time enough to inoculate entire villages at once.

Yet despite cheaper methods, the gradual neutralization of cultural opposition, and growing numbers given immunity, smallpox inoculation was not implemented on a scale sufficiently large either to meet demand or to exercise a significant effect on population size. This had much to do with the spread and lethality of smallpox—it is impossible to state that inoculees would have caught smallpox and died had they not been inoculated—as with regional, urban/rural, and age differences. In England, inoculation was largely confined to the towns and villages of the south and southeast, having limited impact on cities where, despite the London Smallpox Hospital and various dispensing charities, widespread inoculation was always controversial due to its potential to cause a full-blown epidemic.[71] Furthermore, children tended not to be inoculated by these charities yet were at high risk of catching smallpox, which was endemic to large cities. Indeed, the number of smallpox deaths in London actually rose during the period in which variolation was most widely practiced; as the number of adult cases was falling, this represents an appalling toll of children killed.[72] Despite the growing popularity of variolation (sufficient to necessitate its ban in 1840 in favor of the safer technique of vaccination) there exists general agreement that its effect on the size of the population in eighteenth-century Britain was negligible.[73]

Reducing Deaths III: Charitable Foundations

Charities also sought to produce a useful and numerous population, though their immediate objects—from reforming prostitutes to spreading knowledge of mouth-to-mouth resuscitation—differed widely.[74] The quintessential Georgian charity, the Foundling Hospital, exemplified this goal even before its establishment in 1739. In calling for such an institution earlier in the century, Joseph Addison and Daniel Defoe both decried the loss of children, which, "if we consider it only as it robs the Commonwealth of its full Number of Citizens, . . . certainly deserves the utmost Application and Wisdom of a People to prevent it."[75] Such an argument continued to be made during and long after the foundation of the hospital, even by those critical of its policies and constitution.[76]

Brainchild of the shipwright Thomas Coram, the Foundling Hospital was established at the end of the 1730s partly as a result of perceived demographic crisis, Britain's increasingly warlike footing, and broader changes in the culture of charitable giving.[77] Coram sought to provide desperate mothers with an alternative to the abandonment of their children or the questionable mercies of parish care. Those admitted to the hospital were nursed during their infancy in the healthier countryside, inoculated (if necessary) on their return to London, and educated to a trade, domestic service, or the sea. The goal, as stated in the petition that led directly to the hospital's royal charter, was "to enable [such children] . . . to become useful Members of the Common-Wealth."[78] Donors gave generously in its early years, undoubtedly due to a combination of charitable motivations, whereby more traditional religious and sentimental impulses fused with social utility and patriotism.

Indeed, a patriotism that insisted on socially useful charity in the service of the nation at war was at the heart of much mid-eighteenth-century philanthropy. This was no less true for the Foundling Hospital than for charities such as the Marine Society, which was established in 1756 to provide men for mercantile or naval service.[79] When Parliament voted in that year to fund the Foundling Hospital directly from the public purse in return for the automatic admission of all eligible children, commentators lost no time in associating the decision with the imminent outbreak of the Seven Years' War.[80] The linkage of charity and patriotism was also visible in broader culture. For example, the satirical print *A Skit on Britain* (1741), which otherwise concentrates on the misconduct of Britain's politicians during the War of Jenkins's Ear, contains two separate references to "foundlings," as if the new charity were a war measure (figs. 1, 1a, and 1b).[81]

But several factors limited the demographic impact of charities. First, donations were more difficult to obtain after the victory of 1763, which removed the immediacy and gravity of the French threat. At the same time, the discourse on population became increasingly pessimistic and the sheer number of charities competing for donations resulted in the dilution of efforts to effect social change.[82] Second, many charities could not operate on a sufficiently large scale. Some were certainly substantial: over 48,000 men and boys had received support from the Marine Society by 1798, and the Lying-In Charity (established 1757) provided obstetric assistance to women in their own homes and had helped almost 110,000 by 1791.[83] But the numbers for their nonmetropolitan counterparts are far less impressive. Less than a tenth of the number assisted by the London Lying-In Charity, for example, was helped by its Newcastle counterpart over a period lasting twice as long (9,000 in sixty-five years).[84]

Third, good intentions could backfire. The Foundling Hospital was overwhelmed during the period of general admission (1756–1760), when more than 300 infants per month—rather than the anticipated 500 per year—were admitted. Parish officials from as far away as Yorkshire sought to offload the financial burden of unwed mothers by pressuring them to surrender their children to the hospital. Infant mortality, already high at the institution, rocketed. By the end of the century, a total of 18,500 children had been admitted, but 65 percent of these had died in its care. Even those prepared to concede that a foundling hospital might boost population size, such as David Hume, reacted strongly to such figures and denounced general admission as "pernicious to the state."[85]

Finally, some charities had difficulty convincing the public of their moral and ethical legitimacy. This most obviously affected the Magdalen and Lock Hospitals, sanctuaries for penitent prostitutes and sufferers of venereal disease, but even the Foundling was attacked for its potentially deleterious effects on morals. Several authors claimed that it would encourage promiscuity by facilitating the disposal of unwanted offspring, not to mention undermining marriage and with it the prospects for an augmented population.[86] As Coram himself recognized, women were crucial moral arbiters in making the case for and against the hospital, and the diminution of female involvement in its affairs after 1750 made refuting the claim that it promoted immorality no easier.[87]

Hospitals, inoculation, and charities all had *some* effect in reducing mortality. But the scale of the challenge was enormous. An infant mortality rate of 65 percent at the Foundling was undoubtedly better than nearly 100 percent in parish workhouses, but the raising of approximately 6,500 children by 1800 was insignificant among a population of 8.9 million.[88] Various constraints, whether physical, financial, or cultural, unavoidable or self-imposed, ultimately prevented efforts to diminish mortality from exercising any significant influence on population size.

Managing Migration I: Preventing Emigration

If efforts to reduce mortality did not suffice to promote population growth, then the limited and piecemeal control of migration fared even worse. Attitudes to emigration roughly shadowed perceptions of population size: it was tolerated, even encouraged, in times of apparent overpopulation, turning to hostility and obstruction when underpopulation was the predominant concern.[89] But thanks to the consensus behind a large population, by far the most common attitude toward emigration was an anxiety and hostility that smoldered in the background, occasionally flaring up in moments of crisis.[90]

The early 1770s was just such a moment. Thanks to rack-renting, absenteeism, population pressure, and industrial distress, more than 9,000 migrants left Scotland and the north of England for America between December 1773 and September 1775. This wave of emigration produced a flurry of apocalyptic prognostications: in 1773 the *London Evening Post* estimated that at least half a million people had left the British Isles in the past seven years, and "Reflector," writing in the *General Evening Post*, even predicted that this population drain would obstruct efforts to subdue the American colonies should they rebel.[91] Blame for producing this exodus was directed at absentee landlords rather than the migrants themselves, but this had not always been the case. Royal

Fig. 1.1. "Ged Bilchharp" (George Bickham), *A Skit on Britain* (1741). BM 2423. Courtesy of the Lewis Walpole Library, Yale University.

Fig. 1.1a. Detail of figure 1.1.

proclamations of 1635 and 1637, for example, pointed the finger at emigrants both to foreign lands—who implicitly neglected "one of the principall duties of all Our Subjects, to attend at all times the service and defence of their King and native Countrey"—and even to the English colonies in America, the main reason for such migration apparently being "to live as much as they can without the reach of authority."[92]

Despite such hostility, the government was relatively powerless to prevent this outflow. Control of migration of the king's subjects fell within the purview of the Crown, but even at the height of monarchical prerogative power in the 1630s, the restrictions imposed on movement to New England were little more than an irritation. The emigration of skilled workers caused particular concern, resulting in an Act of 1719 that both criminalized the enticement of artisans to move overseas and stripped such workers of British nationality if they failed to return within six months of being recalled by the local British

Fig. 1.1b. Detail of figure 1.1.

ambassador. But it was evidently less than effective, as penalties were stiffened in a further Act of 1750. Serious doubts were expressed about the feasibility of any direct control over movement, especially within British dominions, and it was only once the American colonies had been declared in rebellion in August 1775 *and* it appeared that migrants might be supporting the uprising that emigration from Scotland to America was banned; even this was a controversial move. There had been earlier indirect efforts to discourage migration to the colonies, particularly by suspending grants of Crown land to settlers, but these were weak and of dubious efficacy. Indeed, the constant refrain of witnesses to the 1824 Commons Select Committee that ultimately recommended the repeal of the 1719 and 1750 acts was of their total inadequacy.[93]

Managing Migration II: Promoting Immigration

Attitudes to immigration were altogether more volatile and polarized. In general terms, Whigs and court figures (especially the Crown) looked favorably on immigration, whereas country and Tory politicians were opposed. In the absence of robust migration control, debate centered on "general naturalization" legislation that would make the procedure more widely accessible and do without expensive private acts of Parliament. Among its supporters were Bacon and Locke, whose arguments in favor of naturalization were recycled throughout the eighteenth century whenever such legislation was discussed in Parliament, which was the case at least twice per decade between the 1660s and 1690s, and again in 1709, 1746, 1747, and 1751.[94]

All these measures were fiercely contested, most failed to move beyond the Commons, and only one (1709) became law. Immigration was undoubtedly controversial, not least because such debates were always about more than simply permitting foreigners to reside in Britain. Deeply felt religious, economic, cultural, and political issues played a substantial part in the eighteenth-century immigration controversy.

Before 1688, immigration was inseparably tied to the question of religious toleration. From the earliest populationist tract in favor of encouraging migrants, Samuel Fortrey's *Englands Interest and Improvement* (1663), it was argued that only the toleration of other (Protestant) confessions would make England or its colonies attractive to foreigners. This argument was made repeatedly until the passage of the Toleration Act in 1689.[95] Religious toleration was to remain controversial for a long time, however, and the prospect of large numbers of foreign Protestants in Britain was unnerving for the many who remained convinced that religious pluralism bred political instability.[96]

Furthermore, religious and economic factors were manifestly intertwined. Initial sympathy for religious refugees, especially the Huguenots who fled France after the Revocation of the Edict of Nantes (1685), soon dissipated once they became an economically competitive community. Worse still were those migrants—particularly the "poor Palatines" after 1709 and Salzburg Protestants in the 1730s—whose economic profile was less impressive and who were perceived to be idle and reliant on charity. Many bemoaned "letting in Strangers to eat the Bread out of the Mouths of our own People" and wondered why poverty at home might not be solved before allowing foreigners into the country.[97] Others saw no harm and much good from economic competition: complacent domestic laborers would be forced, by necessity or shame, to work as hard as immigrants. Most agreed that full employment ought to be secured at home, either to attract migrants or to render immigration unnecessary by promoting domestic increase.[98]

It is easy to overstate anti-immigrant sentiment—as Daniel Statt has argued, outbursts of hostility left more archival traces than successful integration—but a generalized xenophobia was clearly part of the popular culture of eighteenth-century Britain, even if there did exist certain exceptions.[99] It could hardly be otherwise during an era in which a "British" identity was being constructed with reference to a wide range of "others."[100] But this xenophobia was at once both deep and paradoxical: it operated on an intra- as well as international scale, yet obvious targets, such as Britain's fledgling black and South Asian communities, seem not to have been singled out for hostile attention. On the other hand, ample evidence testifies to English hostility to Scots, Irish, or Welsh, and even individual towns and cities were unfriendly to outsiders and jealous of their privileges. Bristol, for example, was criticized by Daniel Defoe for its insularity, and a Bristol MP was the most notorious opponent of naturalization in the 1690s, who called—at the end of a vituperative speech in the Commons—for "the Serjeant to be commanded to open the Doors, and let us first kick the Bill out of the HOUSE, and then all Foreigners out of the KINGDOM."[101]

Efforts to encourage migration were thus restricted to the assistance of foreign Protestants and their resettlement in the colonies. Whether to bolster imperial security in Catholic or frontier zones—such as Ireland, Nova Scotia, or Georgia—or to enhance the imperial economy by developing autarky for naval stores (for example), such projects were as utilitarian as eighteenth-century philanthropy. But their failure proved the undoing of efforts to encourage immigration.[102]

The case in point is that of the "poor Palatines." Some 13,000 immigrants from all over southwest Germany poured into Britain over the course of 1709, lured by promises of free land and resettlement in America that had circulated in unauthorized propaganda. Hungry, dirty, and poor, these immigrants were initially encamped around London, but the pressure to find a permanent settlement intensified as ever more streamed off boats arriving from the Netherlands. A massive fundraising drive was initiated, raising over £20,000 in London alone, but generosity was nakedly political: Whigs gave handsomely, but Tories, who worked hard to stoke popular discontent that boiled over into petty acts of violence, gave as little as possible. Eventually some 3,000 migrants were settled throughout England with larger contingents going to Ireland and America.[103]

The Tory reaction was dictated by their intense opposition to the Act of 1709 that introduced a cheap and simple naturalization procedure, which consisted of taking the sacrament, paying a shilling, and swearing oaths of loyalty and against transubstantiation and Jacobitism. Swift spoke for many fellow Tories when he wrote that "some Persons, whom the Voice of the Nation authorises me to call her *Enemies*, taking Advantage of the general Naturalization Act, had invited over a great Number of Foreigners of all Religions, under the Name of *Palatines*; who understood no Trade or Handicraft, yet rather chose to Beg than Labour; who besides infesting our Streets, bred Contagious Diseases, by which we lost in *Natives*, thrice the Number of what we gain'd in *Foreigners*."[104]

The linkage of the naturalization Act and the Palatine crisis was largely unfair—the two were entirely disconnected, and immigrants did not even seek to naturalize until the Act was about to be repealed—but it did resonate with the public, especially when the settlements in Ireland and America failed. This was, in part, the fault of the Tories: having won the 1710 election, they were not prepared to provide the funding necessary to sustain the Palatine settlement in New York, so when its Whig governor, Robert Hunter, ran out first of credit and then of his own funds, the Palatines were left to fend for themselves. The Tories cemented the association of general naturalization and rampant immigration by repealing the Act in 1712, ensuring that it would always be associated with the Palatine debacle, and the question of general naturalization remained politically toxic for a generation.[105]

When revived in 1740, it was applied only to British America, and a bill to enact this quietly slipped through Parliament and onto the statute book.[106] But anything that even resembled a general naturalization measure within Britain was subject to extreme denunciation. The loudest clamor was against the 1753 Jewish Naturalization Act.[107] Ironically, this had nothing to do with general naturalization, as it merely dispensed with the requirement to take the sacrament that had been part of the naturalization process since 1609. Nobody was naturalized by the Act, something only possible by private act of Parliament. The 1740 statute had already dispensed with the sacramental test for Jews in America, so legislators in 1753 did not foresee the opposition that greeted the passage of the so-called Jew Bill into law.

Yet the clamor against it was so intense that it was repealed after only six months on the statute book. Most interpretations argue that the episode was conjured by opposition politicians seeking a *cri de guerre* before the 1754 election. Historians have also focused, quite rightly, on its manifestation of an antisemitism instrumental in the forging of "British" identity, and on religious elements of the controversy, which suggest both that the association of toleration and immigration persisted beyond 1689 and that theological considerations remained crucial in all sorts of political debates until well into the eighteenth century.[108]

Nevertheless, this was first and foremost a controversy about immigration. Electoral politics, religion, and identity were all issues that could only be highlighted once the alarm was sounded that all Jews would be naturalized, making Britain a magnet for the entire community of the Jewish faithful. It did not help that a general naturalization bill, in which there had been talk of including Jews, had only recently been defeated. Heavily laden with antisemitic stereotypes and xenophobic sentiments, tracts opposed to the Act were generally opposed to immigration of any kind. It is a measure of the extent to which hostility to foreign settlement dominated British political discourse that those in favor of the Act often felt compelled to rehearse arguments in favor of general naturalization, however foreign this was to the Act itself.

Matrimony and Reproduction

The only remaining means of securing population growth was to increase births, for which two things were necessary. Monogamous matrimony was to be encouraged and sexual activity regulated to promote productive and eradicate unproductive sex. This section explores the connection between marriage and reproduction, one so ubiquitous as scarcely to require restating. It was exemplified by another controversial statute of 1753, Hardwicke's (Marriage) Act, the debate around which also encapsulates many of the intellectual, social, and cultural trends discussed earlier.

Named after the Lord Chancellor who steered it through Parliament, this Act was intended to resolve the problem of clandestine marriages. Any marriage not conducted in accordance with the Anglican canons of 1604 could be deemed "clandestine," whether it differed in a single minor detail (such as the time of day it was celebrated) or in several of greater importance. The problem with clandestine marriages was not, as has been claimed, the persistence of traditions of informal marriage in which promises were deemed binding marriages, whether made in the present tense (*per verba de praesenti*) or in the future tense with subsequent consummation. Such promises constituted a contract but not a valid marriage. Rather, the problem with clandestinity was that such marriages could violate a sizable number of the canons of 1604 yet remain quite valid. Such marriages were easy to contract (especially at notorious "wedding shops," such as the Fleet prison and its environs), difficult to (dis)prove, and— as with regular marriages—almost impossible to dissolve. Those of propertied minors were particularly grievous, as a family fortune could be undone by youthful infatuation, intimidation, or intoxication.[109]

This problem had not gone unnoticed before 1753, but earlier efforts to legislate had all failed in the Commons, where there existed serious concern that action against clandestinity would detrimentally affect the entire population for the benefit of the landowning elite's narrow self-interest. Anxiety about the overweening power of the landed classes was also writ small, as there were similar fears that restrictions on the ability of minors to marry would place too much power in the hands of parents. All these complaints were made loudly and persistently throughout the debates on Hardwicke's Act.[110]

The controversy surrounding the Act both inside and outside Parliament also encapsulated contemporary attitudes to population, philanthropy, and immigration. All participants in the debate—whether for or against the Act— agreed that "the Wealth and Power of a Kingdom consists in the *Number* of its People."[111] But which "people" was apparently debatable: those in favor of the Act characterized their opponents as improvidently seeking population growth at any cost, heedless of couples' ability to maintain a household; they saw themselves, however, as more prudently seeking the increase only of "mature and valuable [i.e., economically productive] Members of Society."[112] In fact, apart from a few extremists, such as Reverend Alexander Keith (whose chapel conducted over 5,000 marriages between 1744 and 1749), all commentators be- lieved that "increasing the breed of the industrious and labouring sort of people" was best.[113] They even agreed which group was most deserving of en- couragement: sailors, as national security and economic prosperity depended on a powerful navy and sizable merchant marine. This truth was especially potent in the early 1750s, as Britain readied itself for imminent war against France. According to its opponents, the 1753 Act threatened Britain's security because its residence requirements prevented sailors (and other mobile sectors of the population) from marrying swiftly and reproducing.[114]

As we have seen in the case of the Marine Society, the production of more mariners was a central theme of contemporary philanthropy, alongside the pres- ervation especially of women and children so that they could be useful for the nation. These priorities were also reflected in the debates on Hardwicke's Act. Argument raged about whether women would benefit from the Act, which re- moved the (meager) protection provided by the ecclesiastical courts to women who had been seduced into sex by the promise of marriage. Children, too, might suffer if their parents had been married otherwise than in accordance with the statute: such marriages might now be annulled and offspring bastardized as a result. Notwithstanding the insistence of several scholars to the contrary, the judicial interpretation of the Act was never that strict, but this obviously was not known while the Bill was being debated and when it had but lately passed into law. Finally, a key plank of the public discussion of both the Act and sev- eral philanthropic enterprises was the assumption that the poor were funda- mentally improvident. Both sides of the parliamentary debate felt that the poor were apt to run precipitately and heedlessly into marriage.[115]

The controversy surrounding Hardwicke's Act, then, displays in microcosm the broader trends we have encountered: population growth as the ne plus ultra of enlightened social policy, to which end philanthropic activity was overwhelmingly dedicated. Hostility to immigrants, a ubiquitous theme in the political discourse of 1753, also featured strongly among those who fought with increasing desperation against the Marriage Act.[116]

But what the arguments over the legislation show especially clearly is the absolutely fundamental association of matrimony with reproduction. It was rarely—but by no means never—asserted that it was *physically* impossible to reproduce outside marriage, even if extramarital offspring were thought to be weaker and sicklier.[117] But Hardwicke's Act and the surrounding debate did echo the wider sentiment that marriage was by far the best means for producing children, for two crucial reasons. First, the stability of a social order based on the hereditary transmission of property was severely undermined by extramarital sex. Children born of unmarried parents were denied access to property, as only those born within wedlock were deemed legitimate; even the subsequent marriage of unmarried parents did not legitimize their offspring under English law.[118]

Supposedly worse were the results of adulterous sex. All children born within wedlock were legitimate in the eyes of the law—even those not fathered by the husband—so "spurious issue" could end up inheriting a family's property. Hence the stranglehold of the sexual double standard, which sacralized female chastity while excusing as venial an adulterous husband's dalliances.[119] The sexual incontinence of men was hardly inconsequential, but it did not pose the same existential threat as that of women: Samuel Johnson was not alone in reflecting on "what importance to society the chastity of women is. Upon that all the property in the world depends."[120] The Marriage Act sought (largely successfully) to put an end to the bastardization of offspring as a result of prior clandestine unions, and the parental powers it reinforced gave families greater security about the future disposal of their property.

Second, it was exclusively within marriage that the care and upbringing of children could be secured. Only those from a secure and stable marital household were likely to become productive members of society, notwithstanding the charitable projects discussed earlier. In other words, marital procreation was *productive reproduction*, and extramarital, *unproductive*.[121] For the handful who concentrated simplistically on the raw birth rate as the measure of population growth, unproductive reproduction might be satisfactory. But for the vast majority interested in the propagation of "laborious and industrious" people, the stability and material security of a marital unit was essential.[122] Thus, marriage was key to producing a useful population; the raw *quantity* of people was less important than their *quality*.[123]

These ideas were expressed repeatedly across the eighteenth century in a range of genres, especially sermons and novels. Religious texts asserted most strongly and consistently the connection between matrimony and productive reproduction. While many authors acknowledged that marriage also had nonreproductive ends, almost all agreed that these were secondary to the production of offspring. As Matthew Hole wrote in 1719, "The first and great End of marriage . . . was the Peopleing of the World, and the Propagation of Mankind."[124] Filling the ranks of God's elect required a good Christian education, and many interpretations of the most clearly pronatalist biblical verses focused on the upbringing rather than simple production of children.[125] Such was the common interpretation of 1 Tim. 2:15 (a woman "shall be saved by childbearing"): reading "childbearing" straightforwardly as childbirth was unacceptable because it would allow any mother, whatever her circumstances—married or unmarried—to be saved; far better was rendering "childbearing" as child-rearing, for which a marital context was essential. Part of the broader current of debate that emphasized quality alongside the quantity of offspring, the need for the proper upbringing and education of children was painted in stark terms. Authors warned that "a man without education, is worse and more dangerous than a brute" and that failing to educate a child was little better than turning a mad animal into the street.[126]

A heavy responsibility thus fell to the mother, who, in addition to her unfailing fidelity, was charged with the upbringing of her children.[127] By the end of the eighteenth century, such duties had become so entrenched that breastfeeding was obligatory and a mother was empowered to withhold even her sons from her husband, "'till she can resign them, flourishing like the rose-bud, and inviting the warmth of a superior cultivation to open them into a beautiful maturity."[128] This stress on the mother's vital role in educating and shaping children in their early years is a prominent feature of the ethnographic and racial discourses encountered in later chapters.

Religious texts participated fully in broader populationist discourse by highlighting the importance of reproductive matrimony for national power, wealth, and prestige. Authors drew extensively on other elements of populationism, such as the ancients versus moderns debate or the military demand for bodies that fueled philanthropic projects.[129] One clergyman evoked a remarkable scene as he preached before the governors of the London Lying-In Hospital in 1764: "See future armies striving in the womb, pressing to our standards, importuning you to suffer them to guard you in your present possessions, and to lay at your feet the spoils of new Quebecs, other Senegals and future Plassey's. How impolitick then must we be to grudge the necessary expence for recruiting our necessary defence?"[130] Indeed, one of the many ways in which marriage was thought honorable, besides being a divine and prelapsarian institution, was because it was socially useful.[131]

Procreation, whether in the service of spiritual or temporal goals, could never be the sole end of matrimony. Childless marriages were no less valid or important for clergymen, even those who asserted that the production of children was the highest priority of wedlock. Yet it was precisely this emphasis on childbirth that produced the impression that a childless marriage was somehow incomplete, an impression strengthened by repeated assertions that wedlock was the only context in which lawful reproduction could take place. However much it was stressed that marriages without offspring were just as valid or worthwhile, the production of children was an essential telos of matrimony.[132]

This was also the case in popular fiction, and several novels discussed the dangers of childlessness for the state of a marriage. Henry Fielding's *History of Tom Jones, A Foundling* (1749) was typical in stating that "matrimonial felicity" was poisoned by the absence of children ("pledges of love").[133] Lady Delacour's reproductive problems (her first two children were stillborn or died in infancy) in Maria Edgeworth's *Belinda* (1801) were symptomatic of her unhappy, frivolous, and penurious marriage.[134] For Mary Wollstonecraft, the state of childlessness could be potentially liberating: where there existed no children "to reward her for sacrificing her feelings," she claimed in *Maria* (1798), a woman ought to be able to divorce if her marriage was unhappy.[135]

Most novels to address the nature of matrimony adopted the obverse perspective and focused on the delights of children. Offspring are continually described as crowning the happiness of husbands, a joy superior even to marriage itself. The final sentence of Tobias Smollett's *Roderick Random* (1748), for example, has Roderick unable to go to London to secure the fortune of his wife, Narcissa, as "my dear angel has been qualmish of late, and begins to grow remarkably round in the waist; so that I cannot leave her in such an interesting situation, which I hope will produce something to crown my felicity."[136] And in *Pamela* (1740), Mr. B finally gives over his efforts to debauch the eponymous heroine and proposes genuine matrimony.[137] At the heart of this unlikely transformation lay reproductive concerns: he perceived that he could not legitimate any child he might have with her outside wedlock and decided to marry with "a view of perpetuating my happy prospects, and my family at the same time."[138]

The happy prospects of the nation were also invoked in fictional treatments of marriage and childbirth. "I was ever of opinion," wrote Oliver Goldsmith at the opening to *The Vicar of Wakefield* (1766), "that the honest man who married and brought up a large family, did more service than he who continued single, and only talked of population."[139] Another clergyman, Parson Yorick, agreed: "The procreation of children [is] as beneficial to the world . . . as the finding out the longitude."[140] Unsurprisingly, novels were often less conservative than religious texts on this issue, and praise of procreation occasionally went unaccompanied by marriage. Tom Jones, for example, in discovering the philosopher Mr. Square in flagrante delicto with Molly Seagrim, asked, "What can be more innocent than the indulgence of a natural appetite? Or what more laudable than the propagation of our species?"[141] A further example comes from the pen of Benjamin Franklin, whose literary hoax, "The Speech of Miss Polly Baker" (1747), defended extramarital reproduction, particularly in a colonial context: "Can it be a Crime," Baker asks the bench that has just found her guilty for the fifth time of having a bastard child, "to add to the Number of the King's Subjects, in a new Country that really wants People? I own I should think it rather a Praise worthy, than a Punishable Action."[142] She went on to proclaim bearing children even outside wedlock a duty, for which she deserved a statue erected in her honor.

Notwithstanding her brave stand, Polly acknowledged that she "must be stupid to the last Degree, not to prefer the honourable State of Wedlock, to the Condition I have lived in." Fortunately for her, one of the justices was so taken with her defense that he married her the next day.[143] However possible successful reproduction may have been outside marriage—and it was entirely owing to Polly's ability to provide for her children "by my own Industry, without burthening the Township" that her actions were even remotely defensible—the fact remained that matrimony was the only suitable context for child-rearing.[144]

Promoting Productive and Suppressing Unproductive Sex

The association of reproduction and marriage thus ran wide and deep in eighteenth-century British culture. To marry and multiply was to be more than merely obedient to the will of God; it was to act for the benefit of society at large. But a lingering issue was how best to promote productive sex and prevent unproductive heterosexual practices.

Encouraging Productive Sex

One obvious possibility was recourse to the law. Generally more effective in suppressing than encouraging behavior, the public policing of sexuality declined over the first half of the eighteenth century as consensual heterosexual activity increasingly became a private matter. Where sexual regulation persisted (and was in some cases strengthened), the harshest punishments were for anti-reproductive behaviors: sodomy, bestiality, abortion, and infanticide. Incompatibility with pronatalist imperatives was not the sole or even principal basis for their criminality, but it certainly reinforced the view that they were "unnatural" forms of behavior. Nor were irregular heterosexual activities suddenly perceived to be harmless just because they were no longer pursued at law. The masturbation panic of the early eighteenth century is a case in point, but so too were the repeated efforts to introduce statutory penalties for adultery. These bills (introduced in 1771, 1779, 1800, and 1809) reflected the widespread belief that adultery was a serious problem, but their failure to become law equally demonstrates that the time for legislating a solution to such newly private issues had passed.[145]

Legal means were also sought to encourage appropriate reproductive behavior. As "the Number of People is both Means and Motive to Industry," argued Bishop Berkeley in 1721, "it should therefore be of great use to encourage Propagation, by allowing some Reward or Privilege to those who have a certain number of Children."[146] Suggested or enacted measures did not, however, reward procreation so much as marriage (although it was suggested in 1762 that the state should reward parents of multiple births); both public discussion and proposed legislation manifested the principle that marriage and productive reproduction were functionally synonymous.[147] The 1690s saw the passage of legislation to tax bachelors and childless widowers. Although primarily a revenue measure in wartime, an act of 1695 gestured toward the sort of pronatalist measures that had been introduced by Colbert in France. But it was unsuccessful as a tax, complicated to administer, unfair in its distinctions (based on social rank rather than wealth), and it collapsed within a decade.[148]

Later proposals included a marriage loan, additional relief for poor households with several children, and a package of measures contained in a 1751 bill "for granting certain Privileges to poor married Persons, to promote Marriage, for the Increase of his Majesty's Subjects." This would have waived church fees for vital ceremonies (marriage, "churching" of the mother after childbirth, baptism, burial of married persons or their children), released servants from their contracts immediately upon marriage, provided affordable housing, and exempted married people from restrictions on erecting a cottage, from preemptive removal under the settlement laws, from eligibility requirements for exercising a trade, and from obligations to fill parish office or provide highway labor. This bill died in committee but had been prepared by two members of Parliament who fought tenaciously against restrictions on clandestine marriage, suggesting that opponents of Hardwicke's Act were genuine in their pronatalist convictions. Its provisions, especially concerning affordable housing, continued to be recommended in private as a means to boost population.[149]

In the absence of effective or lasting legal encouragements, it fell to wider culture to promote productive sex. In addition to sermons and novels, popular sex manuals and midwifery texts prioritized reproduction and its connection with matrimony. Sex guides such as *Aristotle's Master-piece* (1683) or Nicolas Venette's *Tableau of Conjugal Love* (1687; first English translation 1703) appeared in numerous editions across the eighteenth century and shared an explicitly pronatalist and pronuptial perspective.[150] The valorization of reproductive matrimony was also shared by sex therapists, such as the eccentric James Graham, who lectured widely and whose "celestial bed" supposedly cured impotence and was hired out for fifty guineas a night.[151] As Roy Porter memorably characterized the perspective of sex manuals, within their pages "sex was to obey a value-system and a scheme of activity that . . . prioritized a social economy of procreation."[152]

Midwifery textbooks and courses also betrayed a range of assumptions that situated procreative activity in a marital framework. Most often such texts refer to "men" and "women," but several indicators make it apparent that these are actually "husbands" and "wives." Sometimes the latter terms are explicitly used; at other times they are implicit from disparaging comments made about single mothers and fathers. The most dramatic evidence of the presumptive marital context of reproduction comes from comments made about "man-midwives." Eighteenth-century midwifery texts were written against the backdrop of an ongoing battle between these and female midwives for control of the birthing chamber. Several works written by men sought to convince readers of the propriety and professionalism of male accoucheurs, while those produced by women aimed to defend traditional midwifery. Many commentators—male and female—focused on the sexual threat posed by men-midwives, not so much to *women* in a vulnerable situation as to *wives*. A loud chorus in pamphlets and newspapers sang of the likelihood of adultery from predatory man-midwives and easily seduced wives, and even female authors spoke of the usurpation of a husband's privileges at the hands of the accoucheur.[153]

Discouraging Unproductive Sex

Many midwifery lectures and texts also addressed practical dangers to offspring of extramarital procreation, including congenital venereal disease and the likely weakness or sickliness of illegitimate children. These were weighty considerations that ultimately meant extramarital intercourse was never seriously entertained as a possibility for meeting pronatalist demands, notwithstanding their immense cultural power and the increasing toleration shown to heterosexual delinquency. For obvious reasons, adultery was out of the question and so far from being a serious consideration that at least one divine could jokingly suggest in print that its reproductive potential might palliate this sin.[154] Fornication got no longer shrift, and in contrast with comparatively relaxed attitudes in much popular literature, any attempt to defend it on religious grounds risked serious consequences, as the Scottish cleric Daniel MacLachlan discovered to his cost. His *Essay Upon Improving and Adding to the Strength of Great Britain and Ireland by Fornication* (1735) used a mixture of biblical exegesis, contemporary journalism, patriotism, and philanthropic sentiments to encourage purely procreative sex between single persons. Within weeks of its publication, MacLachlan was arrested and charged with obscene libel in the Court of King's Bench, excommunicated from the Church of Scotland, and—after unsuccessfully attempting to participate in the settlement of Georgia—apparently moved to Jamaica, where he died around 1745.[155]

The eighteenth-century rejection of birth control was in part a matter of logic: if the end of marriage was procreation, then to marry and confute this aim seemed to make a mockery of the institution and to insult God. The most common means of contraception in the eighteenth century was probably coitus interruptus, often discussed alongside masturbation as the "sin of Onan," a reference to spilling seed.[156] Widespread popular knowledge of abortifacients and emmenagogues (substances to promote menstruation) was more often used to terminate than prevent pregnancies, and condoms were rare, expensive, and used principally to avoid venereal disease.[157] Sterilization, although widely practiced on animals and a by-product of alien customs (such as castration to produce castrati or eunuchs), was almost unknown.[158] By the early nineteenth century, radicals sought to popularize the use of the genital sponge as a means of limiting family size.[159] Opposition, however, remained strong, as demonstrated when the young J. S. Mill was detained in a London jail for attempting to distribute pamphlets on contraception.[160] Such an effort, marginal even in the 1820s, was unthinkable a century earlier. Defoe waxed apoplectic about contraception in *Conjugal Lewdness* (1727), in which he accused married women who did not want children of being a short step from lunacy, stated that contraception was murder, and accused "She-murtherers" of calling on Satan for assistance when chemical remedies failed.[161] His utter failure to understand contraception within the context of a marriage demonstrates the power of the connection with procreation. Although characteristic of the seventeenth-century attitudes with which Defoe grew up, this linkage of contraception, abortion, and infidelity was—as the next chapter will show—to fuel planter anger at the alleged prevalence of such practices among slaves throughout the century.

Abortion[162] and infanticide were often conflated, in part because abortion did not exist as a separate statutory offence until 1803 but also because conventional morality considered them synonymous.[163] Indeed, neat divisions between contraception, abortion, and infanticide were impossible in a period when knowledge about conception was hazy.[164] The 1624 English statute against infanticide is perhaps the only in English legal history to presume an accused party guilty unless they could prove their innocence.[165] Within it, the dividing line between marriage and procreation is decidedly blurry: as several scholars have noted, infanticide was a single woman's crime, it being assumed that a married woman would never kill her child.[166] The statute's wording enforces this conclusion, being directed against "any Woman . . . [whose child] being born alive should by the Laws of this Realm be a Bastard."[167] It was reprinted in 1680 by order of London's magistrates and was to be read in church four times annually, suggesting a perception that the law needed restating.[168] Richard Steele certainly felt so when he complained in 1713 that many infanticides went undetected.[169] Furthermore, by the early eighteenth century, increasing numbers of infanticide suspects were deliberately acquitted where evidence was ambivalent or lacking: the law was unpopular among juries and the consequent swathe of acquittals made it so with lawyers and MPs.[170] As attitudes toward unmarried mothers (and other "lewd" women, such as prostitutes) within Britain softened after the first quarter of the eighteenth century,[171] pronatalist imperatives were channeled away from convicting infanticidal mothers toward providing the institutions—such as the Foundling and Magdalen Hospitals—that relieved women's perceived need to kill their children and thereby weaken the nation.[172]

Polygamy, which almost always meant polygyny, was a particular source of anxiety. It was a form of marriage that was common in the extra-European world, had biblical and later theological support, and seemed to be making advances at home among both the high—who frequently debated the topic (in England, the possibility of Henry VIII and Charles II marrying further wives was openly mooted)—and the low, whose bigamous relationships appear frequently in criminal records.[173] Patrick Delany's *Reflections on Polygamy* (1737) is a typical example of eighteenth-century views on multiple marriage, above all by attacking polygamy for its failure to produce offspring.[174] A common line of argument was to explore the marital customs of other societies, especially those that were polygamous, of which the archetype was "Turkey."[175] Polygamy was another supposedly prevalent feature of "savage" (and Islamic) societies across the world, and its African variant was thought to have been transported across the Atlantic to slave plantations in America.

Authors spent much time discussing the marital customs of other societies, simultaneously projecting European concepts onto the non-European world and then reading these back to prove their conformity with nature.[176] Polygamy was believed to produce fewer children than monogamy, so when Europeans visited parts of America and Africa that happened to be thinly populated, they assumed that natives were polygamous.[177] Polygamy was then used to explain the thinness of the population, in a circular reference typical of European racial thought. In his *Observations Concerning the Distinction of Ranks in Society* (1771), John Millar wrote that "the excesses of luxury have been productive of disorders almost equally pernicious [as polygamy]; and instead of introducing polygamy have carried the common prostitution of the women to a height which is not more favourable to the multiplication of the species."[178] Polygamy was therefore *the* benchmark by which perniciousness was measured, and its tendency to undermine pronatalist goals apparently spoke for itself.

Dissenting voices were few, but the most notable was that of Martin Madan, chaplain of the Lock Hospital, who provoked a storm of controversy with his *Thelyphthora* (1780).[179] Madan sought to eliminate prostitution by stating that the marriage bond stemmed from coitus, and men who had sexual intercourse with more than one woman were thus married to each and every one. Madan argued that "wherever, there are the most *married women*, there the increase of the people will be the greatest. *Polygamy*, therefore, as tending to increase the number of *married women*, must certainly tend to *population*."[180] His engagement with questions of population simply amounted to arguing against the current that supposed polygamy to be antiprogenitive.[181] In doing so, he left intact the linkage of marriage and fecundity: polygamy was merely legitimized as a form of marriage.

The story told by this chapter is fundamentally one of continuity. Within a landscape formed by the desideratum of a large population, there occurred gradual tectonic shifts of attitudes to British population size, which progressed from fear, first of overpopulation then of underpopulation from the mid-seventeenth to the mid-eighteenth century, to pessimism and finally Malthusian catastrophism after 1800. But there were also several sudden jolts along the way. One was, of course, Malthus himself, although his ideas must be set against the backdrop of increasing pessimism after 1750 and his own populationist convictions, which were largely ignored by both his opponents and posterity. Another seismic event was the Palatine debacle, which destroyed a fledgling general naturalization policy before many could take advantage of it.

The major discursive earthquakes occurred in the 1750s, a decade that began with a moral panic fueled by two actual tremors that took place in London a month apart. Not merely did the discussion of population take a more pessimistic turn in these years but significant legislation was proposed or passed: yet another general naturalization bill was followed by the crisis of the Jewish Naturalization Act, while the effort to introduce a census came to naught, as did the even less popular attempt to encourage marriage among the poor. Above all, Hardwicke's Act shook marriage to its legal foundations and provided an excellent opportunity to see contemporary attitudes to matrimony and reproduction in the white heat of the legislative process. These attitudes led to the introduction of general admission to the Foundling Hospital, a move that was quickly reversed after overwhelming the charity, and to the incentive to widen access to smallpox inoculation, pioneered by the Suttons during these years.

To meet the demand for physical bodies—a demand that generally grew but that waxed and waned in intensity—eighteenth-century British culture laid great stress on reproduction and its only legitimate institutional form, monogamous marriage. That this remained the case even where marriage could have socially dangerous consequences is the story of the next chapter.

The Limits of Pronatalism

Slavery and Population in the British Caribbean

In 1813 whooping cough hit Waltham, a sugar estate on the northwest coast of Grenada. As is typical with the illness, children were most severely affected and several died over the year. On November 16, the plantation manager, John Fairbairn, wrote to Waltham's absentee proprietor, George Home of Wedderburn, with further bad news:

> I have Meet with a sever loss yesterday a young woman was Brought to Bed of a fine Boy about a Month ago but Never could Recover strength from all the care of the Dr and Myself she Died yesterday a Most unfortunate think [thing] as I am much affraid of the child allso, though I have put it with another Good Woman and have promised her a Good presant if she Brings it up untill its 12 Monthes old and Dewaring [during] that time she is not to be askt for any Worke Worke [*sic*] from the Estate provided the child is keept clean and Louks as well as her Oune—the poor young woman I now talk of was abut 4 years old when I came to the Estate and this is hir second child and as she was Lickly to be a Breeder it Makes the Loss the more Sevear.[1]

Fairbairn goes on to express relief that this woman, named Phillis, was the only valuable slave to have died that year. Phillis's value derived, in part, from the fact that she was an adult female presumably able to undertake field labor. But it also owed much to the fact that she was likely to reproduce. The importance of new enslaved offspring, particularly after the abolition of the slave trade, which came into effect in 1808, is indicated by the care taken of Phillis's child. His adoptive mother was immediately freed from work and would remain so as long as the child was kept clean and as healthy as her own. Had the boy survived (he did not), she was also to have been rewarded with a present.[2]

The complex of issues suggested in Fairbairn's description of this unfortunate case is the subject of this chapter. It will examine reproduction in the discourses of slavery, abolition, and emancipation from the 1780s to the 1830s, to provide an imperial counterpart to the analysis of chapter 1, which established the central importance of population and its natural increase to metropolitan Britain. Natural increase was seen as a sign of a healthy society, and debates over the future of slavery made extensive use of this "principle of population." Both sides agreed that a population's propensity to increase naturally was an indicator of its happiness, and the reverse of its misery. As it became increasingly clear, in the first quarter of the nineteenth century, that West Indian chattel slavery was a demographic disaster, proslavery authors had an increasingly difficult task in reconciling their arguments with the principle of population. A few, realizing this, sought to eschew questions of population altogether.

But anti-slavery campaigners would not let them, and they attacked planters for their inattention toward slave reproduction. This inattention was perhaps understandable prior to 1808, given that it was widely perceived in the Caribbean to be cheaper and easier to buy replacement slaves from Africa than to promote their increase on estates. In North America and in Britain, the movement away from the slave trade and toward promoting reproduction began before the American Revolution. In the one context, prohibiting this profitable branch of British trade served revolutionary ends, whereas in the other it marked the beginning of the broader public sympathy for abolition that was to blossom into a nationwide movement only after the loss of America in 1783. It was because of the growth of organized abolitionism that planters in the Caribbean themselves began to pay at least lip service to reproduction. This was part of a strategy to pass unenforced (and, in many cases, unenforceable) legislation to combat anti-slavery arguments that claimed planters were uninterested in ameliorating the conditions of their slaves. But the general lack of enthusiasm for these measures was clear from the disdain with which resident planters greeted the initiatives even of their own representatives in London. By the eve of emancipation, slaveholders—now realizing that the end of slavery was nigh—began even to repeal pronatalist measures that they had so ostentatiously passed just a few years before.

The question nevertheless remains: why were planters so obtusely opposed to promoting the reproduction of their enslaved workforce? Securing natural increase would have served to neutralize both the damage of abolition and further emancipationist agitation by proving that slaves were well treated. Some saw it this way and worked to encourage their slaves to produce children. Most scholars who have discussed these efforts have concentrated on the unwillingness of planters to make the sacrifices necessary to produce an environment conducive to successful childbearing and -rearing, on their long-standing view that prioritized production over reproduction, and on the resistance of slaves to schemes designed to reduce them to breeding stock.[3] In addition, as this chapter will argue, these initiatives were hopelessly inadequate and unable to act as realistic and achievable incentives, even without opposition from the enslaved population. Besides, economic pressures worked to convince many planters that pronatalism was a low priority. Beyond these limitations there existed two key reasons for slaveowner indifference: racial prejudice and the dangers of Christian matrimony. Racial stereotypes of Africans produced a range of views about slaves that emphasized their ignorance, thoughtlessness, and incorrigible immorality, all of which worked to counteract their supposed natural hyperfertility. Women bore the brunt of these stereotypes, which were also held by many anti-slavery activists, although male slaves were often included. Christian matrimony posed a particular danger to the planter order, not least because it would disrupt the sexual predation of white men on enslaved women. But it also held out the legal threat to admit emancipation through the back door, a threat that carried over into the religious realm. Sympathetic clergy were keen to state that admitting the enslaved to Christian matrimony would leave the rights of planters to their labor unaffected, but planters had always been suspicious of acknowledging slaves as their coreligionists. In addition, the education and literacy deemed fundamental to the adult baptism—and ultimately marriage—of Protestants posed a mortal danger to the plantocracy that its members were unwilling to countenance.

Using a range of primary materials, ranging from plantation records, diaries, and parliamentary papers to pro- and anti-slavery speeches, newspaper articles, and pamphlets, this chapter will situate Fairbairn's comments to his employer within the broader discourse of which they were a part. Most of this material relates directly to Jamaica, as the largest and most important of Britain's Caribbean possessions, but we will revisit Waltham throughout the chapter, not least because the issues discussed below were common to almost the entire region. Indeed, the offer of rewards for thoughtful child-rearing and the valuation of slaves based on their reproductive potential, when tied to the disastrous demographic record of slavery at a time when the institution itself was in a period of existential crisis, helps to indicate the continuities and discontinuities in discourses of reproduction in Britain and its empire.

Slavery and the Principle of Population

The most prominent element linking the discussion of population in metropolitan Britain to that concerning its slave colonies was the "principle of population." This was unsurprisingly of central importance in the abolitionist debate from its first stirrings in the late seventeenth century to the final triumph of emancipation in 1833. As discussed in chapter 1, both "mercantilist" and "free trade" thinkers, before and after the watershed of Smith's *Wealth of Nations* (1776), agreed that the size of a nation's population determined its wealth and power.[4] Particularly important for the debate on slavery was their agreement that a population's propensity to increase naturally was an indicator of its happiness: decrease was a quantifiable sign of the devastation of chattel servitude.[5]

The linkage of population decrease to slavery was of long standing and by no means limited to the abolitionist struggle. It stretches back to the Bible and appears in the thought of such varied figures as Saint Augustine and Jean Bodin.[6] Slavery also featured heavily in the wider population debate during the Enlightenment, not least because this debate was closely related to the "ancients versus moderns" controversy and slavery was a prominent feature of ancient Greek and Roman societies. Montesquieu's argument that tyrannical and despotic governments were the worst for promoting population growth was taken up in the debate between David Hume and Robert Wallace. Although differing on whether ancient or modern nations were more populous, both agreed that slavery (unless unusually mild) exercised a deleterious effect on the birth rate.[7]

Hume and Wallace were not alone in agreeing on the importance of the principle of population: both pro- and anti-slavery activists were united in their belief that a growing population was a happy one. As early as 1786 a group of planters asserted in print that "the increase is the only test of the care with which [slaves] are treated," and one of their representatives in Parliament, George Hibbert, stated in 1807 that he found this principle "to be tolerably exact."[8] Almost a quarter of a century later, Thomas Fowell Buxton, parliamentary leader of the abolitionists after 1823, pointed to this unanimity when he said that "it is a doctrine admitted *by all parties* that, under all circumstances, except those of extreme misery, population must increase."[9] Other anti-slavery campaigners, such as James Stephen and Zachary Macaulay, used the same argument, and by the eve of emancipation, Buxton had made the principle the sole authority in questions of slavery: as he baldly stated before a House of Lords Select Committee in July 1832, "I take it to be the best of all Tests of the Condition of the Negro; I have always considered it so."[10] Consequently, whenever Buxton used arguments based on population in the Commons, the West India interest, spearheaded at this point by the agent for Jamaica, William Burge, raised several objections on points of detail but never questioned this fundamental assumption.[11]

Opponents and defenders of slavery agreed not merely on the principle of population but also that slavery itself was not inherently opposed to population growth. Proslavery authors claimed that the evidence of abysmal treatment, which began to saturate the public domain from the later 1780s, was confined to a handful of bad planters and that "where the slaves are humanely treated and duly taken care of, there is generally an increase."[12] The abolitionist heavyweight Thomas Clarkson, writing around 1815, agreed that following abolition, "the more humane & the more provident [planters] would continue to treat their slaves in something like a rational & prudent, and others in a different manner; and that of course, while there would be, generally speaking, an Increase of slaves on the Estates of the former, there would be a Decrease on those of the latter."[13] His opinion was not unique among abolitionists. The compatibility of slavery and population growth was part of the explicit logic of abolition, as William Wilberforce retrospectively explained: "We trusted, that by compelling the planters to depend wholly on native increase for the supply of their gangs, they would be forced to improve the condition of their slaves."[14] Henry Brougham, later Lord Chancellor in the government that passed the Slavery Abolition Act (1833), went further and expected "an immense natural increase" following abolition, even if the treatment of slaves remained the same.[15] After the success of 1807, abolitionists campaigned for measures intended to reinforce the abolition statute and fulfill its promise to ameliorate the treatment of slaves. Orders-in-council and parliamentary resolutions were passed to provide a pattern of amelioration in Crown colonies (such as Trinidad) that others, with their own semi-independent legislatures, could follow. They also attempted to prevent any contraband trade in slaves that might subvert the aims of abolition, by recommending the passage of slave registration acts. Abolitionists continued to trust in the efficacy of abolition to improve conditions for slaves until 1823, by which time registration returns had begun to show that slave populations were continuing to fall and that colonists were doing nothing effective to arrest the decline.[16]

Nevertheless, although abolition had failed to secure better treatment and natural increase, the consensus behind the principle of population remained unshaken. It was almost entirely unaffected by Malthus's *Essay*, which—as shown in chapter 1—demolished the mercantilist notion that a large population was *invariably* desirable and undid a century and a half of demographic logic in the process.[17] The *Essay* itself studiously avoided modern slavery (and argued that ancient slavery had only a limited impact on population), as did its author until he discovered that proslavery campaigners were using it to support their views. An abolitionist by sympathy, Malthus responded by adding an anti-slavery appendix to the 1806 edition and approached Wilberforce to clarify his position just before a critical debate in the Commons. Wilberforce used this conversation to good effect in the House when countering Hibbert's efforts to co-opt the *Essay* for the West India interest.[18] It is a measure both of the stature of Malthus and of the importance of population for all sides in the slavery debate that Hibbert, undeterred, continued to use the *Essay* to support his arguments.[19]

This overwhelming and bipartisan support for the principle of population forced the defenders of slavery to do battle on demographic terrain. By agreeing that a decreasing population was one in "extreme misery," proslavery campaigners were placed in an impossible situation. Any denial that slave numbers were falling became untenable given the statistical data available via slave registration from the later 1810s. They were thus forced to attribute any "misery" to causes either extrinsic to slavery or that reforms short of emancipation might solve. Occasionally, more perceptive defenders of slavery attempted to tackle the fundamental principle head-on, doing so with growing sophistication as the threat to the institution became increasingly mortal. Soon after 1807, individuals such as Gilbert Mathison used the principle of population to argue against abolition, by asserting that it had directly caused a decrease in the slave population.[20] This was echoed on the plantation itself, as medics such as Thomas Duncan of Grenada made the rather implausible suggestion that the outbreak of diseases associated with despondency, such as pica, or "mal d'estomac," was caused by slaves despairing "because none of their Countrymen are now imported? if so, it may be an Argument for, what I fear never will be re[-]permitted, the Guinea Trade."[21]

Such efforts to blame abolition for any decrease were particularly crude, as they lay open to the counterargument that the cessation of slave imports had merely exposed the previously hidden cruelty of planters. Consequently, the attempt was made to undo the strong connection between population size and misery. Such was the stratagem employed by an anonymous "West Indian" who wrote in the *Jamaica Journal* in the mid-1820s and whose widely quoted essays were collected and republished in 1826. One of the most sophisticated commentators on population, the "West Indian" argued in favor of the principle at the same time as questioning its fitness for the purposes to which it had been put in the slavery debate: there are "few subjects . . . more important than the laws which govern the increase and decrease of mankind," yet "increase in numbers is a *presumption*, not a *proof, of happiness*."[22] The "West Indian" enhanced the scientific credentials of this argument through the judicious use of Malthusian theory. Indeed, stating that "we know the faculty of mankind to increase in a *geometrical* or *compound* ratio," he argued that decrease for a short time following abolition was inevitable, but that "the negroes would descend to 300,000 before they reach the period of permanent increase."[23] The power of this argument lay in the simultaneous acknowledgment of the importance of population increase and denial of its significatory power; it came the closest to successfully undermining population as axiomatic for the debate on slavery. But it failed, and by the 1830s, as statistics proving declining numbers poured into Britain from its Caribbean colonies, proslavery authors recognized the futility of arguing against emancipation on the basis of population. In 1833, one of their number, William Burnley, attempted to salvage the proslavery position by meekly claiming that the argument was not yet decided and that it would be better to drop population from the debate entirely.[24]

A Pronatalist Plantocracy?

Emancipationists could not have disagreed more. For them, as we have seen, population decrease was an immensely powerful argument against the continuation of slavery. Of all the slave colonies in the British Caribbean, only one (Barbados) achieved natural increase before the passage of the 1833 Abolition Act. The statistics from slave registration returns showed that, between 1810 and 1830, the slave population of the region decreased by over 80,000 (just over 10 percent).[25] One question that puzzled contemporaries, and which still exercises historians, is why the Caribbean had such a disastrous demographic record when other slave societies, such as the antebellum United States, had become self-sustaining. Indeed, such was already the case in some North American colonies in the first third of the eighteenth century, and certainly by 1774, when Virginians felt sufficiently secure to call repeatedly for the banning of the slave trade.[26] In general, it appears that the inability of most Caribbean slave populations to increase naturally was due to the exorbitant labor demands associated with heavy sugar monoculture in most of Britain's colonies. Where sugar production had already peaked due to the age and completeness of a settlement, as in Barbados, population had the potential to increase. But the expansion in production elsewhere (prompted, in part, by a fall in the price of sugar in the 1820s) prevented the vast majority of slave populations from becoming self-sustaining, even at a time when planters might have been expected to make this goal a priority.[27]

Planters were surprisingly inattentive to this aim, undoubtedly due to an established culture of buying replacement slaves rather than encouraging the reproduction of their own. The desire to promote slave reproduction had been a hallmark of the early phases of slavery in the Caribbean but gave way in every colony to a culture of "buy not breed" as systems of chattel slavery matured.[28] The ostensible basis for this attitude was the lost labor and expense of an expectant or newly delivered mother and the lengthy period before which her child would become productive.[29] Still, planters who operated in this manner recognized that the attitude was somewhat unpalatable. Witnesses opposed to abolition before the House of Commons Select Committee on the Slave Trade in 1790 uniformly denied that this culture existed; more than one attributed to a single anonymous foreigner the view that it was best to work slaves as hard as possible and then purchase replacements.[30] They stated that it was simply impossible to keep up the numbers of slaves without imports, that planters recognized that it was in their interest to "breed" rather than "buy," and that they earnestly attempted to encourage their enslaved workforce to reproduce, but that, for a range of reasons (discussed below), these efforts were unsuccessful.[31] Other witnesses, less hostile to abolition, described the wide extent of the belief that it was better to purchase than to encourage reproduction, and they were able to cite particular conversations in support.[32]

The dominance of the attitude of "buy not breed" can be demonstrated by the instructions offered by Henry Drax, a Barbadian planter, to his manager in 1679. These instructions circulated widely in manuscript and were printed in 1755. Thomas Thistlewood in Jamaica, for example, read and copied them in 1758.[33] As Drax stated, "to supply the place of those that shall be deseased or Dy you will wantt a yearly Recrute of 10 or 15[. O]r itt may be if by Any Contagious distemper theire Happen a greatt mortality . . . twenty or mor . . . [L]ett all that you buy be Choyce Young Negros Who will be fitt for plantt serwice."[34] He encouraged the purchase of "Coromantee" or Gold Coast slaves as particularly fit for plantation service, and he viewed the deaths of slaves as attritional, as the depreciation of an asset, which must be remedied by the regular purchase of replacements. So, too, did the Tudway family, whose estate in Antigua—Parham—was both indebted and contained a decreasing number of slaves in the early 1720s. Despite the plantation's financial difficulties, proposals to increase the enslaved population centered on purchasing replacements rather than encouraging reproduction, although the adult sex ratio was admittedly heavily skewed toward women, who outnumbered men by almost two-to-one.[35]

Although the culture of purchasing was nowhere dethroned until abolition, the last third of the eighteenth century saw greater attention paid to slave reproduction.[36] This was motivated by a range of factors. Where wartime disruptions to the slave trade combined with slave mortality, plantation officials—such as Parham's overseer in 1759—hoped for natural increase but did little to encourage it.[37] In mainland America, such attention served revolutionary ends. The knowledge that the enslaved population had already achieved self-sufficiency formed the comfortable backdrop to Virginian demands to end slave imports in 1774, which were part of a general nonimportation policy against Britain.[38] Indeed, although the Constitution prohibited Congress from legislating to ban the slave trade until 1808, several delegates at the Constitutional Convention pointed to the redundancy of the trade given the reproductive record of slaves in America.[39] But there appeared no need explicitly to encourage reproduction in the US, even in the face of the abolition of the slave trade and the changed context of antebellum slavery, where the "peculiar institution" had expanded massively due to the demand for cotton. Even here, historians of slavery who have sought evidence of the systematic breeding of slaves have been largely disappointed.[40]

Within Britain itself, reproduction played a central part in early plans for the abolition of slavery and the slave trade. In 1772, the first published British scheme for gradual emancipation recommended the establishment of a colony of free black settlers in Florida. On the assumption that free laborers are more productive, and hence cheaper, than slaves, the flourishing colony would eventually lead to the destruction of slavery in others, forced to make their laborers free in order to compete with their more successful southern neighbor. The inhabitants of this colony were to be imported as children and educated in England until the age of 16. Numbers were limited to keep the plan affordable and attractive, but this mattered little given the proverbial hyperfertility of black people (a racial trope discussed in greater detail below), which would populate the colony in a short time.[41]

Another example is provided by Edmund Burke's model slave code, originally written in 1780 but sent to the Home Secretary, Dundas, in 1792.[42] Burke worked from the principle that it was only by reforming slavery that demand for slaves would be eradicated. Childbirth and marriage are at the heart of his regulations, even if he focuses on the latter as "a principal means of forming men to a fitness for freedom, and to become good Citizens."[43] Marriage itself is defined in reference to procreation: all slaves who have cohabited for at least twelve months "and have a child or children, shall be deemed to all intents and purposes to be married."[44] Burke's code even provides for the forcible marriage of enslaved men to women of reproductive age and for their punishment should they refuse a "competent marriage."[45] Children are central to subsequent regulations concerning the relaxation of labor requirements and compulsory manumission. Every enslaved male over the age of thirty who has served for ten years, is married, and has two children from the marriage is entitled to every Saturday free from labor; once he reaches the age of thirty-seven, he is entitled to every Friday too. But slaves only become eligible for compulsory manumission once they have a third child.[46] Burke may have approached the problem of slavery from the opposite end to the abolitionists, focusing on reforming slavery to reduce demand rather than cutting off supply to impose reform, but both he and they placed reproduction at the heart of their schemes.

In the Caribbean, reproduction became a matter of public attention because of the strident challenge posed by abolitionism. The impact of this challenge is clear from the example of Jamaica. Four separate "Consolidated Slave Acts" were passed by its Assembly, in 1781, 1787, 1788, and 1792. The aim of this legislation was to compile all relevant statutes regarding the management of slaves into one single act, and their treatment of reproduction is telling. The 1781 act, which lapsed in 1784, says nothing about it.[47] A revised act was passed in 1787, by which time metropolitan abolitionist sentiment was increasingly audible in Jamaica. Still, the only attention paid to population and reproduction in this statute was a provision for an annual statement of increase or decrease to be made, along with the reasons for any decrease.[48] The 1788 act followed in quick succession, thanks to louder and more extreme allegations from abolitionists, who were now winning parliamentary time.[49] This improved on the measure of 1787 by offering a tax reduction of twenty shillings per birth for all births in a given year if an estate's enslaved population was deemed to have increased naturally. This tax reduction was to be immediately passed on to the overseer of the plantation as a reward.[50] This statute lapsed at the end of 1791, when the danger from abolition was still profound. It now commanded almost ninety votes in the Commons, and defenders of slavery could not yet rely on the linkage of reform to Jacobinism forged by the excesses of the French Revolution.[51] Consequently, provisions for promoting natural increase became even more extensive in the 1792 act. The overseer's reward per birth was increased to three pounds but was now payable only for the surplus of births over deaths in a population that was naturally increasing. But perhaps because rewarding only the overseer undermined the claim that planters were already benign and gentle, a new clause was introduced exempting enslaved mothers from all work if they had six children alive on the estate.[52] Thus, from saying absolutely nothing on the subject in 1781, Jamaican slave legislation ultimately testified to an escalating concern for reproduction. And, as one witness to the Commons Select Committee testified in March 1791, these changes were directly caused by the growing clamor for abolition.[53]

This legislative activity had its counterpart on the other side of the Atlantic. MPs representing the West India interest, such as Charles Ellis, were keen to popularize the image of planters as caring and attentive to their slaves.[54] Ellis therefore tabled a motion in 1797 to encourage the colonies to adopt pronatalist measures similar to Burke's. Not only were the proposals similar but so was their aim: to eliminate the demand for slave imports, which would therefore allow the trade to disappear without serious consequences for the plantation economies of the Caribbean. Insisting that any legislation ought to be introduced and enforced by the island legislatures themselves, Ellis suggested specific regulations. Among these was the encouragement of reproduction, "by giving annual rewards to the women, in proportion to the number of their children, and by establishing distinctions, such as are likely to be the most gratifying to their vanity, and bestowing on them all those who have families, who are industrious, or, from any other reason, have a claim to peculiar favor—by similar marks of degradation to the profligate of either sex—by holding out encouragement of various kinds, for the purpose of ensuring the strictest attention to the health of the children during the first periods of their infancy."[55]

These proposals were based on a set of racial stereotypes about Africans and those of African descent, to be considered in the following section. But more remarkable than these stereotypes was the indifference of planters even to the proposals of their own representatives in London. Following the success of Ellis's motion, the Duke of Portland, as home secretary, wrote a circular letter to the governors of Caribbean colonies, asking them to place the resolution before their local legislatures and to encourage the passing of measures to ameliorate the condition of slaves.[56] No reply was forthcoming from Jamaica, Barbados, or Dominica, so Portland wrote again in 1798, this time forwarding suggestions and a list of questions that would help to focus ameliorative legislation. These questions were put before the Assembly of Jamaica on November 5, 1799. The governor of Jamaica, the Earl of Balcarres, reported two months later "that this Island in general does not seem to me disposed to reply to the suggestions proposed by Your Grace . . . nor to comply with the orders given to the Governors of the West India islands." By March 1800, the situation was unchanged, and the only information that was forwarded to London consisted of birth and death returns by parish, revealingly thought to be "defective."[57]

The lack of attention paid even to the initiatives of their own MPs and legislatures was typical. Notwithstanding the growing attention paid to reproduction in colonial legislation, many planters showed at best supreme indifference and at worst active hostility to implementing it. Several witnesses to parliamentary committees in the 1790s and 1830s acknowledged that observance of such laws was limited.[58] By the late 1820s, when the demographic situation was steadily worsening in all colonies except Barbados, the Jamaican Assembly actually repealed pronatalist measures. The 1816 Consolidated Slave Act awarded three pounds to be equally divided between the mother, midwife, and nurse of any child surviving at the time of the annual return of births and deaths, if a slave population is deemed to have increased naturally.[59] This was repealed in 1827 ostensibly because, as one commentator explained, "it practically proved to be a source of jealousy and ill-will amongst the slaves on a property; the mother who had lost her child before the age when the premium was received, envying her more fortunate fellow servant, and in many cases inducing abortions."[60] The notion that an enslaved mother would be so violently jealous as to induce a miscarriage[61] was an element of the pervasive racial prejudice that both explains the failure of efforts to promote the increase of population and masks the real reason for the repeal of the 1792 law: fiscal retrenchment.[62]

Resisting Pronatalism

Why were planters so indifferent or hostile to promoting natural increase among their slaves, especially once officially sanctioned imports ended in 1808 and future supplies would be clandestine or via an inter- or intracolonial trade—that is, insufficient? It is important to note that some planters did actively encourage their slaves to reproduce, both by implementing the pronatalist measures that were increasingly a feature of colonial legislation and through their own initiatives.[63] Reproductive capacity could influence the valuation of slaves: in 1788, two Jamaican doctors sought to free a young enslaved male who had been trained as an apothecary and offered "a female Slave who may breed, money, or any eligible mode of setting the poor boy free."[64] Medical writers frequently made recommendations for the promotion of slave reproduction, and the topic was of clear concern to several proprietors.[65] George Home, for example, wrote repeatedly to Fairbairn, to express his concern about the decline in the numbers of his slaves. The slave population of Waltham fell from 186 in 1817 to 156 in 1834, a drop of over 16 percent, which corresponded almost exactly with the overall decline in Grenada's slave population.[66] Fairbairn explained that he did all he could to encourage reproduction: "I am well aware that on this Estate Everry incuragement has been Given to the few Breeding Wiman on it both as a Reward for themselves and to see if it wold not make others wishfull to be Mothers allso."[67] One of these encouragements was the exemption from hard labor of mothers who had six children then alive on the estate. This measure was made law in Grenada in 1798 in direct response to Ellis's Commons motion of the previous year.[68] In 1804, one of the Waltham enslaved, Madeleine, was listed as exempt from hard labor on the basis of having six children alive on the estate.[69]

But the solitary example of Madeleine demonstrates the greatest flaw in ameliorative and pronatalist measures. They were simply hopelessly inadequate. So few slaves had six surviving children that this provision, even where implemented, could not hope to be a realistic encouragement to reproduce.[70] (This did not stop it being misused as a tax break by some Jamaican planters, who, as Sasha Turner has shown, placed orphaned slave children with an adoptive mother unable to work from age or infirmity and then collected the tax relief intended to compensate owners for freeing prolific mothers from labor.)[71] The efforts of Matthew Lewis to promote natural increase on his Jamaican estate were even less effective. He established a reward system whereby the midwife and mother of a baby (if it survived to fourteen days) each received a dollar (roughly 5–6 shillings), in addition to their usual allowance of clothes and provisions. Lewis "also gave each mother a present of a scarlet girdle with a silver medal in the center, telling her always to wear it on feasts and holidays, when it should entitle her to marks of peculiar respect and attention." The more children a mother had, the more medals were to be placed on the girdle, such that precedence went to women with the most. Lewis was aware that this order of maternal nobility was somewhat fanciful, but he allowed himself to be persuaded by his manager that it was a sensible system. When he returned to his estate two years later to find barely a dozen births from over three hundred slaves, he was bitterly disappointed.[72]

Evidence is sadly lacking to decide whether Lewis's slaves were unable or unwilling to respond to his "encouragement." But his experience was typical, as many of his fellow planters found their pronatalist schemes frustrated by opposition from their slaves. Enslaved women resisted efforts to compel them to give birth in plantation hospitals or to stipulate when they should wean their infants; mothers frequently absconded or negotiated with their white masters to secure a degree of parental autonomy. Other measures to improve the perinatal treatment of mothers, such as reducing their workload or working hours, providing new clothing or clean linen, organizing nursery care, or offering food or cash incentives, were patchily implemented even where mandated by legislation. Planters sought to incur a minimum of expense or disruption to production, maintaining a long-established perspective that viewed enslaved women first and foremost as units of production and only secondarily as prospective mothers.[73]

Resistance by the enslaved to pronatalist initiatives began long before pregnancies came to term. Slaves knew of contraceptive plants and practices, such as delaying the weaning of their infants. This was frequently cited by planters and doctors as a deliberate contraceptive technique, and slaveowners often offered rewards for weaning within a certain number of months (usually 12).[74] Evidence of infanticide is much rarer, but abortion appears quite frequently.[75] In July 1767, three months to the day after raping her, Thomas Thistlewood recorded that Mountain Lucy, one of his slaves, took contrayerva (*Dorstenia contrayerva*) to induce a miscarriage, a practice that was reputedly common.[76] In October 1804, another of Waltham's enslaved women, Jeanne (a creole) was listed as "convalescent from miscarriage—which she herself brought."[77] Somewhat unusually, the prevalence of abortion formed the centerpiece of the testimony of the medic Sir Michael Clare before the House of Lords in June 1832. More typical, however, was the reason he offered for enslaved women aborting pregnancies: pregnancy would impede their amorous pleasure.[78] Clare was asked specifically whether he might not infer, "from the Frequency of Attempts to procure Abortion, a Disinclination on the Part of the Mother to bear Children to an Inheritance of Slavery?" He responded, "I do not think they care about it; they care to get rid of the Child, that they may have no more trouble with it."[79] Janet Schaw and Edward Bancroft, both writing about interracial sex in the 1760s and 1770s, agreed: "As even a mulattoe child interrupts their pleasures and is troublesome, they have certain herbs and medicines, that free them from such an incumbrance."[80] We know very little about slave women's reasons for terminating their pregnancies, and the comments by Bancroft, Clare, and Schaw are singularly unenlightening, except as an insight into their own attitudes and those of the broader class of planters.

Beyond the obstacles to natural increase imposed by the ineffectiveness and impracticality of pronatalist measures or active slave resistance, there are several reasons why slaveholders were indifferent or hostile to a goal that they nevertheless recognized was redolent of good plantation management. Perhaps most obvious, and certainly the most studied, is the economic motive, which took two forms: pessimistic and optimistic. For those who chose to do so, it was possible to see a series of economic problems that justified working slaves as hard as possible, discouraging activities (such as reproduction) that led to increased costs and the loss or suspension of labor, and buying replacements where available. Before 1807, the American war and lingering hostility thereafter, natural disasters (hurricanes), and the wars against France disrupted production and intermittently closed markets in Europe and America for essential supplies and colonial produce. After abolition this pessimism was fed by the disturbances of wartime, which produced competitors both internal and external to the British Empire (such as Trinidad and Cuba, respectively), the emergence of apparently more ethical sources of sugar in India (which was the supplier of choice for those who boycotted slave-grown sugar), and a serious fall in the price of sugar in the 1820s (to its lowest since the mid-eighteenth century). Those who subscribed to the pessimistic view worked an ever-decreasing number of slaves hard to produce an ever-greater volume of sugar in the hopes of turning a profit. As slaves succumbed through exhaustion and maltreatment, their population fell, resulting in a vicious cycle of increasing labor and decreasing numbers.[81]

As a wide and growing field of scholarship has proven, slavery and the slave trade were not in decline at the turn of the nineteenth century.[82] Those who refused to be bowed by the developments just described held a more optimistic view of their economic future. Confident that they could continue to make large profits on their plantations, such individuals were happy to continue to follow accepted practice and buy their slaves rather than rock the boat with initiatives to encourage reproduction. Thus, whichever perspective one cared to take on the economic state of the Caribbean colonies—whether teetering on the brink of disaster or looking forward to a bright and prosperous future—encouraging reproduction was in few planters' economic interest, at least before 1807.[83]

But planter apathy and ineptitude survived abolition, for two other reasons that have been less well studied. Their general hostility to slave marriage will be discussed in the following section. But just as important was the existence of a complex of racialized assumptions about black women. For one thing, they were assumed by all sides to be naturally hyperfertile, much like Africa itself.[84] Lewis reflected this attitude when he wrote in his diary, "I really believe that the negresses can produce children at pleasure; and where they are barren, it is just as hens will frequently not lay eggs on ship-board, because they do not like their situation."[85] This was accompanied by other stereotypes asserting that black women could give birth with ease and quickly resume work as uncomplicatedly as livestock.[86] It is therefore unsurprising that, given this assumption of hyperfertility, it was black women themselves who were most vociferously blamed for their failure to reproduce sustainably.

Plantation managers offered many reasons for the lack of natural increase, including absentee proprietorship, the unhealthy situation of certain plantations, the prevalence of (infant) disease, and the unbalanced sex ratio. But by far the most common and persistent were the supposed promiscuity and polygamy of slaves.[87] The physician and poet James Grainger stated bluntly in 1764 that "black women are not so prolific as the white inhabitants, because they are less chaste."[88] This explanation was thriving fifty years later, when Fairbairn wrote to Home: "You justly observe that there is something in a state of slavery that prevents the Natural incriss [increase]. I can Give one and I am convinced the Greatest Reason, which is the too Early, but above all, the promiscuous intercourse between Male and Female, which we cannot prevent."[89] Promiscuity and polygamy were frequently seen to act in conjunction to prevent population growth, as in the testimony of several witnesses to the 1832 House of Lords committee or the writings of proslavery Captain George Henderson.[90] As the doctor John Williamson wrote of Jamaican slaves in 1817, "polygamy is one of the most prominent and injurious circumstances connected with their present situation. Increasing population is incompatible with such a condition in any country; and other evils, emanated and diffused by this source, may be comprehended."[91] Polygamy, long associated with low rates of population growth and attributed to Africans as a racialized characteristic, was a common feature of explanations of the failure of slaves to reproduce.[92] There was, however, occasional flexibility on this point where dictated by tactical necessity: Edward Long, for example, argued that the slave trade did not depopulate Africa partly because of widespread and productive polygamy.[93]

Both sides in the debate on slavery roundly condemned sexual incontinence, differing only in their views on its ultimate cause. Proslavery writers placed the blame squarely with slaves themselves whereas abolitionists and emancipationists indicted the system of slavery. This difference was often more apparent than real, however, and even anti-slavery authors could write in terms that blamed the slave. In 1825, the planter and geologist Henry de la Beche wrote of his difficulty in convincing slaves to marry: "Their African ideas lead them to prefer polygamy and promiscuous intercourse to marriage, as it exists in Christian countries."[94] In like fashion, an anti-slavery pamphlet of 1830 stated clearly that "the ordinance of marriage is scarcely known among [slaves]; while the most unrestrained licentiousness and profligacy of manners, as well in their intercourse with each other, as with the Whites, is indulged and encouraged."[95] The broad agreement on the existence, extent, and resultant degradation of slave immorality inexorably led to the slave being blamed, even if their behavior was partly attributed to the poor moral example offered by whites.[96] The tendency of anti-slavery argument, to find fault squarely with the system of slavery rather than with the slaves themselves, was more of an ideal than a consistently held position.

Blame did not stop at the factors that might prevent conception, for even once this had taken place, planters were apt to accuse fathers, mothers, and midwives of mismanagement resulting in miscarriage, stillbirth, or neonatal death. Women, in their capacity as mothers or midwives, bore the brunt of such criticism. Enslaved mothers were widely believed to be more careless of their offspring than their white counterparts.[97] Among other things, planters complained of mothers exposing their children to cold (by going on night walks to a partner on another plantation), of their not doing so (by omitting to plunge them in cold water, believed to protect against tetanus), of slave huts being too smoky, and of general negligence.[98] Lewis told of a slave mother who received a reward for ensuring the health of her child until it was ten days old, at which point she went to a dance on a neighboring estate, abandoning the child, which subsequently perished.[99] The emphasis in this anecdote is on the unnatural behavior of the mother, whose care for her infant was a result merely of avarice. In keeping with the broader view of black people as intellectually backward and bound to their sensory impressions—a perspective that fed into the bestialization of Africans— music and dance were seen to cast a dazzling spell on the average slave.[100] Both this and her avariciousness were based on powerful racial assumptions.

Midwives fared little better, as attitudes toward them were also based on existing prejudices against female midwives in Europe.[101] Seen as superstitious, ignorant, and dangerous, the position of the female midwife was even more precarious in slave societies, where her reputation depended greatly on her skin color.[102] European notions of the sexual looseness of midwives, a stock-in-trade for eighteenth-century erotic writing, blurred with stereotypes of African immorality to produce, even in abolitionist writings, the viewpoint that enslaved midwives were procuresses.[103] Worst of all were African midwives, who were supposedly tied to "barbarous" traditions, as Lady Nugent's description of her own lying-in in 1802 made clear: "The old black nurse brought a cargo of herbs, and wished to try various charms, to expedite the birth of the child, and told me so many stories of pinching and tying women to the bed-post, to hasten matters, that sometimes, in spite of my agony, I could not help laughing, and, at others, I was really in a fright, for fear she would try some of her experiments upon me. But the maids took all her herbs from her, and made her remove all the smoking apparatus she had prepared for my benefit."[104]

In offering his testimony on abortion before the House of Lords, Clare directly blamed midwives and cited one case of an African midwife administering abortifacient herbs.[105] Far better than African midwives were black creole women, and better still were women of color: according to Mathison, cases of lockjaw were unknown where the midwife was an "intelligent Creole, who, by living in families of white people, had learned a cleanly judicious method of managing infants."[106] Male midwives were best of all, even if they were of color. For example, Jonathan Troup, a Scottish medic in Dominica, wrote positively about Joseph, a French mulatto man-midwife who delivered quadruplets in February 1789. Troup spent some time recording the details of the case in his diary and unsuccessfully attempted to convince Joseph to write an account of the delivery. Troup's curiosity was almost entirely self-interested: the authority of more senior doctors rankled with him, and he sought to make a name for himself in Britain by communicating the details of the case to scientific celebrities such as Sir Joseph Banks. Nonetheless, his attempt to hijack Joseph's professional success bespeaks an esteem for the man-midwife absent in all other discussions of plantation midwifery.[107]

This example of a mixed-race man-midwife also highlights the ambivalent position of men in discourses of slave reproduction. Doctors and planters appeared as the providers of medical care and childcare facilities demanded by ameliorationists or as the selfish individuals withholding these in the name of profit, an accusation some colonial doctors did not scruple to level at planters.[108] The image of fathers in these discourses was largely drawn on racial lines, as demonstrated by the example of John Stewart. His *Account of Jamaica* (1808) argued both that African fathers sold their own children to European slave traders and that white fathers took paternal care of their children by slaves.[109] In other words, African fathers sold their children into slavery where white fathers released theirs from it. White men were attacked for their moral laxity, but even some abolitionists claimed that this was due to immorality spreading from the slaves to their masters.[110] Attitudes to white men were fundamentally equivocal, praising and damning the planters and their staff in equal measure. The only significant role given to male slaves in the discourse of reproduction, however, was as a treacherous paterfamilias.

Disrupting Marriage

Planter hostility to encouraging reproduction was also based on the principle established in chapter 1: that successful procreation depended on monogamous Christian matrimony. The implicit connection between monogamous marriage and procreation was so strong and ubiquitous throughout the discourse on slave reproduction that commentators seldom felt the need to explicitly state that one entailed the other. Wilberforce, for example, was astonished in 1823 that planters had not encouraged marriage among their slaves, "because the owners of slaves are powerfully called upon by self-interest, no less than by religion and humanity, to make the attempt to promote it."[111] A generation earlier, Sir George Young, a naval officer, was explicitly asked, "How do you know that due attention is not paid by the planters to the rearing of children?" He just as explicitly answered, "I found no encouragement given to the blacks to marry."[112] Fairbairn claimed to offer such encouragement, blaming the failure of his pronatalist measures on the slaves themselves: "I have often told some of the most stedey of our Negroes bothe the men and the weman that if they wold Marrey and Live together as man and wife, that I wold pay the parson for doing so and that I wold help then to make a Good House and do many things for them that wold not do for them otherwise, the answer constantly is that Marriage is only fit for white pipole not for slaves."[113] He was not alone in attributing the absence of monogamous matrimony to the opposition of slaves.[114] Few, however, recognized that the hostility of enslaved women to Christian matrimony may have been partly due to the loss of independence they would suffer under coverture.[115]

As discussed here and in chapter 1, polygamous marriage was perceived to be detrimental to the production of offspring, and slaves were assumed to be incorrigibly polygamous and promiscuous. The consequences of the lack of monogamy were drawn by Thomas Cooper in 1824: "It is, doubtless, in part owing to this cause, and to the universal profligacy of manners prevailing among Blacks and Whites, that the Negroes in Jamaica are a very unprolific race."[116] Why, therefore, did planters not at least try to encourage monogamous marriage? Some did, and many authors offered schemes and sober reasoning to convince the rest.[117] Burke's slave code provided for forcible marriage, for example, and the author of *A Plan for the Abolition of Slavery* (1828) suggested placing a tax on single slaves to encourage owners to pair their slaves off.[118] Although premiums were only generally awarded in the case of natural increase, a few commentators, such as Captain Henderson, suggested that rewards should be payable on marriage and that planters should establish a household for each married couple.[119]

But in general, slaveowners opposed the monogamous Christian matrimony of their slaves in both practical and theoretical terms. Other than refusing their consent to slaves who wished to get married, practical opposition came from the massive disruption of slaves' relationships caused by white sexual predation on the enslaved community.[120] Thistlewood, for example, meticulously recorded in his diary over 3,850 separate acts of sexual intercourse with almost 140 women during his time in the Caribbean (1750–1786). The traumatization of enslaved females at living under a regime of systematized rape is suggested by Thistlewood's comments in his diary entry for September 27, 1753: "Last night Sus[ana]h piss the bed again, makes 3 times, will bear no more."[121] Even proslavery fiction did not shy away from portraying the sexual activity of white men and enslaved women, although it suggested that the latter enjoyed a greater measure of control over their lives and persons than was ever the case in reality.[122] Thistlewood himself often paid his sexual partners, which may have been an attempt to convince himself that a mutual, consensual exchange was taking place. Whether coercive or not—and noncoercive sexual relations between slaves and their white masters must have been vanishingly rare—it was universally acknowledged that sexual relations between these groups took place on a scale sufficiently large to prompt remedial legislation and to act as a focus of anti-slavery propaganda, especially after 1823.[123]

More theoretical opposition to slave marriage came from legal discourses, as Katherine Paugh's work has made clear. Various legal cases in Britain amply demonstrated the principle that slavery and marriage were incompatible, a view that was held by most clergymen in the Caribbean, either explicitly or in practice.[124] The Anglican Church was the only that could legally solemnize marriages, and its clergy were sympathetic, if not warmly enthusiastic, to the interest of their white slaveholding congregations.[125] Parish registers in Barbados, for example, contain no slave marriages before they were explicitly permitted by legislation in 1826, and in Jamaica over a fourteen year period (1808–1822) only 3,600 slave marriages were solemnized among an enslaved population of 330,000.[126] Legal officers of the Bahamas stated in 1816 that slavery and marriage were incompatible because of the legal status of the slave: as *res* rather than (however inferior) *persona*, the slave could not form a contract, such as marriage. Were slaves to obtain the consent of their master to form such a contract, their admission to the category of personhood would ensue, thus forming a case for their emancipation. As such, abolitionists were as ready as slaveowners to agree on the incompatibility of marriage and slavery, but with obviously different ends in sight.[127]

Even disregarding the issue of possible freedom, marriage and slavery were difficult to reconcile because the absolute power of masters over slaves competed with that of husbands over wives. Spouses could be sold to distant plantations—even at the height of amelioration and despite the possibility that some slaves opted for a church-sanctioned marriage to prevent such separations—making the maintenance of family life almost impossible.[128] (Slaveowners, though, recognized that they would then have to contend with the inevitable resentment and absenteeism this caused, so their freedom of action was, to the limited extent dictated by pragmatism, constrained.[129]) In slave societies themselves, the legal power of enslaved husbands was nonexistent, and that of free husbands to enslaved wives so circumscribed as to be practically so.[130] In Britain, however, the situation was different. Legal efforts against slavery here began with assisting runaway slaves to prevent their owners removing them to the Caribbean. On occasion these runaways were married, and, as shown in Paugh's analysis of the case of *Hylas v. Newton* (1768), the rights of husbands over their enslaved wives were respected over and above those held by a slaveowner. Indeed, the mere fact of marriage to a British spouse was used as an argument to demonstrate the incompatibility of slavery and marriage, as holding the husband (as breadwinner) in slavery would ipso facto enslave his British wife and family.[131]

Ironically, given these legal moves in Britain, government legal authorities responded to the Bahamian argument of 1816 by stating that marriage and slavery were perfectly compatible in the colonies and that planters had no right to prevent their slaves from marrying. The exchange had been prompted by a letter from a local clergyman, John Stephen, who was keen to Christianize and marry slaves. He was not alone: upon translation to the See of London, which position included ecclesiastical responsibility for the West Indies, Beilby Porteus wrote a *Letter to the Clergy of the West India Islands* (1788). He sought to encourage the instruction of slaves in the principles of morality and religion, and advocated the establishment of schools to do so. He attempted to allay the fears of planters by pointing to the fact that Christianity teaches "servants, [to] be obedient to them that are your masters."[132] With such safeguards in place, it would conceivably have been sensible for planters to allow their enslaved workforce the consolations of religion and marriage in order to combat one of the anti-slavery camp's strongest arguments: that planters had done nothing to civilize or Christianize their slaves.[133]

But they did not do so for one further, and curiously overlooked, reason. For slaves to be legitimately married as Christians would have entailed their baptism. Notwithstanding the need to demonstrate that they were doing all they could to civilize their slaves, baptism was a step too far for many planters. For one thing, if a slave was a baptized Christian, the traditional justification for their inability to give evidence in court—that they could not understand the gravity of a sacred oath to tell the truth—was no longer valid. Furthermore, the interior logic of Christianity, which preached the equality of all men before

God, was dangerous to inculcate in an enslaved population. But to these broader Christian issues may be added another unique to Protestantism, which possibly helps to explain, if not the higher incidence of slave marriage in the Catholic colonies, then certainly the greater legal protections for their slaves' right to marry.[134] Protestant baptism entailed access to God's word in print. The right and duty of the individual believer to contemplate scripture was of paramount importance in Protestant confessions, and clergymen generally coupled their demands for slave baptism and marriage with those for slave education, especially after *c.* 1810.[135] Planters, on the other hand, looked with horror at any proposal to introduce literacy among their slaves, which clearly irked the missionaries of Moravian Church. In a statement to the Privy Council in 1789, they made clear that they had taught basic literacy to native children in America and Greenland, "but the Situation of the Children of the Slaves in the West Indies puts it out of the Power of the Missionaries to act herein according to their Wishes."[136] Opposition to literacy also existed within Britain, where many members of the elite feared that it would lead to revolutionary sentiments, as in the 1640s.[137] This opposition, both at home and in the colonies, meant that Porteus's proposals fell on deaf ears.

In 1808 he tried again, confident that the passage of abolition and the consequent need to encourage slave reproduction would force planters to look more favorably on the issue of the marriage and thus education of the enslaved. He explicitly argued that slaves would be taught only to read, not to write, thus obviating the possibility of clandestine communications between slaves that could potentially instigate rebellion. His confidence may have been bolstered by the appearance in the 1790s of proslavery texts that also recommended marriage, baptism, and the teaching of reading.[138] Porteus made the further claim that reading was a preventive against revolution. Indeed, mischievous publications "*will* find their way among them; and if they are not able to read themselves, they may and certainly will hear them read by others; and then being incapable of reading any thing in confutation of them, they of course receive them as undoubted facts, and are thus easily and fatally imposed upon by wicked and designing men. Whereas if they are capable of reading what is alleged on the other side of the question, they may and probably will escape the snare that is laid for them."[139]

Porteus blamed the French Revolution and Irish Rebellion of 1798 on the gullibility of the illiterate and praised the English peasantry for its generally good literacy, such that Painite doctrines held the field for only a short while. Even this impassioned defense of his scheme for slave education was unsuccessful, as planters never lost their suspicion of reading or indeed of slave Christianity, which was implicated in the rebellions of 1816, 1823, and 1831–1832.[140] Thus, on the basis of their own sexual opportunities, of the legal danger of inadvertently freeing slaves by simply giving them permission to marry, and of the threat posed by a religious education that entailed literacy, planters were unremittingly hostile to the single greatest measure that they and abolitionists both believed would produce natural increase.

As debates over slave registration and the future of slavery itself were heating up in the aftermath of the Napoleonic wars, George Home found himself increasingly exasperated. The octogenarian responded to Fairbairn's explanation of the inability of the Waltham enslaved to increase in number (because they were incorrigibly immoral) with the hope that Caribbean legislatures would do something soon:

> Hitherto the Proprietors of Slaves have paid little attention to these Improvements and while the Slave trade continued, and the manners of the old were preserved and perpetuated by new Importations all such improvements were Impracticable. The case is different now and the Legislators of the different Islands cannot in my opinion apply their Labours so beneficially for their Country or themselves as in Endeavouring to form a body of Laws to promote the Instruction and improvement of the Negroes in the knowledge and practice of Morality and Religion, they may still continue in a State of Slavery greatly improved no doubt, but still subject to their masters in the same way as their peasantry still are to their Landlords in many parts of Europe.[141]

He thoroughly agreed with the assessment provided by Fairbairn and argued, as did so many abolitionists, that moral instruction of slaves was necessary. In addition, he sided with the government legal authorities of 1816 and clergymen, such as Porteus, who asserted the compatibility of slavery and marriage. Ultimately, however, the stubborn refusal to view slave relationships as valid ultimately proved a large part of the undoing of slavery. Home's own recommendation bears this out: "I have heard it suggested that much good might be done by passing a law in the different Islands to authorise the Proprietor of a Female Slave to purchase her husband at a valuation by Publick valuators named by the Assembly, The Transfer being always made with the Consent of the husband, but this I am afraid might be looked upon under our free Government as too great an incroachment upon private property."[142] Given the tiny proportion of slaves married by Christian rites, surely the only that would have been viewed as legitimately married under this proposal, Home's suggestion was yet another initiative that was characterized by hopeless impracticability. But it does point to the importance of marriage for efforts to promote slave reproduction. And this, as chapters 1 and 6 of this study also demonstrate, was part of the wider nexus of ideas making the production of new beings a central concern in the culture and policy of Britain and its wider empire in the long eighteenth century.

PART TWO

THEORIES OF RACE AND REPRODUCTION, 1600–1850

CHAPTER THREE

Gentes *and Genitals*

Sex in Enlightenment Racial Theory

During the twilight hours snatched from the numerous calls on his time, Charles White, a man-midwife from Manchester, put the finishing touches to his own amateur discussion of racial diversity. Published in 1799, White's *Account of the Regular Gradation in Man* was a forceful restatement of the older notion of a "great chain of being," in which all natural phenomena, living and inert, were arranged hierarchically with humans at the top. He drew the inevitable conclusion for a racial scientist of his era—that Europeans were at the pinnacle of creation—but his final paragraph was nevertheless revealing of the sexual reasons for drawing this conclusion:

> Ascending the line of gradation, we come at last to the white European; who being most removed from the brute creation, may, on that account, be considered as the most beautiful of the human race. No one will doubt his superiority in intellectual powers; and I believe it will be found that his capacity is naturally superior also to that of every other man. Where shall we find, unless in the European, that nobly arched head, containing such a quantity of brain . . . ? Where that variety of features, and fulness of expression; those long, flowing, graceful ringlets; that majestic beard, those rosy cheeks and coral lips? Where that erect posture of the body and noble gait? In what other quarter of the globe shall we find the blush that

overspreads the soft features of the beautiful women of Europe, that emblem of modesty, of delicate feelings, and of sense? Where that nice expression of the amiable and softer passions in the countenance; and that general elegance of features and complexion? Where, except on the bosom of the European women, two such plump and snowy white hemispheres, tipt with vermillion?[1]

Alongside such commonplaces of naturalistic and racial thought as the equation of a large brain and "intellectual powers," or the identification of an erect posture with humanity, White referred to explicitly sexual characteristics: beards, breasts, and nipples. The use of sexual features in texts on human varieties was hardly rare and shadows the discussion of human diversity in reproductive and embryological theories that we will encounter in chapter 4. But White's rhapsody in praise of European beauty and majesty, which reaches a crescendo in its euphemistic paean to breasts, suggests a deeper significance to the presence of sexual features in discussions of human diversity.

Indeed, enlightened writers on human difference were fascinated by sex, which permeated both the conceptual apparatus of such categories as "species" and "race" and the source materials on which these concepts were based. Dealing with each of these in turn, this chapter uncovers the depth of influence exercised by sex and reproduction on all aspects of racial theory. This marks a significant departure from existing work on racial thought in the early modern era, even where it explores race's links with sex and gender. Scholars since Foucault have recognized the close and important ties between these categories, but a sustained examination of their mutual influence, even under the guise of "biopower," has gone largely unattempted. The closest most researchers have come is to interpose "metaphor" or "analogy" as a mechanism governing their association, ultimately isolating sex and race from each other as they sought to explain a bilateral relationship.[2]

The reality was, however, far messier, and the following pages explore at length the conceptual and practical dimensions of sex's involvement in racial thought. The first section below addresses the concept of "species," a category whose very meaning was defined by sexual reproduction. Indeed, it is arguably the case that "race" owed its power, if not its existence, to the overreliance of species on sex for its definition. Substantial limitations to the sexual definition of "species" became increasingly clear over the course of the eighteenth century, driving leading naturalists toward other means of defining "species" that weakened interfertility as the guarantor of human unity and strengthened the case for multiple human species and origins.

This ultimately led to the emergence of modern, scientific concepts of race. The process by which these emerged is charted in the second section, which explores key areas of the discourse on human difference—the monogenist-polygenist debate; scriptural ideas about difference; climate and its problems; and efforts to move beyond climatic approaches toward more sociological and biological models—to highlight the importance of the conceptual and physiological dimensions of sex and reproduction. The place of heredity and sexuality in these developments will be examined alongside the persistent influence of issues relating to broader questions of speciation, such as degeneration and domestication.

The final section examines sex as a racial characteristic—that is, a pervasive datum used in theories and discussions of human difference. This involves physical sexual features and physiological processes as exemplified by White's raptures on European beauty, but it also incorporated considerations of social, cultural, and religious factors, which were critical registers for the discourse on human diversity. The treatment of sexual customs, rites, and traditions was an important counterpart to the analysis of physical features and helped to forge connections between the body and mental or moral characteristics. Considering polygamy, promiscuity, and virginity makes it clear that sex was a powerful lens through which human moral and intellectual diversity was brought into focus before becoming associated with physical features, such as breasts, genitalia, beards, parturition, and menstruation.

Species: From Logic to Biology

The part played by "species" in accounting for the variety of the world's peoples, flora, and fauna was seldom straightforward. Not only are different, often competing definitions of "species" possible but there were two major registers in which the term was used before the nineteenth century, one logical, the other biological.[3] The current dominance of the biological usage of "species" is a development particularly associated with the Swedish botanist Linnaeus (Carl von Linné, 1707–1778). It was he who developed both the Latin binomial nomenclature and the associated hierarchy of kingdoms, classes, genera, and other categories. At the heart of Linnaean botany was an emphasis on the sexual reproduction of plants that scandalized the scientific community and led to a rich seam of botanically themed erotica. Furthermore, as Londa Schiebinger has shown, Linnaeus's choice of mammary glands as the defining characteristic of mammals (over and above the other five features they all share) was directly related to eighteenth-century gender politics, particularly the Europe-wide campaign against wet nursing that was an integral part of general population-ist policy. Sex was, therefore, at the heart of the conceptual apparatus of the modern life sciences at their very inception.[4]

This should come as no surprise. Indeed, the concept of species has been associated with reproduction at least since Aristotle. Long before the emergence of a purely biological notion of species, Aristotle explained that reproduction takes place to make "such a class [*genōs*] of things as animals . . . eternal in the only way possible. Now it is impossible for [that which comes into being] to be eternal as an individual . . . but it is possible for it as a species. This is why there is always a class of men and animals and plants."[5] The Epicurean Lucretius, in *The Nature of Things* (*c.* 50 BC), was even more explicit in framing a generative conception of species: "Nothing can be created from nothing. . . . For if things came out of nothing, all kinds of things could be produced from all things, nothing would want a seed. But manifestly none of these things takes place, since all things grow, little by little, as is proper, from a fixed seed, and in growing preserve their kind."[6]

The generative tradition continued into the Middle Ages and Renaissance, as figures such as Frederick II (1194–1250), Albertus Magnus, Nicholas of Cusa, Marsilio Ficino, Francis Bacon, and William Harvey all discussed species along-side reproduction. Bacon, for example, spoke of experiments to create new "kinds" in his utopian *New Atlantis* (1627): "We find means to make commix-tures and copulations of different kinds; which have produced many new kinds, and them not barren, as the general opinion is."[7] In Bacon's Bensalem, not merely were genuinely new "kinds" created—a possibility that had been hotly debated but was generally deprecated by orthodox Christians—but their fertil-ity was noteworthy and, as we shall see, dangerous.

Meanwhile, the logical usage of "species" remained current until well into the nineteenth century. Its use as a logical category also dates from antiquity and, in the most general terms, signifies a group of individuals, itself part of a broader genus: "The individual man belongs in a species, man, and animal is a genus of the species."[8] All species and genera contain characteristics (differen-tia) that distinguish them from other species and genera. For example, the spe-cies "table" might have differentia such as "flat-surfaced," "four-legged," and so on, to distinguish it from other species within the genus "furniture."

The use of these terms by Linnaeus and other naturalists to describe the natu-ral world neither immediately nor entirely replaced their application as logical categories, which continued to feature even in scientific texts. This makes for a sometimes confusing diversity in the meaning of "species," as both logical and biological uses might be present in the same work. Take, for example, the *Lectures on Physiology, Zoology, and the Natural History of Man* (1819) by the physician Sir William Lawrence (1783–1867).[9] Early in this text Lawrence defined species in an exclusively biological sense, "as a collection of all the individuals which have descended one from the other, or from common parents, and of all those which resemble them as much as they resemble each other."[10] When later discussing the mixed offspring of "Europeans" and "Negroes," however, he stated that "in obedience to that principle by which the properties of the offspring depend on those of the parents, we have the power of changing one species into another by repeated intermixture."[11] Given that Lawrence also wrote that "no difficulty prevents us from recognising the unity of the human species," he was clearly using both the natural and logical registers of "species."[12] It is essential that studies of racial thought—a field in which the discussion of terminology (especially "species") plays a central role—recognize the persistence of non-naturalistic uses of such terms; texts can otherwise be misread with unnerving ease.

Nevertheless, the biological use of species did become predominant over the course of the eighteenth century and with it the generative conception that linked species identity to a reproductive community. This generative idea came to be formulated in terms of the "interbreeding criterion." This stipulated that two animals are of the same species if they can produce fertile offspring with each other; the infertility of animals such as mules demonstrated that their parents were of separate species. The first concrete expression of this principle was given by the English botanist John Ray (1627–1705), who reckoned "all Dogs to be of one *Species* they mingling together in generation, and the breed of such Mixtures being prolifick."[13] This idea had been expressed before (among others by Bacon, as we have seen), but Ray was the first to produce an idea of species uniquely tailored to living creatures rather than as a catch-all classificatory concept.

It was, however, to the Comte de Buffon (Georges-Louis Leclerc, 1707–1788) that the interbreeding criterion was most frequently credited. Author of the most popular work of science of the eighteenth century (including in translation), the *Histoire naturelle, générale et particulière* (1749–1767), Buffon was enormously influential and remained the standard authority on natural history until the middle of the nineteenth century. His use of the term "species" varied somewhat over the course of his career, and it is not entirely unreasonable to suggest that he deliberately misused Linnaean terminology to undermine what he considered an excessively abstract and artificial system. But he was a strong advocate of the interbreeding criterion and made it the foundation of his own concept of species, over and above morphological and other similarities. As he wrote in 1753, "It is neither . . . the number nor the collection of similar individuals, but the constant succession and renovation of these individuals, which constitute the species. . . . Species, then, is an abstract and general term, the meaning of which can only be apprehended by considering Nature in the succession of time, and in the constant destruction and renovation of beings."[14] Buffon's diachronic and generative concept of species proved of lasting importance and was directly cited and employed by many others, including Kant. But there were two principal drawbacks with the interbreeding criterion that shaped Buffon's later discussion of species and led several naturalists to favor other definitions of the concept.

The first was the lengthy time lag involved in experiments to determine species identity. Since proof that two animals are of the same species could be provided only by the fertility of their offspring, investigators were forced to wait throughout the pregnancy of the female and the immaturity of her progeny before its fecundity could be tested. Related to this is the fact that such an experiment entailed two instances of sexual intercourse *and* the creation of another being. This was of little moment when the experiment concerned nonhuman animals but was of enormous moral significance in the case of perhaps the most sought-after clarification in eighteenth-century zoology: whether "orang outangs" (common chimpanzees, *Pan troglodytes*) were human. No less a figure than Jean-Jacques Rousseau pointed to these problems when he wrote that "if the Orang-Outang or others did belong to the human species, there would be one way in which the crudest observers could satisfy themselves on the question even with a demonstration; but not only would a single generation not suffice for this experiment, it must also be regarded as impracticable because what is but an assumption would have to have been demonstrated as true before the test to confirm the fact could be tried in innocence."[15]

As a means of getting around such problems, naturalists began to pay attention to the "repugnance" that animals of different species exhibited toward sexual congress with one another. The reluctance or resistance of animals to their forced pairing became a proof presumptive of separate species identity, and interestingly one over which animals could exercise their own agency.[16] It was not an infallible—or even reliable—measure, however, leading to the second and more fundamental issue with the criterion.

This was its inability to cope with instances of fertile hybrid reproduction. According to the criterion, any two animals that can produce fertile offspring must be of the same species, but instances came to light in the eighteenth century of prolific young born to parents of manifestly different species.[17] Fertile mules were the most common example, and Buffon wrote at length (in the 1760s and

1770s) about the relative fecundity of crosses of horses, donkeys, mules, and hinnies. He acknowledged that the fertility of mules was rare, but that it existed at all forced Buffon either to admit horses and donkeys to the same species or to acknowledge that species were variable and that the interbreeding criterion applied only to genera. As early as 1727, one German naturalist had expressed the view that the interbreeding criterion was best applied only to wild (rather than domesticated) species, and by 1766 Buffon had organized the quadrupeds described in the first fifteen volumes of his *Histoire naturelle* into genera of interfertile species; within each genus, all the species were related and one of them was the common ancestor of the rest.[18] In the same year, Linnaeus made explicit his belief (in gestation since the 1740s) that new species could be created by hybridization, underlining the importance of genus over species in his work.[19] Among other widely described mixtures were those of dogs and foxes, various avian species, bovines and equines, and—most notoriously—humans and apes. Not all these mixtures produced young, let alone fertile offspring, but there were enough such tales in circulation to convince naturalists like Johann Friedrich Blumenbach (1752–1840) to abandon the interbreeding criterion as a test of species and focus instead on morphology.[20]

Here was one way in which the reliance on sex in a diachronic, reproductive test of species had left open a yawning gap that concepts of race could fill. If interfertility was no longer an adequate definition of species, then external appearance would have to suffice. But doing away with one of the strongest proofs of a single human species and adopting another that, given the diversity of human appearance, could be readily used to argue for separate species (as was recognized by several contemporaries), was clearly a gift for racist authors.[21]

Yet while a belief in separate human species could easily morph into a conviction that there were distinct human races, they were not necessarily the same thing. Some of the most powerful racial theories of this era emerged from *within* a strongly reproductive concept of species that insisted on the unity of humanity as a species. For all its limitations, the interbreeding criterion did not disappear, and several of its adherents put up a stout defense. The anatomist John Hunter (1728–1793), for example, also wrote about fertile mules but explained this fecundity as simply a result of "monstrosity in the organs of the mule which conceived, as not being a mixture of two different species, but merely those of either the mare or female ass."[22] Even those who had turned away from the criterion found themselves trapped within a reproductive paradigm: in the fifth (1797) edition of his *Handbook of Natural History*, for example, Blumenbach outlined his reasons for using morphology to determine species but still justified his use of "Gattung" as the German translation for "species" by referring to the related verb *sich gatten* (to reproduce).[23] Nevertheless, it was Kant, whose commitment to a generative concept of species never wavered, who produced the most powerful articulation of race. So while some notions of race were tunneling under the reproductive foundations of human unity, others were invited inside by the very defenders of the interbreeding criterion. Exactly what these concepts of race were, how they developed, and the role played in such discourses by sex and reproduction are discussed in the following section.

Reproductive Sex and Racial Thought
Generating "Race"

The intellectual career of race owes its byzantine nature to the same sort of terminological ambiguity that characterizes "species." The chief difficulty facing any historian of "race" is locating this "moving and fuzzy target," and many have focused their efforts on examining the shifting vocabulary that expressed what might loosely be termed "racial" ideas.[24] Although frequently illuminating, there is an inherent teleological drawback to this approach, which traces developments from their consequences rather than causes and often appears to assume the completely synchronized development of both the content of an idea and the vocabulary with which to express it. Thus tracing familiar uses of "race" and its cognate terms runs the dual risk of identifying a concept only after it has been developed and used (possibly for a long time), and ignoring or underemphasizing relevant ideas that were expressed in an alternative vocabulary. At its worst, this approach would—to borrow an example from another field—entirely ignore Joseph Priestley's "dephlogisticated air" and unduly concentrate on Antoine-Laurent de Lavoisier's later isolation of *oxygène*.[25] Furthermore, examples abound of translators putting the word "race" in the mouth of an original author. Blumenbach seems unduly to have suffered from this, at the hands both of his mid-nineteenth-century translator into English, Thomas Bendyshe, and one German translator of his *De Generis Humani Varietate Nativa*, Johann Gottfried Gruber, whose text Blumenbach advised his students to burn.[26] The following discussion of racial ideas therefore concentrates on surveying the range of ideas and themes used in eighteenth-century writings on human varieties, and in particular on the part played in these by sex.[27]

That said, it is significant that the term "race" came to signify these ideas, especially because it was freighted with reproductive meaning. One of its oldest connotations is "lineage," denoting a genealogical relationship be-tween members of a community, whether on a smaller (family, tribe) or larger (*ethnos*, nation) scale. This was, especially at the familial level, the principal meaning of "race" in English throughout the eighteenth century. As the term came to be adopted by naturalists, via French and German usages and their translations, it acquired a more specific set of connotations, eventually coming to be synonymous with nation in the early twentieth century.[28] Before then, its implications were strongly reproductive; among the definitions of the term listed by Samuel Johnson in 1756 were "a family ascending," "family descend-ing," "a generation; a collective family," and "a particular breed." In some cases, it was even used to refer to a specific generation (in the modern sense), as when the lawyer and rake William Hickey was rebuked by the intoxicated host of a dinner party in 1769: "What's got into the present race? There is not an ounce of proper spirit about them."[29] Well into the nineteenth century, dictionary definitions of "race" continued to emphasize its genealogical rather than clas-sificatory connotations, and the seventh edition (1842) of the *Encyclopædia Britannica* defined it cavalierly as "RACE, in genealogy, a lineage or extraction con-tinued from father to son."[30] The term was, of course, being used in sophisticated ways by this time, so the continuity of its vernacular, reproductive meaning is certainly significant.

Furthermore, it is noteworthy that "race" also functioned in several languages as a catch-all term to be employed when other natural, anthropological, or zo-ological categories did not suffice. For instance, Georg Forster, writing in the mid-1780s against Kant's use of "race," suggested that "it would be a task for a person with nothing to do, to unravel in what senses each writer may have used this word. Of the travel writers who have recently depicted the inhabitants of the South Sea islands, I may say that they seem to have taken refuge in the word 'race' only where it was inconvenient to say 'variety.'"[31] This sense of "race" as a catch-all term to be used in default of others that were perhaps more precise but somehow inapplicable in a given situation, was also present in English half a century after Forster wrote.[32] In the third edition of his *Researches into the Physical History of Man* (1836), James Cowles Prichard wrote that

> the instances are so many in which it is doubtful whether a particular tribe is to be considered a distinct species, or only as a variety of some other tribe, that it has been found by naturalists convenient to have a designation applicable in either case. Hence the late introduction of the term *race* in this indefinite sense. Races are properly successions of individuals propagated from any given stock; and the

term should be used without any involved meaning that such a progeny or stock has always possessed a particular character. The real import of the term has often been overlooked, and the word race has been used as if it implied a distinction in the physical character of the whole series of individuals.[33]

Such a versatile yet equivocal use of "race" in both English and German shows that the term was being employed in radically different ways by Forster, Prichard, and the writers they were describing, on the one hand, and modern "racialists" and their precursors, who have attracted so much scholarly attention, on the other. This is to deny neither the overall trajectory of "race" from a descriptive genealogical to a prescriptive biological category nor the importance of tracing the use of terminology in the history of this idea. But it is a reminder of the slipperiness of our quarry, whose development over the long eighteenth century we will now consider.

Monogenism, Polygenism, and Inheriting Difference

At its heart, the early modern discussion of human diversity consisted of a debate between two broad perspectives that took opposing positions on the biblical story of creation detailed in Genesis. Supporters of monogenesis[34] ("one creation") held firm to the Mosaic account, whereby all humans stem from Adam and Eve and later from Noah's children as the only survivors of the Flood. This perspective enjoyed several distinct advantages: it was impeccably orthodox, making it also the majority viewpoint; it was consistent with intellectual tradition stretching back to Augustine, if not further; it sat more easily with developing concepts of species because it raised no awkward questions about human interfertility; its emphasis on genealogical descent accorded with traditional models of political authority and social order; and the belief in common ancestors suggested a shared humanity that *could* (but by no means necessarily *did*) support humanitarian projects. By insisting on a single act of creation and a common ancestral pair, monogenetic ideas relied on secondary mechanisms—such as curses or climate—to account for human diversity.

Divine action gradually gave way to more naturalistic explanations of phenotypical diversity, which emphasized incremental change over time as each generation developed still further characteristics inherited from its forebears. As a result, sexual reproduction became an increasingly prominent element in monogenetic thought. It was not absent even from theological accounts: for the mark inflicted on Cain following his murder of Abel (Genesis 4:15) to function as an explanation of physical differences, it had to have been both inherited *and* carried by one of Noah's daughters-in-law, for example. But reproduction's significance grew noticeably when climate became the chief agent of monogenetic change. This was arguably the case by the mid-seventeenth century, as Robert Boyle, Thomas Browne, and other Fellows of the Royal Society pioneered arguments against the attribution of skin color to a divine curse, especially the "Curse of Ham." This had become an increasingly prominent fable whose association of Africans, blackness, and slavery coalesced only in the early modern era. Yet while it was frequently and extensively debunked over the course of the eighteenth century, in sermons, learned essays, and the popular press, it persisted into the nineteenth, especially in the antebellum South.[35]

Polygenesis, the second perspective, was a more straightforward idea that had its origins in theological controversy which, in the sixteenth and especially mid-seventeenth centuries, suggested that other humans were created before or alongside Adam.[36] As with the shift from providential to naturalistic monogenesis, by 1700 pre-Adamism had begun to range beyond pure theology into naturalistic polygenism. This argued that the present diversity of humanity was too great to be consistent with conventional readings of the Bible: monogenetic change was so imperceptible that a large number of generations must have been necessary for all the observable differences between people to have emerged, and the earth—at a mere 6,000 years old, according to traditional biblical chronology—had simply not existed long enough for monogenism to be convincing.[37] Polygenists instead explained global diversity by a greater or lesser number of separately created peoples and by their sexual intermixture.

The principal drawback—or attraction—of this school of thought was its heterodoxy, which discouraged many from expressing their views publicly. Indeed, the danger it posed to Christian teaching was profound, even existential. Admitting separate creations would not only make a nonsense of the Flood (which would thereby not have affected all people) but would also exempt those of non-Adamic descent from the foundational and guiding principle of Judeo-Christian belief, original sin.[38] Just as disturbingly, polygenism threw the possibility of a coherent moral science or even a meaningful concept of human nature into doubt: separate human species implied distinct human natures, meaning that a morality founded on one of these natures would not be automatically valid for all.[39] The religious and philosophical stakes could therefore not have been higher, making it remarkable that polygenism garnered as many supporters as it did.

Beyond the seductions of heterodoxy, the likeliest reason for its appeal was its simplicity. Polygenesis cited no complicated intermediary factors, such as climate, to explain human difference, so it was vulnerable to attack on a smaller number of fronts than monogenism. Its simplicity rested in part on a common sense understanding of sex, reproduction, and heredity. By attributing all human diversity to originally distinct people and their sexual intermixture, polygenism demanded only the ability to notice human diversity and to acknowledge how children resembled their parents.

Polygenism left questions of heredity largely to individual intuition. Practically the only discussion of the issue offered by polygenist authors was to attack monogenist ideas on heredity, a tendency apparent from their earliest writings. For instance, two early examples of naturalistic polygenesis from the late seventeenth century both emphasized that the color of white and black people was innate and "could not proceed from any Accident [i.e., external influence], because when Animals are accidentally Black, they do not Procreate constantly Black ones (as the *Negroes* do) . . . but a *Negro* will always be a *Negro*, carry him to *Greenland*, give him Chalk, Feed and Manage him never so many ways."[40] Later, more sophisticated, polygenist accounts were more thorough in their dismissal of monogenetic mechanisms but provided just as little detail on how the original differences between peoples could be transmitted or blended. We will see in chapter 4 how Edward Long sidestepped this entire issue by asserting that white and black peoples were of different species, a popular strategy given the state of flux surrounding "species" in the second half of the eighteenth century. Other authors, such as Henry Home, Lord Kames (1696–1782), Charles White (1728–1813), and Julien-Joseph Virey (1775–1846) all argued against the possibility that acquired characteristics could be inherited, the latter even claiming that original differences were so strong they could be perpetuated even within the same family.[41]

At the same time, polygenists differed on the precise role played by sex and reproduction. Some, including Long, White, and Virey, held to the interbreeding criterion and used supposed cases of sterile mixed offspring to assert the separate species identity of their parents. Others questioned the criterion, which allowed them to claim that black and white people were of separate species without the need to deploy questionable anecdotes about infertile children.[42] William Bourke, for example, speaking at the Edinburgh Society for Investigating Natural History in November 1793, argued that Buffon "has laid down his arbitrary law, and then says that all mankind are of the same species, because they generate together, & their progeny generated. This, certainly, is forcing facts."[43] But Bourke made no mention of mixed children, leaving open the possibility that they might both exist and be fertile. Other polygenists avoided talking of different species of humans altogether, choosing instead to emphasize distinct origins by a variety of terms, including "race." For example, Alexander Robertson, a medical student at Edinburgh, stated clearly in 1798 that "there are different races of men, the progenitors of each of which were originally created in those climates, in which Providence intended their progeny to live."[44] Yet against these rather technical differences must be set the broad polygenist consensus that human diversity was an original condition, furthered by sexual intermixture, which owed nothing to the complex intermediary mechanisms on which monogenesis relied.

Climate

The foremost of such mechanisms was undoubtedly climate. In the premodern world this provided an explanation for every kind of natural and cultural phenomenon, remaining until the nineteenth century the paramount cause of human diversity for orthodox thinkers. It had exercised this function since Hippocrates, whose *Airs, Waters, Places* (*c.* 400 BC) overshadowed every subsequent account of the environment's influence on the human body.[45] This text also demonstrated that a prerequisite for climatic/environmental explanations of diversity was an associated and relatively advanced concept of heredity.[46] By the eighteenth century, most explicitly environmentalist accounts of human difference contained a discussion of reproduction and heredity. Few relied solely on the heat of the sun to explain skin color, especially after encounters with people in Africa, America, and Asia who had diverse pigmentations yet lived at similar latitudes.[47] Instead, an extended web of factors, which came to include humidity, topography, local flora and fauna, soil, and wind patterns, was gradually developed to add considerable complexity to environmental explanations of human difference.[48]

Probably the most sophisticated variant of climatic theory was contained in the first few volumes (1749) of Buffon's *Histoire naturelle*, which described how climate and geography directly affect vegetable life that, via the food chain, influenced animals and finally humans through an operation Buffon named "intussusception." This was his term for a process of assimilation by which "organic molecules" (*molécules organiques*), the basic building blocks of all living things, are absorbed by the human body in digestion and distributed throughout the body. Excess organic molecules from every part of the body are later collected to produce the male and female seed necessary for reproduction. (Until puberty, all organic molecules are required for growth, so a superfluity not only signifies but constitutes sexual maturity.) Crucial for these processes is the body's "internal mold" (*moule intérieure*, a force the Newtonian Buffon likened to gravity), which controlled the development of an organic body from the moment of its conception. The internal mold was affected by the factors that he believed could produce variations (climate, food, customs, and time), but these effects were limited and reversible.[49]

This elaborate theory was both ambitious and powerful. Organic molecules and the internal mold could account for all the observable differences between peoples, the heritability of acquired characteristics, and the connection between climate, nutrition, and diversity. Furthermore, it offered a compelling explanation of how and why climate exercised an influence on human populations that was transmissible across generations. But it was highly speculative, involving concepts and mechanisms that were at best unverifiable or, at worst, incomprehensible. None other than Albrecht von Haller admitted that he was completely baffled by the interior mold, and even some of Buffon's admirers treated his account of nutrition and reproduction as "Ingenious but a mere Hypothesis."[50]

Beginning in the seventeenth century, environmentalist explanations of human difference came under pressure that increased steadily after 1700.[51] One reason was the wealth of data pouring into Europe from travelers in various parts of the world (especially the Pacific after 1760) that described people who did not appear as predicted by existing climatic models. But the most serious threat to these ideas was the dawning realization that bodily change might not be reversible, which led some to doubt whether climate could effect significant physical alterations at all.

The classic case was that of black Africans. Almost all commentators believed that humans were originally white and that the pigmentation of sub-Saharan Africans had developed completely within the fewer than 6,000 years since creation.[52] (Even this window was too generous, as Herodotus explicitly referred to "black-skinned . . . Ethiopians" in the fourth century BC, meaning that their emergence must have taken only 3,600 years.[53]) If climate alone was responsible for the total transformation from white to black in a few dozen centuries, then, it was asked, surely the movement of sub-Saharan Africans to cooler climes within the previous three would have produced some signs of physical change?

This question preoccupied naturalists throughout the century, but a definitive answer remained elusive. In 1766 Buffon suggested relocating some black Africans from Senegal to Denmark to observe the effects of climate, but so far from being tried, the question was posed again by the Glaswegian John Anderson in 1774.[54] The answer offered thirteen years later by Samuel Stanhope Smith (1750–1819)—"the features of the negroes in America have undergone a greater change than the complexion"—was so ambiguous that in a later edition of his *Essay*, Smith angrily rebuked those who thought he had claimed "that the negro complexion had hitherto become sensibly lighter in America."[55] In any event, Smith's contemporaries were clear that there had been noticeable change in skin color, but it was a result of sexual intermixture, not climatic influence; this had produced nothing that could compare with the speed, completeness, and self-evidence of the difference exhibited by the offspring of interracial couples. Interracial sex demonstrated both that rapid change was possible and that climate failed to produce it.[56]

Climate was challenged even where the influence of sexual intermixture was arguably negligible. The debate over Jewish characteristics, for example, reached no firm conclusions on whether appearances changed following movement from one climate to another. But all sides of that debate assumed Jewish endogamy: for those who thought Jewish appearance altered with the climate, the lack of sexual intermixture confirmed that climate alone was responsible; for their opponents, endogamy ensured the preservation of a characteristic complexion in a foreign environment.[57] Where intermarriage between Jews and members of host societies was acknowledged, prejudice was used to shore up a leaky argument. White, for instance, relied on antisemitic stereotypes to qualify this threat to his polygenist convictions: "The Jews have gained proselytes in every part of the world where they have resided, and they are at liberty to marry those proselytes. But the truth is, that the Jews are generally swarthy in every climate."[58] The ambiguity of the Jewish example meant that it could support diametrically opposite conclusions, even within different editions of the same work.[59]

The crisis of climate had significant ramifications. Peaking at roughly the same time as abolitionist activity became more prominent (late 1760s to early 1770s), it contributed to a resurgence in polygenism. Two important judgments of the 1770s, in England (*Somerset v. Stuart*, 1772) and Scotland (*Knight v. Wedderburn*, 1778), rendered slavery in Great Britain legally unenforceable, outraging the West India lobby and leading to the first polygenist accounts of human difference to be published in Britain since the 1730s. The shadow cast over these texts by sex, species, and the interbreeding criterion is unmistakable. Their authors, Samuel Estwick and Edward Long, both made the argument "that human nature is a class, comprehending an order of beings, of which man is the genus, divided into distinct and separate species of men," and that "all other species of the animal kingdom have their marks of distinction: why should man be universally indiscriminate one to the other?"[60] The idea that the unity of humankind somehow violated the "Great Chain of Being," undermined the analogy of humans and animals, and implied an imperfection in

God's creation was shared by other authors, such as White, Virey, and the German scholar Christoph Meiners.[61]

One result of abolitionist sentiment flexing its muscles while climate floundered was therefore a resurgent polygenism. Undue stress should not be laid on this development: polygenism still remained an unorthodox and minority viewpoint, even if it was more readily expressed in private than in public.[62] Furthermore, polygenesis and proslavery did not automatically go hand in hand. Some polygenists were opposed to slavery while certain monogenists were in favor: a telling example is provided by Lords Kames (polygenist) and Monboddo (monogenist), who respectively judged for and against Joseph Knight's liberty in 1778. Monboddo exemplified the proslavery standpoint that defended the institution on the basis of tradition and therefore looked skeptically at novel racial ideas.

More significant than the impact on polygenesis was that on monogenist thought, which henceforth proceeded along two separate but intersecting paths. One of these concentrated on human society, the other on the human body or, as Silvia Sebastiani has put it, on the natural history of humanity and the natural history of man.[63] Neither of these strands was entirely new, but they had hitherto been in climate's shadow; now they began to predominate, and what may be termed "post-climatic monogenism" did not so much jettison climatic explanations as make them serve newly dominant sociological or biological narratives. What is more, both forms of post-climatic monogenism continued to manifest the importance of sex and reproduction in their accounts of human difference.

Post-climatic Sociology

This is less immediately obvious for the sociological "natural history of humanity" that played a significant part in the intellectual development of racial thought, particularly in Scotland. It was here that stadial theory—the idea that all human societies proceed through several defined stages as they approach modernity—reached its fullest flower. Each stage was characterized by a predominant mode of subsistence, from hunting and scavenging to domesticating livestock, and thence to organized arable farming before finally trading as commercial societies for staple and luxury foodstuffs. This theory was based on a universalist conception of human nature that was frequently honored more in the breach than the observance: stadial models were teleological and ethnocentric, consistently placing modern European society as the inevitable (if not always desirable) end point of development. Stadial authors were also frequently pessimistic because a universal idea of human nature left frustratingly little room to relativize cultural "progress." The Scottish historian William Robertson, for example, stated categorically that "the intellectual powers of man in the savage state . . . *cannot* acquire any considerable degree of vigour and enlargement."[64] The problem for such thinkers was not so much that certain societies were less advanced than modern Europe—this would be remedied in the fullness of time—but that they did not follow the European pace or pattern of development.[65]

Sex was nevertheless vitally important both within stadial thought and where pessimism had taken hold. For stadial theorists, the state of "relations between the sexes" (more precisely, the treatment of women by men) was deemed a reliable indicator of the position of a given society on the path to civilization. As William Alexander stated confidently in 1779, had the history of the earliest societies been "entirely silent on every other subject, and only mentioned the manner in which they treated their women, we would, from thence, be enabled to form a tolerable judgment of the barbarity, or culture of their manners."[66] Most authors agreed that women were subjugated by their men in "savage" societies and that their piecemeal emancipation was both cause and consequence of social development.[67] In ways that echoed the heterosocial aspects of "politeness" earlier in the century, the complementarity of the sexes was viewed as a quintessential sign of civilization.[68] But it had its limits: homosocial man was rude and uncouth, but while the company of women might refine him, when surrounded solely by women he became effeminate. Sexual politics was, therefore, at once a constitutive element, an index, and a limiting condition of stadial theories of social development.

Sexuality was also crucial, particularly in the writings of pessimists. For all stadial theorists, the sexual neglect of women in the earliest stages was a clear sign of their oppression.[69] But the lack of sexual activity among "savages" was a sign of cultural stasis that threw into doubt the inevitability and universality of social progress. For writers such as Cornelius de Pauw, whose *Recherches philosophiques sur les américains* (1768) was excerpted and translated into English, the perceived asexuality of Amerindians was proof of their incorrigible effeminacy and inherent degeneracy. As he wrote, "The lack of inclination, the scant heat of Americans for the [female] sex undoubtedly demonstrates the want of their virility and the deficiency of their reproductive organs: with difficulty does love exert on them hardly half of its power: they know neither the torments nor the sweetness of this passion, because the most ardent and most precious spark of nature's fire is extinguished in their tepid and phlegmatic soul."[70] Deficient libido and genitals were only the tip of the iceberg, as De Pauw then claimed that syphilis was caused by the corruption of Amerindian blood, which also produced lactating breasts and a general lack of body hair (especially pubic and facial) among Amerindian men.[71] He was the most outspoken of such pessimists, but there were several who tied together the perceived sickliness, thin distribution, and apparent asexuality of the decimated native population of the Americas, and associated these with physical degeneracy.[72]

Degeneration

The more explicitly biological strand of post-climatic monogenism—Sebastiani's "natural history of man"—also adopted this new concept of *irreversible* degeneration as a response to the crisis of climate. Degeneration had its origins in climatic and religious thought: Buffon's claim that American flora and fauna were degenerate variants of Old World species was influential, and it seemed that only degeneration could explain why several of the world's peoples, all of which were descended from Noah's sons, were so primitive.[73] "Degeneration" could therefore be interpreted as a neutral, descriptive term that simply indicated movement away from a primordial form, an interpretation supported by Buffon's rejection of the idea that degeneration could produce new species.[74] But it almost always implied degradation, prompting not only Buffon to reject De Pauw's application of it to people, but also Thomas Jefferson's angry defense of American nature against both authors.[75] A significant factor in explaining both the power and clearly negative connotations of degeneration was the frequently applied and readily understood analogy with disease. Even Buffon wrote in this manner: "Varieties have been perpetuated from generation to generation, as deformities and diseases pass from fathers and mothers to their children, . . . and have been confirmed and rendered permanent by time and the continued action of the same causes."[76]

Other authors also used degeneration less analogically and made direct connections with disease or systemic flaws to explain differences. Blumenbach's account of skin color, for example, concentrated on the morbid excess of carbon in the body that would "precipitate" in contact with atmospheric oxygen to produce dark pigment. While this was common to all humans, it particularly affected those with an excess of black bile or whose liver function was overstimulated by a hot climate. Despite claiming he did "not wish to insist too much on the analogy of jaundice with national tints of the skin," he proceeded to do exactly that.[77] Blumenbach also used his carbon thesis to explain why parts of pregnant women's bodies darkened, as they had both their own and the child's share of carbon to respire. Likewise, positing menstruation as another mechanism to expel excess carbon, "a similar blackness is also observable in women who never menstruate."[78] For Blumenbach, blackness was linked to climate but also strongly to bodily dysfunction along axes both of sex and race. On the other hand, he also used "degeneration" in its neutral sense to characterize all changes from an original type.[79] Indeed, it is arguable that his chief contribution to reproductive theory—the "formative drive" (*Bildungstrieb*)—was fundamentally a means of explaining degeneration.

Yet it was Kant's writings on race, which produced a complex technical vocabulary of human difference, that show degeneration at its most sophisticated. For him, "race" described a group of people whose characteristics were maintained across generations irrespective of physical location and who, when coupled with people from another "race," consistently produced visibly mixed offspring. All humans shared a common "ancestral form" (*Stammgattung*) that possessed "germs" (*Keime*) containing the predispositions for all possible racial features. As humans dispersed throughout the world, these germs developed in a unidirectional process to produce the different races. Climate, food, and other traditional explanations for human varieties could not directly produce permanent change so much as stimulate the unfolding of predispositions. Above all, this was a one-way process: once ur-humans had migrated throughout the earth and their germs had unfolded in response to local conditions, these characteristics were then set and unchangeable. Kant believed that only one human feature was so consistently transmitted that it could form the basis of racial classification. The diversity of skin color led him to posit four races that differed from the ancestral form: "high blondes" (Northern Europeans), "copper-reds" (Native Americans), "blacks" (from Senegambia), and "olive-yellows" (Indians).[80]

This diachronic theory privileged natural history (*Naturgeschichte*) over the mere description of nature (*Naturbeschreibung*), and it centered on "race" (*Race*) precisely because of the term's genealogical connotations. Kant carefully distinguished "race" from "degeneration" (*Ausartung*) because, unlike Blumenbach, he associated the latter with the creation of new species, a phenomenon Blumenbach was not prepared to admit. But Kant's was clearly a theory of degeneration, both in the sense that it involved departure from an original archetype and by implying that this movement was—for all non-white races, at least—a negative development. Kant's anthropological writings and lectures were awash with derogatory references to black Africans or Amerindians, at least until the 1790s, when he appeared to reevaluate his ideas both on individual races and on race itself.[81]

Before then, however, Kant had to face several critics, the most important of whom was Georg Forster (1754–1794), erstwhile companion of his father and

Captain Cook on the latter's second voyage (1772–1775). Forster attacked several different aspects of Kant's ideas, but the most powerful criticism—as Kant himself recognized—was against the one-way, once-only development of the "germs": if nature could produce one germ in the original "ancestral form," then surely it could produce several to enable man to adapt to a range of environments? Kant defended his ideas on the basis of empirical circumstances: people tended to remain within the climatic zones to which they had adapted and they did not change color when they strayed beyond these zones, meaning that unless one was prepared to admit multiple creations or several human species, all humans had a single original color that then changed within different groups and could not now be altered. Furthermore, Kant thought his theory offered the best explanation for an issue that had troubled climatic thought—namely the uniformity of Amerindian complexions across a range of climates.[82]

Domestication

Forster was apparently oblivious to two important features of his attack on Kant. The first was its irony, as it contained manifestly racist statements in a manifesto against the concept and language of race. He exposed the moralistic hypocrisy of monogenists, who—blind to the cruelty prevailing within a monogenist paradigm—argued that conceding multiple creations or human species would condemn black Africans to cruel treatment: "Let me rather ask, whether the thought that blacks are our brothers has anywhere, even once, meant that the raised whip of the slave driver has been lowered?"[83] Yet he also speculated that black people had been separately created and argued that there existed an instinctive mutual sexual repulsion of black and white people, which could be overcome only by "lust and concupiscence."[84]

Second, in talking of reason overriding instinct, Forster unwittingly invoked "domestication." This set of ideas concerning the physical and behavioral alteration of animals in proximity to humans could also be readily applied to humankind itself. Among the changes produced by domesticating animals were reduced ferocity; a diminished fear of people; changes in color, size, shape, and proportion; and the suppression of the natural "repugnance" felt by animals of different species toward mating with one another. As we have seen, the "interbreeding criterion" came under increasing scrutiny throughout the century, making "repugnance" arguably the chief mechanism by which species boundaries were maintained; its suppression resulted in—and was even deemed necessary for—the production of varieties within and hybrids between species. Forster's comments therefore not merely suggested that white and black people were of different (zoological) species but invoked domestication by referring to the corrupting power of human reason, which had broken down their mutual "repugnance."

Such an application of domestication to humans was not unprecedented. Rousseau described civilization as self-imposed domestication and claimed that changes produced in domesticated animals were less dramatic than those imposed on humans.[85] Belief in an arcadian state of nature was no prerequisite to interpreting social and political development as domestication. Thomas Jarrold, speaking in 1811, argued that "what takes place in an animal on its being domesticated, is I apprehend a full illustration of the constitutional, or in other terms, the physical change which passes upon a nation in its progress from the barbarous to the civilized state of society."[86] More outspokenly racist was Virey, who communicated his pessimism that black people could ever be domesticated by explicitly—and somewhat sarcastically—stating that "black animals . . . appear to be less degenerated by domesticity."[87]

The undoubted popularity of this metaphor did not, however, make it unambiguous. Some, such as Rousseau, saw domestication as a fundamental corruption of primitive innocence, a viewpoint reinforced by the frequently asserted link with degeneration.[88] Viewed negatively, domestication and degeneration apparently produced "unnatural" hybrids and defective or "monstrous" offspring, and had a deleterious effect on fertility. Another phenomenon with close links to domestication was slavery, both within naturalist discourses (where domestic animals were described as enslaved) and discourses of slavery (where slaves could be perceived as domestic animals or slavery as a process of domestication).[89]

But for many authors writing on humans and animals, domestication signified social progress or human control over nature. Indeed, much as it was possible to view degeneration in a neutral or even positive light, so too could the assertion that domestication *is* degeneration have positive connotations. Blumenbach stated that the only domestic animal more degenerated than the pig is the human and that "the difference between him and other domestic animals is only this, that they are not so completely born to domestication as he is, having been created by nature immediately a domestic animal."[90] Yet it was precisely because of these characteristics that domestication, according to Samuel Stanhope Smith, produced "the greatest perfection of [man's] form, as well as of his whole nature."[91] Ultimately, therefore, domestication both showed how the sociological and biological strands of post-climatic monogenism were intertwined and offered one prominent example of the overlap between discourses of race and species.

Sex as a Racial Characteristic
Sexual Customs: Polygamy, Promiscuity, and Virginity

The relationship between sex and racial thought shaped not only the concepts but also the data on which theories of human difference were based. Customs, rites, and traditions were central components of accounts of foreign and exotic people, remaining at least as prominent as descriptions of physical bodies until late in the eighteenth century. These cultural and social elements were essential for the construction of newer forms of racial thinking, not least because they attested to characteristics—"savagery," for example—that could be signified only indirectly by the physical body. The boundary between the physical and the cultural was often blurred because customs left physical traces, such as the flattened noses of a whole host of "savage" peoples, attributed to parents crushing the noses of their infants in adherence to supposed fashion. As with language (discussed in chapter 5), customs and traditions provided a guide to the interior mental life of others and exposed the unseen and insensible intellectual characteristics to which physical features were increasingly becoming a convenient shorthand.

Sexual customs were especially significant in this process. They were powerfully tied to the production of new bodies and the use of existing ones and thus bound the physical and cultural in a uniquely intimate way. Sexual customs were also governed by religious rites and traditions, particularly marriage. Like language, religion—and marriage—was something that all human societies shared, making it an ideal basis for comparing the world's peoples: Giambattista Vico's *New Science* (1725), for example, identified religious belief, marriage, and the burial of the dead as the three features shared by all cultures. Correspondingly, the discussion of sexual practices took place in (at least) two registers—the naturalistic and religious—that often overlapped.[92]

This is exemplified in one of the most important works of comparative religion of the eighteenth century, Jean-Frédéric Bernard and Bernard Picart's *Ceremonies and Religious Customs of the Various Nations of the Known World* (1723–1737, English translation 1733–1739). This work proved enormously influential in furthering the cause of religious toleration during the Enlightenment by highlighting the common features and shared origins of religious practices across the world. Bernard and Picart aimed at "making all religions comparable," and they used a wide range of travel writings and descriptions of foreign societies to provide an authoritative account of global religions.[93]

In examining marital customs around the world, Bernard and Picart devoted substantial attention to polygamy (polygyny), which they discussed in both religious and naturalistic terms. They highlighted polygamy as an element of almost all non-Christian faiths yet did not spare Christian Europe, making oblique references to "Polygamy of another Nature," which anticipated Martin Madan's view of prostitution-as-polygamy in his *Thelyphthora* (1780).[94] Furthermore, polygamy in the Abrahamic faiths was explicitly linked to that in African, South and East Asian, and American creeds to support, if not a common ancestry of all religions, then at least their agreement in certain particulars.[95] They also drew on nature and climate to explain polygamy, beginning with its Jewish manifestations. In an echo of the wider discussion concerning Jewish features, they argued that many Jewish customs, including polygamous matrimony, varied according to local circumstances. In addition, while sticking to the familiar line that polygamy was an ineffective means of producing population growth, they claimed that a society's acceptance of polygamous marriages was often tied to the importance it laid on populationism. And they even accepted that in some circumstances, such as where women greatly outnumbered men, polygamy might "follow the Order of Nature."[96]

They were not alone, as several other authors who sought to account for polygamy on naturalistic grounds identified several apparently plausible reasons for the custom. The most common such explanation was an unbalanced sex ratio, to which was occasionally added the supposedly shorter reproductive window for women in warm climates. But for all the apparently "natural" reasons why societies might permit polygamy, European writers discovered or devised more that condemned the practice. According to these authors, polygamous marriage more often impeded than promoted population growth, and it was clearly contrary to nature given both the obvious balance of the sex ratio[97] and the physical and moral consequences of polygamy: men with several wives became progressively emasculated and exhausted, even to the point of death; those unable to wed turned to sodomy; and wives who enjoyed only infrequent sexual contact were likely to indulge in infidelity or same-sex activity.[98]

The presence of polygamy in Christianity's own past and the ongoing debate about its potential legitimacy prevented a clear moral distinction from being drawn between Christians and the rest of the world. Other sexual practices were discussed in less ambivalent terms. Promiscuous fornication was roundly condemned and strongly associated with extra-European peoples, despite the prevalence of such behavior within Europe. The execration of this practice was based on a mix of religious and natural factors: to the biblical prohibition announced in Deuteronomy 22 was added the naturalistic argument that promiscuity impeded fertility. Promiscuous individuals and social groups within

Europe were certainly attacked for this behavior, but it was not viewed as a fundamental characteristic of European society.

Elsewhere, it was a different story. Promiscuous sex was most consistently associated with black Africans and peoples from the extreme north. Bernard and Picart stated baldly that "some of the Natives of the *Golden Coast*, are addicted, notwithstanding they are indulg'd in Polygamy as well as their Neighbours, to strolling abroad, and lying with Strangers."[99] Some acknowledged that Europeans in Africa were partly responsible for African promiscuity, but it remained a popular explanation for the failure of slave populations to increase naturally.[100] The expectation that extra-European societies were characteristically promiscuous was not disappointed when Europeans began to chart the Pacific, as Johann Reinhold Forster (Georg's father) explained: "There is hardly a country to be found, where the young unmarried females are allowed such a latitude as at O-Taheitee [Tahiti] and its neighbourhood in admitting a variety of young males, and abandoning themselves to various embraces without derogating from their character."[101] Amerindian peoples were the notable exception to this general picture of exotic sexual abandon, due either to their "nobility" or incapacity.

Yet notwithstanding this widespread and sustained vision of an extra-European world unburdened by sexual morality, marriage remained the lens through which sexual customs were viewed. Take, for example, Monboddo's unpublished comments on a people originally described by Buffon who "live all in herds of Thirties and Forties, & mix together, promiscuously in the same manner as a Herd of Oxen does. And tho' in almost every other Nation which we have now discovered there is something like Marriage; it is in many countries so loose an Institution, as to savour very much of the original freedom of Copulation."[102] Promiscuity, which even in Europe was sometimes interpreted, only half jokingly, as a form of ersatz polygamy, was not the only sexual practice to be attacked because of the damage it inflicted on traditional matrimony. Masturbation was condemned for wasting the vital energy of men, thereby jeopardizing their chances of producing healthy offspring, and same-sex behaviors were linked both to climate and to the sexual frustrations of polygamy or an unfaithful partner. Promiscuity and incest may have been permitted as necessary for the original peopling of the world, but such were the unique circumstances of the era of Genesis and hardly applicable to later times.[103]

Not all the (sexual) customs and practices of the extra-European world manifested radical difference. The cult of virginity, for example, which Bernard and Picart identified in various cultures throughout the world, was at the heart of European ideals of chastity, fidelity, and virtuous marriage. The veneration of virginity in other cultures demonstrated that these ideals were widely shared. Significantly, this was a moral quality with an associated physical manifestation (the hymen). There were undoubtedly skeptics: several doubted the hymen's existence, Buffon and Blumenbach both felt the signification of a moral quality by a physical feature was illegitimate, and the counterfeiting of a "maidenhead" occurred repeatedly in both erotic and mainstream literature of the eighteenth century.[104] Yet the hymen remained the most prominent example of the association of morality with specific corporeal features, an association that became increasingly powerful over the eighteenth century.

Maternal Influence: From Cause to Catalyst

Physical characteristics became moral signifiers most clearly when bodily features directly resulted from human action that manifested savagery or perverse custom, if not downright immorality. Mothers were thought especially prone to such behavior, and accounts abounded of women physically manipulating their children's features in order to produce a particular skull or nose shape.[105] Buffon expressed the consensus view that "this capricious practice of altering the natural figure of the head is very general among savage nations."[106] It was admitted that features might be misshapen by accident rather than design, as when the stock flat noses and thick lips of black Africans were supposedly produced by the repeated pounding of a child's face while carried on its mother's back.[107] Deformities might also be attributed to the mother's imagination, as we will see in chapter 4. But accidents were never entirely blameless, as the source of mishap could just as easily be attributed to negligence as intention.

Some authors ceased to blame mothers exclusively for these physical features, arguing instead that savage custom merely accentuated existing characteristics. Blumenbach wrote in 1795 that "the natural conformation of the nose can only be exaggerated by this violent and long continued compresson of the nose when soft, but can in no way be made thus originally, since it is well known that the racial face may be recognised even in abortions."[108] Rather than creating the flatness of a nose, maternal action now merely exaggerated an extant feature.[109] In a move that echoed the demolition of the maternal imagination's explanatory power, the responsibility for facial features was removed literally from the mother's hands over the course of the eighteenth century.

Flesh Versus Bone: Female Sexual Characteristics

Nevertheless, mechanical explanations of facial features—especially those involving the repeated collision of an infant's face with its mother's back—were tied to a set of persistent stereotypes about the sexual characteristics of exotic women. Children could be strapped to a mother's back for extended periods because her allegedly massive breasts enabled the child to suckle from over her shoulder, allowing her to work uninterrupted.[110] Pendulous breasts came to be associated with black African women in particular, but they were part of earlier descriptions of other peoples, constituting a generic marker of savagery. They were even attributed to men in Mandeville's *Travels* and to Irish women in the seventeenth century: William Lithgow had written in 1632 that the breasts of Irish women were so large that they would make an ideal money bag.[111] This observation was also made of Khoisan ("Hottentot") women in southern Africa, as noted by Lawrence, whose critique of it extended only as far as acknowledging its "evident air of exaggeration."[112]

Opinion was divided on the cause of large breasts. Some, such as Johann Reinhold Forster or William Webb, treated them as a matter-of-fact corollary of the environment.[113] Others linked them to the process of childbirth.[114] White was more racial in tone: "Long flabby breasts . . . are not the effect of relaxation in a warm climate, but are found with people of colour in the frigid as well as torrid zone. No European white woman, however, in any age or climate, was ever known to have a breast of such a description. The African, therefore, in this particular approaches to the simia."[115] White argued for the innateness of pendulous breasts when most other commentators thought they were a product of climate, custom, and manipulation, some even arguing that European breasts were artificially small because of fashion.[116] The discussion on breasts shows that opinion was divided on extrinsic or intrinsic causes of physical features, even within the work of an individual author. Blumenbach, for example, discussed techniques employed to increase breast size at the same time as he pointed to the innateness of other characteristics, such as noses.

This dualism, in which certain features remained climatically or culturally determined while others became more inherently fixed in the structure of the body, was representative of thought on human difference in the wake of climate's crisis and is a reminder that the emergence of biological race was no smooth or linear development.[117] Female sexual features and processes exemplify this mixture of intrinsic and extrinsic sources of characteristics with unusual clarity. Certain aspects of pregnancy and childbirth, for example, came to function as intrinsic signs of difference by being linked to more permanent bodily structures, like the skeleton, while menstruation remained bound to extrinsic factors such as climate and social rank.

Pregnancy and childbirth had long been discussed in terms redolent of racial thought, particularly among exotic women. We have already seen how Blumenbach used the darker skin—especially nipples—of pregnant women to illustrate his carbon thesis of pigmentation. He was not alone in accounting for skin color with reference to expectant mothers, nor indeed in referring to the common trope that exotic women found childbirth easier than Europeans.[118] The myth of easy childbirth among women in warm climates dates from Aristotle and was successively applied to almost every female population encountered by Europeans over the centuries.[119] It was a popular sign of difference because it manifested the broader stereotype that exotic peoples were able to bear physical pain to a far higher degree than Europeans (which became a staple of accounts of Amerindians), and it was a means by which biblical proof of radical difference could be adduced.[120] Easy childbirth suggested an exemption from the curse pronounced on Eve (Genesis 3:16) that both implied polygenetic origins and justified urgent conversion, often via enslavement.[121]

Easy parturition was, however, no monopoly of exotic women. It was also related to social rank, as women from the lower orders supposedly gave birth with ease and produced milk in sufficient quantity and quality to make ideal wet nurses.[122] But giving birth easily within Europe meant that it could not be a climatic phenomenon, so theorists posited other causes. Oliver Goldsmith associated birthing pains with a greater delicacy of women or their being "enfeebled by luxury or indolence."[123] The man-midwife Robert Bland wrote more generally in 1794 that "this facility of bringing forth, is not occasioned by the warmth of the climate" but rather to the level of civilization.[124] The abolitionist James Ramsay agreed, arguing that advances in civilization produced physical changes in the mother's body that facilitated childbirth.[125]

Dissociating easy childbirth from the climate opened the way for alternative biological explanations, which ultimately came to rest on the wider pelvis of the female skeleton. Londa Schiebinger has shown both how descriptions of the female skeleton were powerfully influenced by prevailing gender norms and that these texts and images emphasized a broader pelvis in order to naturalize the woman's primary function as mother.[126] A similar process was at work among those authors most committed to demonstrating that the differences between Europeans and non-Europeans were intrinsic. White, Virey, and the comparative anatomist Samuel Thomas Sömmerring—who was an important source for White—all argued that the pelvis of non-European women was wider (or the heads of their infants smaller) to explain easy childbirth. A few commentators rejected this notion (Blumenbach, Camper, Lawrence) or sat on the fence (Prichard), but it was nonetheless possible to attribute the phenomenon to the (semi-)permanent structures intrinsic to the human body, making it an ideal sign of racial difference.[127]

Such was not the case with menstruation. The link between its cessation and pregnancy was abundantly clear, but its exact purpose and causes remained a mystery.[128] Menstruation was no unequivocal sign of pregnancy, sexual maturity, or even of being female. Bleeding while pregnant was often interpreted as a continuation of menstrual flow, it was possible to become pregnant without having first undergone menarche, and the periodic bleeding of men—whether from groin, anus, nose, extremities, or elsewhere—was viewed as "male menstruation" until well into the eighteenth century.[129] Furthermore, menstruation could not even function as an exclusive marker of humanity. There existed a general but not universal consensus among naturalists and medics that it was unique to humans, and while many more denied than asserted that higher apes menstruate, all those who addressed the topic thus testified to its popularity.[130]

Further challenges to the uniquely human status of menstruation came from the fact that some adult women between the ages of menarche and menopause apparently did not menstruate. Amerindians, Brazilians, and Lapps were all identified as amenorrheal peoples, something that, again, came to be more disputed than accepted but which was evidently a widely held view.[131] Menstruation also intersected with other aspects of the discussion on human difference. For Blumenbach, amenorrhea produced darker skin, and other authors thought different human varieties particularly prone to menstrual disorders. In the 1780s, menstruation was believed to promote the acclimatization of European women in hotter climates, and Jewish men were thought to menstruate (and lactate), although such views were more common before 1720.[132]

Amid this range of viewpoints on menstruation's role in understanding human difference, three broad areas of agreement may be identified. First, the importance of diet for proper menstruation was stressed. The amount of menstrual blood could, it was argued, vary with the amount of meat in a woman's diet, and "obstruction of the menses" could be due to an improper regimen, perhaps too high in milk and fish. Related to diet was social rank, particularly as indicated by a luxurious or sedentary lifestyle. Such women often suffered from menstrual disorders and bled copiously, in contrast with those whose lifestyle was more frugal and who worked physically. The final and most important point of consensus was on the influence of the climate. Almost every medical writer on menstruation made a point of emphasizing that hot climates promoted early menarche and menopause, and that menstrual flow in these climates was more copious than in colder regions. Some writers—Buffon, for example—thought that a hot climate produced less menstrual blood, but the details of climatic influence are less significant than the unanimity on its existence.[133]

In marked contrast to the large chorus stressing the powerful influence of climate on menstruation, there were few, if any, voices that sought to interpret menstrual differences as intrinsic to the human body or based on its semipermanent structures, such as the skeleton. This had much to do with the continuing mystery surrounding menstruation, as well as its strong basis in the traditional humoral, fluid economy of the body, which had long highlighted the body's porosity and its susceptibility to external influences. Unlike easy parturition, therefore, menstrual differences could not be readily transformed into intrinsic biological phenomena but rather remained tied to climatic, dietary, or other factors.

Some of these factors were cultural and moral. It was argued that excessive lust, for example, stimulated menstruation and produced not only larger breasts but also the quintessential racial/sexual characteristic—the "Hottentot apron."[134] It is difficult to exaggerate the attention paid by writers on human difference to the enlarged nymphae (or labia) of particularly Khoisan women of southern Africa. It featured in the earliest descriptions of this region, in works such as Olfert Dapper's *Naukeurige Beschrijvinge der Afrikaensche Gewesten* (1668), to an extent that rivaled the concentration on large breasts.[135] In the late eighteenth century, one Edinburgh commentator reported that this feature assisted menstruation—it "is said to be useful at certain period[s] w[hic]h are much less troublesome to the fair Sex here than in Europe"—and some even went so far as to claim it provided a defense against rape.[136] Blumenbach minimized the significance of this organ, stating that in Africa "the nymphæ are a little more turgid and prominent, a defect the less to be astonished at in that country, because it is certain that it sometimes occurs in this."[137] Others felt that these organs were the direct result of living in a hot country.[138]

Irrespective of its cause, this organ was clearly associated with an insatiable and animalistic sexuality. Some authors made this linkage explicit. Buffon stated that Khoisan women have no shame: "All the women who are natives of the Cape are subject to this monstrous deformity, which they uncover to all those who have the curiosity or audacity to view or to touch it."[139] In fact, the first thorough examination of this organ was conducted by the zoologist and comparative anatomist Georges Cuvier who dissected Saartje Baartman, the "Hottentot Venus," on her death in Paris in 1815.[140] Lawrence explained away the "apron" by stating that "a considerable development of these organs is more common in warm climates, and has been noticed in the Negroes, Moors, and Copts, among whom it has been the practice for females to be circumcised."[141] Even if not exclusively "Hottentot," these organs were still thought common to peoples whose aberrational sexual appetites, behavior, and customs had already attracted comment.

Male Sexual Characteristics: Genitals, Beards, and Hair

The same applied to male genitals, which, like all sexual organs, were minutely examined by enlightened investigators into human differences. This was not a novel development: not for nothing are the male gonads called "testes" after the Latin for "witness" (or "proof," of one's manhood), and the myth that the semen of black men was not white, which persisted into this period, dates from Strabo.[142] Abnormal sexual organs were increasingly becoming representative not just of individual monstrosities but of entire classes of people. Black African males, for example, supposedly possessed prodigious genitals that signified both masculinity and, paradoxically, the effeminacy implied by a sexual appetite out of control.[143] The threat posed by the endowment of African males to white European manhood was clearly expressed by the usually cautious Blumenbach, who wrote that "it is said that women when eager for venery prefer the embraces of Negroes to those of other men."[144] The other extreme was no less pathological: we have seen how De Pauw's description of Amerindians drew a direct link between hairlessness, small genitals, and a lack of virility.[145] Other groups, like "Hottentots," might voluntarily emasculate themselves by removing a testicle in order to run faster or further.[146] Sexual characteristics were therefore doubly laden with both gendered and racialized connotations. Indeed, the implicit genital ideal was much like the optimum climate: moderate.

This is perhaps why male genitals of all sizes, not just the extremes of large and small, were subjected to such scrutiny that they became precise markers of identity. Charles White, for example, devoted sustained attention to the *frænum præputii* (the membranous fold that attaches the foreskin to the underside of the penis) as the apparent proof that black peoples were an intermediate group between Europeans and apes.[147] But it was in discussions of skin color that male sexual organs were most frequently mentioned.[148] As body parts that almost all cultures tended to shield from view, they were useful indicators of the role of climate as a direct cause of skin color. In 1775, Blumenbach described how the scrotum of a black man was considerably darker than the rest of the skin, "for it is well known that some parts of the human body become more black than others, as, for example, the genitals of either sex, the tips of the breasts, and other parts which easily verge towards a dark colour."[149] Virey used the example of dark areolae to prove that "such a conformation is not the effect of heat alone (though it contributes much to it) but the natural conformation of those races, in whatever climate they live."[150] The color of male genitals was used as a more stable marker than overall skin color, which could change. White wrote that "we are informed that the children of negroes, when first born, are of the same ruddy colour as European children; but that the *scrotum* and the *glans penis* are black; and that they have a black or brown thread or circle on the extremity of the nails."[151] For such authors, genitals and fingernails were mobilized as markers of identity when even skin color proved unreliable.

They performed the same function even for those who took a diametrically opposed view of the causation of skin color. The army surgeon, John Hunter (1754–1809), for example, produced a doctoral thesis in 1775 that was a strongly environmentalist account of human differences.[152] In asserting that the sun directly caused skin color, Hunter claimed that all children are born of one color before turning into their particular varieties. White did so too, he and Hunter sharing the same method of proof: the genitals again provided the key to skin color. In Hunter's case, those parts of the body shielded from the sun and air do not lose their original white color, "as is observed in those blacks who have the gland covered with the prepuce."[153]

Beyond genitals, the male sexual characteristic that most attracted comment was the beard. Despite being somewhat out of fashion in eighteenth-century western Europe, beards were a marker of both sexual difference and male virility. Mention was unfailingly made of the asexuality of eunuchs, Amerindians, and others who supposedly had no facial hair. The claim that Amerindian men naturally lacked beards attracted much criticism, yet it continued to be made even by respectable naturalists into the nineteenth century. Most of those who rejected this idea pointed instead to the existence of cultural practices of plucking facial hair, one author remarking that this practice promised a lucrative market for tweezers.[154] Nonetheless, the volume of texts that addressed this subject testified to the longevity of this myth and the power of the emasculation it signified.[155]

But this was not the only reason why beards were a popular topic. Beards, and hair in general, formed a key aspect of the discourse on human variety because, on the one hand, the hair of sub-Saharan Africans seemed distinct from that of other human societies, and, on the other, hair was closely related to skin color. African hair was almost uniformly described as "woolly," even among the minority of writers who denied that it actually constituted wool. They claimed rather that it was genuine hair because it did not exhibit the same seasonality as the development of wool and because sheep in warmer countries also tended to be more hairy than woolly. Their opponents also made comparisons with animals, but to argue against the influence of climate, that such hair was a sign of savagery, and—in one extreme example—that Parliament had therefore decreed that "negroes" should not be considered human beings under the law.[156]

Woolly hair was also frequently discussed by all sides in relation to black skin. One Jamaican clergyman wrote that the combination of black skin and woolly hair was the Mark of Cain, while a Scottish judge tried to assert his own color-blindness by arguing that "the wisest nation was a nation of Black men with wooly [*sic*] hair."[157] The association of hair and skin was so strong because the color of both was seen to derive from the same cause. Francis Moore in 1738 described the people he encountered in Africa by stating that they, "by the Intenseness of the Heat and Burning of the Sun, are of a Black Colour, and have their Hair curling."[158] Oliver Goldsmith in the 1770s differentiated between peoples by their hair, an approach echoed by Blumenbach, who argued that hair and skin color were environmentally acquired but also that there was a remarkable correspondence between the hair and the whole constitution and temperament of the body (for example, he directly associated black hair and lunacy).[159] Others were more focused in their assessment of hair. Prichard, a staunch monogenist, treated the question anatomically and agreed that the basis of skin color was identical to that of hair.[160] For him, hair was dissociated from inner temperament and anchored more precisely to skin color, which operated as a primary racial signifier.

Interest in hair was arguably due ultimately to the ambiguity of skin color. De Pauw's *Recherches philosophiques* contains a chapter devoted to skin color, of which by far the largest section focuses on Africans rather than Amerindians. He concluded that "the Americans of the north, exposed to the inclemencies of . . . all the changes of the seasons have a heavily tanned face, but it would be much less black if they did not rub medicines and grease into it."[161] Many authors said the same about "Hottentots," who were supposedly not as naturally dark as they appeared, and the discoloration from "filth" and "nastiness" was a factor that complicated the discussion of skin color in John Anderson's account of human varieties.[162] Where pigment could be counterfeited, investigators sought certainty in a closely associated and potentially less falsifiable characteristic.

This search for reliable markers of identity is characteristic of the discourse on human difference. The sustained and sophisticated interest in sexual features failed, however, to produce unambiguous indicators of identity. This was perhaps a problem that stemmed from the very foundations of this discourse, in the overreliance on sex within contemporary understandings of "species." This concept was in a state of flux for most of this period, as counterexamples to the interbreeding criterion repeatedly came to light, forcing the redeployment of the criterion from species to genus, thereby permitting naturalists to speak of different human species. This may not have automatically entailed different human races, but the breadth and slipperiness of that term—"race"—certainly did nothing to foreclose the possibility.

Sex was also at the heart of all strands of the discussion that were eventually to be entwined around the idea of race. A basic notion of hereditary transmission was key to polygenist ideas, but more especially to monogenists, who relied on complex intermediary mechanisms to account for the emergence of phenotypical differences. The chief of these mechanisms, climate, remained powerful throughout the period but encountered growing difficulties over the course of the century, leading to a split in monogenist thought whereby the environment no longer ruled but rather served newly dominant sociological and biological narratives. Sex remained vital within these threads, however, whether as a multifaceted entity within stadial theory or as a causal or catalytic factor in degeneration and domestication.

It was as racial data, as characteristics that might be exhibited on the body, or within the society, culture, or religion of peoples around the world, that sex and sexuality were most clearly visible in these discourses. The broad shift from providential to naturalistic bases of racial ideas was mirrored in the discussion of polygamy in Bernard and Picart's *Ceremonies and Religious Customs*, which dealt with the phenomenon in relation both to theology and the environment. Morally laden sexual practices, such as promiscuity or virginity, often had associated physical features (whether anatomical, such as the hymen, or pathological, like venereal disease), much as did "savage" fashion, manifested in the facial features of newborns. The blame laid on mothers for deforming their infants became less intense over the period as researchers became more familiar with unborn children.

Yet uncertainty surrounding the permanence or reliability of markers of identity remained, and sexual features did little to assuage these anxieties. Some characteristics or processes could be associated with more permanent bodily forms like the skeleton, as was the case with parturition. But many others, from menstruation to abnormal genitals or bodily and facial hair, could not be relocated to such structures and so remained either biological mysteries or associated with individual or communal morality.

CHAPTER FOUR

Ex Ovo Omnia

Embryology, Sex, and Race

In Houyhnhnmland, the problem and its solution were reproductive in nature. Three months before Gulliver's sojourn abruptly ended, his "Houyhnhnm Master" attended the Grand Assembly, a representative council in which Houyhnhnms assembled every four years to conduct the affairs of state. Befitting creatures for whom "the Meaning of the Word *Opinion*, or how a Point could be disputable" was incomprehensible, this assembly decided matters summarily and did not—could not—engage in debate.[1] Except on one perennial issue: "Whether the *Yahoos* should be exterminated from the Face of the Earth." One supporter extended his argument beyond the Yahoos' incorrigible turpitude, filthiness, and ugliness, to adduce their origins:

> He took Notice of a general Tradition, that *Yahoos* had not been always in their Country: But, that many Ages ago, two of these Brutes appeared together upon a Mountain; whether produced by the Heat of the Sun upon corrupted Mud and Slime, or from the Ooze and Froth of the Sea, was never known. That these *Yahoos* engendered, and their Brood in a short time grew so numerous as to overrun and infest the whole Nation. . . . That, there seemed to be much Truth in this Tradition, and that those Creatures could not be *Ylnhniamshy* (or *Aborigines* of the Land) because of the violent Hatred the *Houyhnhnms* as well as all

other Animals, bore them; which although their evil Disposition sufficiently deserved, could never have arrived at so high a Degree, if they had been *Aborigines*, or else they would have long since been rooted out.[2]

The sophistication of Swift's satire is visible in this piling of one contradiction upon another: Houyhnhnms are incapable of debate yet engage endlessly in the same one; the Yahoos are quintessentially aboriginal, having been spontaneously generated by the action of sea or sun on (Houyhnhnm)land, yet are cast as outsiders; and their continued existence, itself due to Houyhnhnm dithering on the question of their eradication, is taken as proof of their foreign origin given the "violent Hatred" that ought long ago to have led to their deracination.

The question remained, what is to be done? The answer had been unwittingly provided by Gulliver, which his master proceeded to convey: "He added . . . That, among other things, I mentioned a Custom we had of *castrating Houyhnhnms* when they were young, in order to render them tame; that the Operation was easy and safe; that it was no Shame to learn Wisdom from Brutes, as Industry is taught by the Ant, and Building by the Swallow. . . . That, this Invention might be practiced upon the younger *Yahoos* here, which, besides rendering them tractable and fitter for Use, would in an Age put an End to the whole Species without destroying Life."[3]

The common thread linking all the deliberations of the Houyhnhnm Assembly concerning the Yahoos—from the statement of the problem to the speculative account of their origins to the proposed solution—is reproductive. The "odious Qualities" of the Yahoos, a result of their nature rather than of any moral failing, clearly motivated calls for their destruction, but only because their reproductive success had allowed them to multiply beyond a tolerable limit.[4] Their proposed castration enabled the continued exploitation of their labor while securing their gradual extinction, a suggestion that had real-world counterparts. In one Dublin broadsheet from 1725, the proposal to castrate thieves was supported by a Barbadian precedent, "where the Planters wisely consider'd that many of them would be Losers by putting all the Guilty to death; whereupon a great Number of the rebellious Slaves were ordered, like some wild vicious Cattle, to suffer the Punishment of Castration."[5]

Crucially, the "Tradition" that claimed the original Yahoos were spontaneously generated was a clear statement that they were an inferior form of life. Although only finally laid to rest by Pasteur, the idea of "spontaneous generation" was by the seventeenth century confined to explaining the origins of "lower" creatures, such as insects or crustaceans. As long as this idea persisted, the possibility of human spontaneous generation continued to be addressed, even if overwhelmingly satirically or by negation.[6] An example from the same year that *Gulliver's Travels* was published (1726) speculated that "Peter the Wild Boy," a feral child found in Germany and brought to Britain in that year, might have "originally sprung up out of the Ground like Corn, or Asparagus."[7] But the Houyhnhnm Assembly clearly had no time for such jeux d'esprit. As much a reproductive as a legislative or executive council (the redistribution of Houyhnhnm children to the childless was one of its key tasks), the assembly embodied an association of generation and human diversity that was devastating in its clarity: a persecuted population was yoked to a confected origin story that stigmatized this group as a lower form of life, and its eradication was to be secured by destroying the ability to reproduce.

While Houyhnhnmland was mercifully fictional, the nexus of racial and reproductive thought was all too real and is the focus of this chapter. Its principal claim is that racial theorists were ultimately embryological theorists, and vice versa. Put simply, it was not possible prior to the mid-nineteenth century to make an argument about the causation of human phenotypical differences *without thereby* adopting a position on embryological development. Equally, embryological thought both established the parameters of contemporaneous racial theorizing and made substantial contributions to these ideas.

This argument is made in three sections. The first surveys embryological thought from the mid-seventeenth to the early nineteenth centuries. Emphasis is laid in this section both on sources and ideas of importance in racial discourses—the use of non-European populations, animals, and ideas arising from them (such as telegony), for example—and on the embryological research of key racial theorists, such as Buffon, Blumenbach, and Kant. The second section examines "racial embryology," the principal terrain shared by these discourses. After briefly analyzing the case of John Atkins, whose writings in the 1720s and 1730s exemplified how racial thought was thoroughly dependent on reproductive theory, the section moves to consider the crucial datum shared by both bodies of thought: the fact of mixed-race fecundity. That parents with different phenotypical characteristics can produce fertile offspring was proof both of the unity of the human species and of the equality of parental contributions to the fetus. This single datum was of immense importance for both discourses and had implications in a wide variety of more or less allied fields, from hybridity to heredity, the most important of which are discussed here. Finally, the chapter turns to the thought of two influential authors on racial and reproductive topics active in the mid-eighteenth century, Pierre Louis

Moreau de Maupertuis (1698–1759) and James Parsons (1705–1770). This section uses their work to guide the discussion through several outstanding topics that also embodied the strong ties between racial and embryological thought: albinism, the power of the mother's imagination to shape her fetus, and hermaphroditism.

Embryological Debate from Generation to Reproduction

The Sources of Embryological Thought: Genre, Genera, and Geography

Ideas about generation were expressed in a multitude of forms in this period. Learned texts, from individual letters published in scholarly journals to richly illustrated and expensive multivolume treatises, were the most significant vehicle for discussing theories and observations on generation. But contributions were also made in sermons and theological works, in cheap print and the popular literature of sexual advice, and generation formed an important stock-in-trade for pornographers and satirists throughout the period. These genres did not simply parrot or popularize knowledge that first appeared in more exalted forms, but they often playfully recombined old and new ideas with other aims in mind than the elucidation of nature's secrets.[8]

Furthermore, many texts on generation written in Latin or other foreign languages were translated into English, and there was much cross-fertilization between texts and genres. Some works, such as Thomas Browne's *Pseudodoxia Epidemica* (1646) or John Jones's *Medical, Philosophical, and Vulgar Errors* (1792), set out to attack "vulgar" beliefs, in the process testifying to their durability. To be sure, some reproductive ideas were less popular among mainstream scientists than among the wider population by 1800: the ability of a mother's imagination to physically alter her unborn child is one example. But this idea continued to be addressed in university medical courses and was difficult to eradicate entirely from a field in which empirical observations were few and difficult to obtain.

Concrete data for researchers working on generation came from numerous sources, two of which are highlighted here. One was nonhuman animals. These had always featured heavily in embryological studies, but examples drawn from birds and poultry, rabbits, fish, amphibians, insects, snails, larger game, and livestock became especially important in early modern investigations. One reason for this was expediency: domesticated animals were familiar and readily available, meaning they could be continuously observed and that observers could be more confident in their ethological findings than with less well-known species.[9] Experimenting with larger animals was rare: William Harvey's association with Charles I gave him the unique opportunity to work on deer—to the extent that Maupertuis accused him of committing a "scholarly massacre"—at a time when few theorists could afford to study even livestock.[10] Indeed, the comparative absence of direct examples drawn from cattle, sheep, horses, or pigs in theoretical work on generation may have contributed to the reluctance of working breeders to consult this literature.[11]

A second source of embryological data came from peoples outside Europe. This, too, achieved a new prominence in the early modern period, due in this case to greater exposure to the extra-European world, the humanist rediscovery of ancient texts, and the circulation of both sorts of evidence in print. Indeed, the geographical range of ethnographic data used in the discourse on generation was broadly in keeping with European expansion and exploration across the period: examples involving Amerindians and sub-Saharan Africans (both in Africa and the Americas) predominated in the seventeenth and early eighteenth centuries, while data from the Pacific was increasingly deployed after 1775. Furthermore, the urge to use "exotic" data in debates on generation was stimulated by the matters in dispute, to which this chapter now turns.

The Course of Embryology, 1640–1840
THE ANCIENT AND MEDIEVAL HERITAGE

Developing in three overlapping phases between 1640 and 1840, the debate on generation was profoundly influenced by the legacy of ancient embryological thought. The key ancient writers, Hippocrates, Aristotle, and Galen, all agreed that the fetus was formed after fluids from both parents mixed in the uterus. The various parts of the fetus then developed sequentially in a process termed "epigenesis" until all that remained was for the complete fetus to grow. This consensus was tempered by some significant differences, principally concerning the role played by each parent. Aristotle privileged the male contribution, arguing that the father alone contributed the true "semen" that shaped the fetal form, while the mother provided only matter through her menstrual blood.[12] For Hippocrates and Galen, both partners produced semen/seed, meaning that the mother provided both matter and form, and that female orgasm held a higher status in their embryology than in Aristotle's.[13]

Differing parental contributions also shaped accounts of sex determination and the resemblance of children to relatives.[14] For Hippocrates, the sex of the fetus was decided by competition between the parents' seed: each parent produces both strong (male) and weak (female) seed, and if they both contribute a stronger seed, the child is male, a weaker results in a female, and if they produce different kinds of seed, sex is determined by whichever prevails in quantity.

Aristotle, in denying any formative role to the mother, relied instead on the influence of heat and cold. These were also crucial for Galen, who, despite adhering to a two-seed theory, thought that the location of the fetus within the uterus exercised an important role in sex determination.[15]

Of what was semen composed? One venerable theory, based on pre-Socratic thought, was that seed (σπερμα, *sperma*) were distributed throughout the world, absorbed by humans and other animals through breathing and eating, and were concentrated in their bodies. This theory of panspermism surfaced periodically over the centuries but was never as popular or influential as Hippocrates's argument that semen derived from fluids that flowed throughout the body. This idea, known as "pangenesis" because it posited the collection of seminal matter from all over the body, remained the most convincing explanation of hereditary phenomena until the nineteenth century, notwithstanding its rejection by Aristotle.[16] Ironically, his belief that semen was produced from excess nutrition by a refinement ("decoction") of the blood was often cited to support pangenesis, given the presence of blood throughout the body. This idea of semen as particularly refined blood, whose loss occasioned physical exhaustion, explained both the physiology of sexual climax and the dangers of immoderate sexual activity, especially masturbation. What all these theories had in common was a powerful link between nutrition and generative matter: you and your offspring are what you eat, and what you eat is shaped by its environment and climate.[17]

Beyond establishing its general parameters, the ancients bequeathed four enduring aspects of reproductive thought that shaped its links with racial discourses. The first was the use of "exotic" peoples and their sexual intermixture as decisive evidence. Aristotle, for example, argued that pangenesis could not explain the resemblance of children to their grandparents, since "there was at Elis a woman who had intercourse with a blackamoor [Αἰθίοπι]; her daughter was not a black, but that daughter's son was."[18] Second, the power of the mother's imagination was stressed in several ancient sources. A third aspect concerned "spontaneous" or "equivocal" generation, which pointed to the action of the sun on decaying matter. This was quite in line with regular reproductive theory, which also emphasized the causative power of heat, but it struggled to account for the constancy of animal form. Finally, the ancients influenced formal elements of the discourse on generation by their frequent use of analogy and metaphor. These often involved animals or other mechanical, chemical, and culinary processes such as cheese-making, brewing, or distillation.[19]

The principal medieval addition to this ancient heritage resulted from the integration of Aristotelian ideas and Christian theology. While Aristotle had argued that his three souls—vegetative, sensitive, and rational—progressively developed in the fetus, medieval commentators such as Albertus Magnus rejected the automatic emergence of the rational soul and made its implantation a divine prerogative. The status of the fetus thus differed before and after "ensoulment," usually thought to occur with its first perceptible movements, with implications for the legality and morality of abortion, for the transmission of original sin, and for "monstrous" or deformed offspring.[20]

EMBRYOLOGY AND THE ENGLISH REVOLUTION, C. 1640–C. 1670

These ideas survived into the seventeenth century, when the study of generation in England was overshadowed by civil war and revolution. Many scoured the natural world for signs that might make sense of the chaos engulfing Britain, although their task was complicated by the inevitable disruption. The royalist Kenelm Digby produced his synthesis of Aristotelian and mechanical embryology while a prisoner of Parliament in 1642–1643, and the London crowd destroyed William Harvey's papers while ransacking his house early in the Civil War. Devastated by this loss, Harvey, in alluding to the "intestine troubles" of his day in the *Exercitationes de Generatione Animalium* (1651),[21] reflected a wider despondence that led many to seek solace in older beliefs.[22]

Despite the difficulties, the study of generation did not stand still. Digby largely toed the Aristotelian line (except concerning pangenesis) in his *Two Treatises* (1644), while Nathaniel Highmore, whose *History of Generation* (1651) was one of the first studies to use microscopic evidence, maintained that the various parts of the fetus emerged all at once and subsequently grew, an idea termed "metamorphosis."[23] Both agreed that embryogenesis began with the physical mixture of generative matter, whereas Harvey denied that the two seeds came into physical contact as he found no semen in the does he dissected. Not recognizing that they had been killed too early in the rutting season, before they could copulate, he concluded that the male's contribution was incorporeal (or "spirituous").[24]

This was just one of several influential assertions made by Harvey, who reinforced them using two powerful metaphors. The first exploited the double meaning of "conception" to establish an analogical link between mental and uterine conceptions that became a fruitful source of literary imagery and strengthened beliefs in the power of the mother's imagination to shape her unborn child. The second used the concept of contagion to show how a father could influence his offspring without physical contact between seed and egg. Following Girolamo Fracastoro, whose *De contagione* (1546) argued that contagion was itself a process of generation, Harvey reinforced an association between reproduction and disease that was to prove important in thinking about heredity.[25]

Also of lasting importance was Harvey's conviction that all generation necessarily involved "eggs." Emblazoned on the frontispiece to the *Exercitationes* (fig. 4.1), the motto "ex ovo omnia" proved popular not because it was based on empirical observation (the mammalian ovum was not discovered until 1827) but because it corresponded to a perception of nature that demanded absolute consistency: the Aristotelian conviction that nature uses the same means to the same end ensured that since birds, fish, and frogs all procreated using eggs, so too must humans and other "quadrupeds."[26] This belief was so powerful and ubiquitous that it not only thrived in the absence of proof but led many to deny that unanticipated discoveries, such as spermatozoa, played any part in generation.

It also arguably fueled ideas such as telegony, the hypothesis that offspring sometimes inherit characters from a previous mate of their dam. Telegony is usually associated with nineteenth-century biology and eugenics: Darwin was a supporter, and the idea became a mainstay of racist thought in the US South and Nazi Germany, where the Nuremberg Laws codified the notion that if an "Aryan" woman became pregnant by a "Jew," all subsequent children she might have (with any partner) would bear traces of this prior liaison.[27] Earlier expressions of telegony were very rare, yet remarkably Harvey offered two. The first was based on his interpretation of avian embryology. To fertilize their eggs as they are produced, hens can store sperm in the body for a prolonged period (in chickens, up to 2 weeks; in turkeys, up to 10). The spermatozoa donated by individual treadings are stored discretely, and evidence suggests that the sperm of the most recent partner is used first in conception; thus the chicks of an earlier partner might be produced *after* those of a more recent one.[28] Early moderns were aware of these facts—Harvey's mentor, Fabricius, believed chickens could store sperm for up to a year, and Julius Caesar Scaliger recognized the "last in, first out" mechanism of semen deployment—and this alone could account for some awareness of the possibility of telegony. Harvey's own views took this further. He acknowledged the prolonged fecundity of hens while denying that they stored sperm or that egg and sperm had any physical contact. Instead, conception occurred after the entire body of the female was fertilized in coitus, making telegony far less predictable than the finite and serial mechanisms suggested by Fabricius and Scaliger, and far more difficult to confute.[29]

Fig. 4.1. Frontispiece to William Harvey, *Exercitationes de Generatione Animalium* (Amsterdam, 1651). Courtesy of the Wellcome Collection, London.

Fig. 4.1a. Detail of figure 4.1.

Hence, perhaps, his explicit discussion of telegony in viviparous species, as reported by John Aubrey. In Aubrey's manuscript play, *The Country Revell* (1671), Harvey is named as the source of the following: "He that marries a widdowe makes himself Cuckold. *e[xempli] gr[atia]* . . . [if] a good Bitch is first warded w[i]t[h] a Curre, let her ever after be warded w[i]t[h] a Dog of a good straine, & yet she will bring curres, as at first, her wombe being first infected with the Curre. So—the children will be all like the first Husband (like raysing up children to y[ou]r brother.) So—the Adulteress (though a sinn in her) the children are like the husband."[30] Although the language here is of "infection" and "sin," this hypothesis is arguably more generous to unfaithful wives than the usual explanation given for the likeness of illegitimate offspring to cuckolded husbands. Instead of the mother's deliberate and manipulative use of her imagination to mold the features of the unborn child, Harvey makes resemblance a result of unconscious processes. In so doing, he provides valuable evidence to support Harriet Ritvo's conjecture that telegony had a long history of popular acceptance before its first explicit scientific discussion in 1820.[31]

Beyond Harvey's innovations, other topics of embryological thought continued to overlap with the discussion on human difference. Nutrition, the handmaid of climatic influence, featured prominently. Numerous texts discussed the effect of diet on fertility and offered recommendations for food appropriate to provoke desire and ensure well-formed seed.[32] On this basis, Highmore warned against transplanting flora and fauna from their original habitat: "They either degenerate, or die, wanting their proper aliment; but seldome or never propagate their kinde."[33] But nutrition and climate could be used to combat prejudice, such as when Browne rejected the canard "that an unsavory odour is gentilitious or national unto the Jews." He argued that a disagreeable odor is not characteristic of or heritable by particular nations but can be produced by diet, before stating that Jews were unlikely even in this manner to smell, given their temperate diet and moderate sexual habits.[34]

Evidence originating from outside Europe was also deployed—for example, concerning hereditary phenomena: Digby described his privileged access to Muslim women when investigating the polydactyly of an Algerian family.[35] Another example concerned ease of parturition, which Harvey discussed using examples from Ireland and the New World, alongside "country women and such as are used to heavy work."[36] By contrast, Nicholas Culpeper's consideration of complexion and its relation to infertility in his extremely popular *Directory for Midwives* (1651) made use of no evidence from antiquity or beyond Europe. His argument, reprinted throughout the eighteenth century, that husbands and wives hoping to procreate ought to differ in complexion, serves as a useful reminder that "complexion" most often referred to humoral constitution and not skin pigmentation in early modern Britain.[37]

MECHANISM AND PREFORMATION, C. 1670–C. 1760

The learned consensus behind epigenesis did not survive the Restoration. Changing ideas about the universe exposed a serious flaw that Harvey himself recognized: "The chief problem of all [is], what is there in generation that . . . constitutes the parts of the chick in the egg in an established order by means of epigenesis, and produces a univocal creature like to its own self?"[38] How, in other words, did a homogeneous mass of fluids develop over the course of pregnancy into a fully formed child? Traditional answers to this question involved a governing agent—whether God, entelechy, or a soul—that directed the development of complex matter.

By mid-century these had become unsatisfactory. Foremost among the reasons why was the newer, mechanical model of the universe in which natural processes were governed by laws operating on inert, uniform matter. This reconceptualization of nature enabled major advances in physics and astronomy but struggled to account for vital phenomena. Efforts to produce a mechanical account of epigenesis failed, most spectacularly in the case of Descartes, whose posthumously published *La formation de l'animal* (1664) became notorious. He sought to explain embryogenesis using a range of analogies, particularly fermentation, and when these proved inadequate to the task Descartes was forced to invent increasingly exotic particles to shore up his rickety edifice.[39]

Unable to accommodate epigenesis within a mechanical framework, researchers abandoned it in favor of models that posited a preformed fetus inside one parent's seed that could simply unfold once implanted in the womb. Preformationism in the seventeenth century was really two theories. One, derived from botany, held that the fully formed plant or animal could be observed in miniature in its seed. The second theory, which dominated by 1700, extended the first to its logical conclusion by stating that *every* generation was preformed inside its parent, they inside theirs, and so on.[40] Souls were neither created following conception nor inherited from one parent but have rather *preexisted* since the Creation. Preexistence theory—known at the time as "evolution" (*evolutio*)—was also named *emboîtement* (encasement) from the image evoked by each creature existing inside its parent like a set of Russian dolls.[41]

One of preformationism's principal attractions was that it reflected mainstream views on God's involvement in the world, hence its advocacy by churchmen such as Malebranche. Few devout Christians would have denied that God could intervene in worldly affairs if he wished—miracles would be otherwise inexplicable—but few imagined a radically interventionist God, "a continual busybody," in the words of the late Peter Reill, "pulling every string, demonstrating incompetence as well as irrationality."[42] Preformationism provided for a single divine creative act and absolved God from involvement in individual cases of monstrous or defective births or of sinful sex that produced children, whether from fornication, incest, or rape. In addition, the semipresence in the Garden of Eden of all beings ever to exist provided a powerful support for the doctrine of original sin and bolstered Calvinist predestination.[43]

Which parent bore the preformed seed? Preformationists divided themselves into ovists, who argued for the mother's egg, and animalculists, who championed the father's sperm. Ovism was the oldest and most popular variant: it fit more neatly within established patterns of natural knowledge, enjoying the advantage of a wide range of analogues from seeds to bird's eggs. Investigations of chicken eggs were particularly important, and the superb empirical studies of Fabricius ab Aquapendente (1621 posth.) and Marcello Malpighi (1672) served to make the early formation of the chick a principal battleground for later attackers and defenders of ovism.[44]

Without eggs to examine, ovists studying mammals focused on other aspects. Working on rabbits taught Regnier de Graaf that copulation left a visible trace on the female testicle, the corpus luteum that marks the site of an earlier ovarian follicle. The number of these corresponded with the embryos he found in the uterus and oviducts, so he believed these follicles were genuine ova, fertilized by male semen while still part of the ovary. The opinion that female testes produced eggs and that these were fertilized by male semen *before* leaving the ovary had become commonplace among researchers by 1700, thanks to de Graaf and later studies of matters such as ectopic pregnancy. But analogies drawn from rabbits could be deceptive: these animals ovulate in response to coital stimulation, and the belief that women did the same was unsettling, reinforcing the Hippocratic/Galenic idea that female orgasm, and hence female desire, was central to the process of generation.[45]

Even more unnerving was the possibility that women might ovulate, indeed procreate, without the involvement of men. Lust-induced ovulation was discussed in melodramatic terms by the anti-masturbation tract *Onania* (c. 1712) and more soberly by later researchers such as Blumenbach, who lectured and wrote on the existence of corpora lutea in virgins. But while parthenogenesis was a topic of (entomological) study and a staple of reproductive satire, genuine belief that it might occur in humans was rare since male semen was deemed necessary to stimulate the unfolding of a preformed seed. Despite prioritizing the mother's contribution, ovism clearly entailed no real enhancement in the status of women as it continued to reserve to the male the privilege of bestowing the spark of life.[46]

This was explicit in the case of animalculism, which emerged after Antoni van Leeuwenhoek discovered spermatozoa ("animalcules") in 1677, and which won some high-profile supporters, including Leibniz and Boerhaave, the foremost physician of the age. Animalculists acknowledged that women and even eggs were essential for procreation, but many reduced them to mere vessels that housed the unfolding seed.[47] Some were skeptical of preexistence, such as Jean Astruc (physician to Augustus the Strong of Saxony), who supported animalculist preformation because he thought it could best explain how children resemble their parents: while ovists struggled to show how the momentary and possibly nonphysical interaction of semen and ovum could communicate the father's characteristics, animalculists could rely on other mechanisms that came into play at conception or during pregnancy. Astruc's own view was that the animalcule containing the future child was partly remolded by the mother, as it had to squeeze through an opening into the ovum where it could develop.[48]

Yet animalculism had some serious drawbacks that drastically curtailed its popularity. The idea that male semen exercised a purely occult power—the *aura seminalis*—had taken root by the 1680s after gaining powerful support from Harvey and later ovists.[49] Persuading contemporaries that animalcules might play a part in generation was no easy matter, especially given their close and lasting association with parasitic worms, which directly influenced the nineteenth-century coinage *spermatozoa* (literally, "animals in the seed").[50]

Far the greatest challenge came from the early recognition that of the millions of sperm in every drop of ejaculate, only one would develop into a child. What of the remainder? The destruction of millions of preformed humans with each emission offended against belief in a benevolent God and mainstream perceptions of the simplicity and uniformity of nature.[51] Answers to this objection ranged from Leeuwenhoek's metaphor of an apple tree (every pip is capable of becoming a tree, but plenty of them do not grow) to Astruc's hairsplitting insistence that sperm contained merely a *potential* organism, which only became a full, ensouled fetus after implantation in the ovum. Notwithstanding such efforts, animalculists were never able satisfactorily to respond to this problem.[52]

To these particular difficulties can be added those faced by preformationism as a whole. Preexistence involved spectacular degrees of miniaturization, as each generation was sized in proportion to its immediate predecessor as a sperm or egg is to an adult human. Projected back 6,000 years to the Creation, or forward to all future generations, the levels of minuteness involved were eye-watering and hardly credible.[53] More troublesome was the problem of explaining modifications to the preformed seed that would be necessary to account for the resemblance of the child to parents or more distant relatives, and for birth defects or "monstrosities." Until it began slowly to lose credibility after the first quarter of the eighteenth century, the maternal imagination offered a wide-ranging answer to these problems. But the desideratum remained a convincing mechanical explanation that was not forthcoming given the unpredictability of these phenomena.

VITAL FORCES: NEO-EPIGENESIS, C. 1760–C. 1840

The final drawback with preformation was so severe that it ushered in a new embryological paradigm. Preformation was unable to explain the regeneration of the freshwater polyp (*Hydra vulgaris*), a creature whose regenerative powers were first investigated by the Swiss naturalist Abraham Trembley in 1740. He cut a polyp in two and observed both parts regenerate into smaller but complete polyps. It retained this ability even when diced up into dozens of pieces, and Trembley showed that a polyp could survive even if turned completely inside out. This phenomenon demonstrated that the orderly sequence of preexisting seeds could be disrupted by arbitrary violence without apparent effect, and, worse, it seemed to entail the division of the animal "soul." It was this last aspect that proved most troubling because the production of multiple complete souls from one individual implied either that redundant souls existed throughout the parent organism, that matter had its own creative powers, or that God miraculously intervened with each regenerative act. All these possibilities were unpalatable given the epistemic constraints of a mechanical universe based on nature's order and frugality, so even those preformationists who made strenuous efforts to incorporate regeneration, like Charles Bonnet, remained unable to explain the apparent division of souls.[54]

Regeneration did not immediately or completely overturn preformationism. It remained debatable whether the polyp was an animal or vegetable: if the latter, its disruptive potential could be contained, although several animal species with regenerative powers (such as the crayfish) were swiftly identified. Principally, however, many continued adherents to preformationism because they were committed to a mechanical universe and, in the words of Jacques Roger, "preferred ignorance to the abandonment of clear ideas. Other minds would be needed to look for other systems."[55]

What came to replace preformationism was a novel form of epigenesis that resolved Harvey's conundrum by using a new conception of organic matter subject to vital forces. Much of the impetus for this development came from the work of Newton, in which a mechanical universe was subject to occult forces, such as gravity, whose effects could be observed, described, and mathematically predicted if not explained. It is therefore unsurprising that the first major works in the life sciences to embody this new paradigm were produced by French Newtonians such as Buffon, whose theory of generation we encountered in chapter 3. Buffon invoked occult forces in the shape of the "interior mold" to direct embryonic development and employed a modified form of panspermism in which male and female semen were composed of "organic molecules." His was no epigenetic theory, being instead a sort of metamorphosis where the form of the future organism was determined immediately at conception by the interior mold. Its unpopularity in certain quarters owed much to the materialism implied by its mix of panspermism and occult forces.[56] Other Newtonians were more authentic epigenesists—for example, Maupertuis, who based his model of fetal development on the example provided by the formation of crystals.

This sort of iatrochemistry was an advance on the mechanism that had failed so dismally to explain epigenesis. Although it, too, fell short of offering a convincing physiology of generation—chemical analogies were scarcely better than mechanical—it helped to establish that living matter might contain intrinsic properties or be subject to hidden forces, such as irritability or attraction. As such, chemical analogies paved the way for more convincing models of epigenesis without functioning as such themselves. This has an analogue in the intellectual career of the towering figure who did so much to establish attraction and irritability as vital forces, Albrecht von Haller. He began his career following in Boerhaave's footsteps as an animalculist preformationist. Thanks to Trembley's experiments, he supported epigenesis between *c.* 1740 and the late 1750s, arguing in 1747 that an "attractive" force served to form the embryo. In 1757, Haller publicly converted back to preformationism, this time as an ovist but in line with the theory of preformation advocated by Bonnet, which was more sophisticated than simple *emboîtement*. Haller's change of heart was principally due to his observations of the developing chick, but it was also because his critical engagement with Buffon led him to reject the idea that a simple "force" of nature could consistently direct the development of the human form.

Haller maintained this viewpoint even when confronted with more sophisticated forces, such as the "essential force" (*vis essentialis*) described by the young scholar Caspar Friedrich Wolff in his dissertation *Theoria generationis* (1759). Wolff sent Haller a copy of his work in the hope of winning him back for epigenesis, but it sparked a lengthy controversy between them that was never resolved. Wolff's theory was based on the ability of animal and vegetable fluids to solidify: each plant or animal is produced from the secretion of fluids that then solidify into structures in a serial order, and as each part solidifies, it acquires vessels and vesicles—that is, becomes "organized"—from the movement of fluids into it. This process occurs throughout the lives of plants, but in animals it takes place only in the embryo. For Wolff, the *vis essentialis* is the only explanation for the movement of fluids from the soil and throughout the plant; in the chick embryo, the growth that occurs before a visible heart has developed must also be due to the "essential force." The controversy between Wolff and Haller went unresolved partly because the two spoke across each other but also because the differences between them were predominantly philosophical.[57]

While more sophisticated, Wolff's *vis essentialis* was not fundamentally different from other simple and unidirectional forces, like attraction or irritability, that fell short of explaining epigenetic development.[58] What was required was a genuinely teleological agency that could consistently produce the regular species form, while allowing for limited variation in addition to hereditary and accidental phenomena, such as resemblance to parents or birth defects. This is precisely what Blumenbach provided in the early 1780s as he developed his theory of the "formative drive" (*Bildungstrieb* or *nisus formativus*). He conceived this as an organic version of a Newtonian force that "continues to act through the whole life of the animal, and that by it the first form of the animal, or plant is not only determined, but afterwards preserved, and when deranged, is again restored."[59] Blumenbach's account did not actually explain very much, but he did make some specific empirical claims. One was that embryogenesis could only occur once the "unorganized matter of generation" was placed in its proper destination, be that the womb, egg, or soil. The drive would fail to operate in the case of missing or incompatible generative matter, thus preventing both spontaneous generation and hybrids arising from wholly mismatched species. But the drive, which he named as the "chief principle" of generation, growth, nutrition, and regeneration, might change direction (*Richtung*), which could explain both monstrosities and the regular degeneration of species into varieties. Indeed, in the second (1781) and third (1795) editions of his *De Generis*, Blumenbach listed the *Bildungstrieb* alongside climate and nutrition as a cause of the production of varieties and subspecies.[60]

The impact of this theory in Britain and Germany is hard to exaggerate. The second (1791) edition of *Über den Bildungstrieb* was the only one of Blumenbach's reproductive or racial works to be translated into English and published before the mid-nineteenth century, and his ideas had a profound influence on the next generation of German biologists and on Kant, especially on the second part of the *Critique of the Power of Judgment* (1790). As several historians have shown, Kant and Blumenbach never quite understood each other's projects—Kant understood the *Bildungstrieb* as a purely heuristic device rather than a teleological cause fully resident in nature—but they certainly appreciated each other and were often lumped together by their supporters. Between them, the long-standing controversy over the process of generation was largely resolved, and despite the slow death of preformationism (particularly in France), the epigenesis they both championed soon became scientific orthodoxy. The *Bildungstrieb* and its account of generation was part of the medical curriculum at Edinburgh by the turn of the century and went from strength to strength, garnering support and attracting substantial empirical backing from the likes of Karl Ernst von Baer, whose *Entwicklungsgeschichte der Thiere* (1828) announced his discovery of the mammalian egg to the world.[61]

Racial Embryology: Race Mixing, Heredity, and Hybridity

The Preformationist Impediment to Racial Theorizing, 1670–1740

The prominence of major racial theorists—particularly Buffon, Blumenbach, and Kant—in this final, vitalist phase of early modern embryology testifies to the close links between racial and embryological thought. It also highlights their apparent absence in earlier periods, which is puzzling given the important work done in both fields in the late seventeenth and early eighteenth centuries. Indeed, Malpighi's anatomical research laid the essential foundations for subsequent discussions of both ovism (in *De formatione pulli in ovo*, 1672) and skin pigmentation (in *De externo tactus organo anatomica observatio*, 1665), and his identification of the epidermal "rete mucosum" or Malpighian layer as the seat of skin color shaped the discourse on race for the following two centuries.[62] The classificatory schemes of William Petty (1677) and François Bernier (1684) were also produced at this time, alongside speculative accounts of the peopling of the (New) World by John Toland (1695), for example, all of whom devoted space to reproductive matters in their writings.[63]

Why, then, are there so few examples from these years of work that integrated embryology and the study of human origins? The answer lies in the fundamental incompatibility of preformationism with the two prevailing theories of human origins, monogenesis and polygenesis, both of which relied on an embryology that allowed for transformative change. Monogenists insisted on the role of secondary factors—principally climate—in producing difference, while polygenists insisted that differences were original. Both schools, however, held to the view that bodily changes, once produced, could be spread or perpetuated by sexual intermixture. Neither the inheritance of acquired characteristics nor the admixture of parental features was compatible with preformationism, and while this embryological paradigm was in the ascendant, the prospects for integrating theories of race and generation were dim.

A clear example of how ill-suited preformationist embryology was to serving theories of racial difference is provided by the foremost advocate of polygenesis in early eighteenth-century Britain, the naval surgeon John Atkins. He visited West Africa in the early 1720s while surgeon on board the *Swallow* and wrote up his experiences on his return to England. The discussion of skin color in his *Treatise on [. . .] Chirurgical Subjects* (1724) set a pattern followed in later works, in which generation was treated as a marginal aspect of human difference. He began by acknowledging Malpighi's discovery of the seat of skin color but highlighted its limitations as an explanation of diversity. Atkins then tackled the monogenist orthodoxy head-on, listing five reasons why climate was an unconvincing engine of change. Only one of these mentioned generation, and then as a secondary factor: "No *European* totally changes by length of Cohabitation with them ["Negroes"], and in Generation begets a Mulatto Race, which ever remain so."[64] Having, as he believed, eliminated the monogenetic explanation, Atkins embraced the only remaining alternative: "From the whole," he concluded, "I imagine that White and Black must have descended of different Protoplasts, and that there is no other Way of accounting for it."[65]

Atkins repeated this discussion practically verbatim throughout the 1730s in the various editions of his *Navy Surgeon* (1734, 1737) and *Voyage to Guinea* (1735, 1737).[66] In 1742, a substantially revised edition of the *Navy Surgeon* was published that featured a radical revision of his views on human diversity alongside serious reflections on generation. Most significantly, Atkins removed his argument for different "protoplasts," abandoning polygenesis. His was no full-throated conversion to monogenism, however: he continued to enumerate the weaknesses of climatic accounts, and his new explanation for diversity—"the Soil each is bred and nurtured in"—could just as easily support polygenesis as monogenesis.[67] The difference was made by the place of mixed-race couples in his thinking, as racial mixture went from being casually invoked to vitally important. Indeed, Atkins based his argument for soil almost entirely on its ability to overpower differences arising from the "Mixture of different Nations and *Complexions*." England's soil was therefore able to stamp its diverse population with shared characteristics that distinguished them from their closest neighbors.[68] Mixed-race children drove him to abandon preformation because he observed that they were so constantly the median between their parents in both likeness and color that neither ovism nor animalculism could be correct. His newfound religiosity and ongoing hostility to the slave trade certainly contributed to Atkins's change of heart, but it was also a product of considering generation in depth for the first time. His example shows that once preformation was renounced, either theory of human origins could be advanced; while embraced, neither was tenable.[69]

The Mixed-Race Datum (1): The Equality of Parental Contributions

Atkins's example gets to the very heart of the nexus of racial and reproductive thought that is the subject of this book. A vital bond linking these discourses was their shared evidence base and at its center was sexual mixture. More specifically, it was one particular datum: the fact that mixed couples produce children that are fertile and consistently combine their parents' features and color. The mixed-race datum was used in theories of generation and of race to prove one or both of two things: that the parents contribute equally to the fetus, and that black and white people belong to the same biological species. Only a handful of authors followed Blumenbach in employing the datum in both ways, but by examining its uses alongside the ideas it was used to combat or support, we can appreciate the breadth and depth of the links it forged between these discourses.[70]

As one of the strongest proofs of the equal role of parents in generation, the mixed-race datum provided vital support for two-seed embryologies.[71] Their resurgence after 1740 was partly due to the datum and helps to explain its prominence in embryological debates. Examples that used black-white couples to illustrate hereditary phenomena remained common, and the Aristotelian tale of the mixed couple from Elis was recycled frequently in early modern texts.[72] Only later appeared the specific use of the mixed-race datum to prove the equality of parental contributions (Leonardo da Vinci was among the first to deploy it), and later still emerged the abstract concept of "heredity," which completed its migration from the legal to the biological sphere only around 1830.[73] Between these developments the perennial interest in hereditary phenomena waxed and waned, but it intensified steadily from the mid-seventeenth century for several reasons. Political upheaval led to a prioritization of the smooth inheritance of legitimate authority, the economic and social dangers of spurious offspring inheriting wealth haunted patriarchal society across the period, and the growing fashion for using mathematical reasoning in discussions of social and political problems produced an interest in predicting hereditary characteristics.[74]

In the following century, thinking about the hereditary assumed two principal forms. One was the discussion on hereditary disease that got underway after 1770. Conducted mainly by physicians, it was highly pragmatic in orientation and avoided the minefield of speculations about generation, although the mere existence of illnesses that could be inherited from either parent did suggest that both parents contributed to the fetus.[75] Attention was focused on aspects neglected by embryologists such as homochrony, the tendency for a hereditary disease to appear at a similar stage in the life of parent and child. Interpreting homochrony was especially challenging because it involved illness that did not break out immediately but remained latent. Many medics explained latency by arguing that what was transmitted was not the disease itself but rather a predisposition to it that produced the full-blown disease when later triggered by "causes occasionelles."[76]

For all they avoided generation, debates about hereditary disease overlapped with those on human diversity. Historians have only recently begun to examine the important relationship between race and disease in this era, and few have yet focused on hereditary disease.[77] The belief that black people were immune to illnesses such as yellow fever was not uncommon even before the Philadelphia epidemic of 1793–1794, and several physicians claimed this immunity was heritable by mixed-race offspring.[78] Equally common was the idea that blackness entailed greater susceptibility to certain diseases, such that some (for instance, yaws) were thought unique to black people.[79] Writers as varied in their racial views as Blumenbach, John Anderson, Edward Long, and Benjamin Rush believed that black skin was at least a symptom of infirmity, if not a disease in its own right. Yet most commentators went no further than to assert an analogical link between hereditary disease and blackness, arguing that skin color was transmitted by parents *in much the same way* as hereditary illness. Indeed, the strategy for explaining the latency of hereditary disease—inherited predisposition triggered by an external stimulus—was very close to the widespread opinion that all babies are born with a ruddy colored skin that attains its proper coloration only after exposure to the elements.[80]

The second form of thinking about the hereditary concerned the resemblance of offspring to parents and other relatives. Many participants in this discussion, which featured the heaviest use of the mixed-race datum and engaged more intensively with theories of generation, were themselves reproductive theorists.[81] It was a perplexing topic and remained so until Mendel. While there was little doubt that children typically blended the features of their parents and often resembled more distant relatives, it was impossible both to predict the nature and extent of these similarities and to explain why mixture almost always stopped at the sexual organs, which resembled those of only one parent. It was a moot point whether intellectual or moral traits could be inherited, and if so, whether this was limited to the parents or could extend to wet nurses, whose milk might convey such characteristics.[82]

Many relied on pangenesis to explain similarities, especially before preformationism became dominant, but others—Harvey, for example—were skeptical, mindful of Aristotle's critique of pangenesis and its difficulties in explaining resemblance to distant relatives.[83] Animalculists found succor in phenomena that undermined pangenesis, such as Jewish boys born with a foreskin or Amerindian men growing beards when their fathers had eradicated theirs.[84] But pangenesis remained an unresolved issue by the time epigenesis underwent its eighteenth-century renaissance, leading some to adopt similar explanations to hereditary disease: lecturing in the late 1770s, John Hunter stated that any likeness of children to their grandparents was the result of "a latent hereditary Disposition."[85] Others attributed similarities to the maternal imagination, a faculty whose unpredictability fit well with the vagaries of resemblance.[86] At the same time, similarities between parents and children were mobilized repeatedly to attack preformationism or its individual variants, especially animalculism.[87] The most popular examples for this purpose featured parents whose differences were so conspicuous yet so unmistakably mixed in their children that biparental influence on the fetus was incontestable. Significantly for what follows, the two most popular examples almost invariably appeared side by side after 1750: the mixed-race datum and the quintessential animal exemplar, the mule.

If the general debate on resemblance pitted preformationists against epigenesists, the more particular controversy over the inheritance of acquired characteristics divided the parties along the lines of racial theory.[88] Such a division was inevitable given the reliance of monogenesis on this mechanism to explain the persistence of racial features. Polygenists like Lord Kames and Charles White scornfully rejected the idea while their opponents—especially Lord Monboddo and Samuel Stanhope Smith—put up an equally strident defense.[89] These battle lines were maintained by ordinary students and members of the public who contributed papers on human variety to two of Edinburgh's learned societies, the Royal Medical Society and the Society for Investigating Natural History. Polygenist speakers such as the Antiguan Alexander Macpherson (1785), William Bourke (1793), and Alexander Robertson (1798) rejected the inheritance of acquired characteristics, while it was supported by monogenists like Richard Millar (1786), John MacFazdean (1788), and William Webb (1794).[90]

The exception to this rule was James Cowles Prichard (1786–1848), whose denial that acquired characteristics could be inherited was almost as unwavering as his lifelong commitment to monogenesis.[91] Both positions were evident in his paper on "the Varieties of the Human Race" delivered to the Royal Medical Society in its 1807–1808 session, in which Prichard argued that changes introduced artificially (e.g., circumcision) or from the action of the climate could not be inherited. The only heritable changes are those produced by "domestication" (among humans, "civilization"), and he firmly rejected the idea that black skin was a result of hereditary jaundice or any other disease. Prichard developed and maintained these views without recourse to any specific embryological theory until the second (1826) edition of his *Researches into the Physical History of Mankind*, when he broadly followed Kant and Blumenbach.[92] Prichard was not uninterested in reproduction before this date—he had been apprenticed to a man-midwife in 1802 and attended James Hamilton's lectures on midwifery in 1806—but he left little evidence of his early embryological leanings, and the claim that he was preformist before 1826 is entirely conjectural.[93] He gave ground on the inheritance of acquired characteristics only at the very end of his life, when the inroads made by a resurgent polygenism forced him into sacrifices to shore up his beloved monogenesis.[94]

The Mixed-Race Datum (2): Hybrids, Mulattoes, and the Unity of the Human Species

A novel feature of this resurgent polygenism was the claim that different human races constituted separate species. Earlier polygenists had stopped short of this mark: Atkins (1724) wrote of "different protoplasts," Francis Lodwick (c. 1675) of different "Originalls of Mankind," and John Toland (1695) "against the Propagation of all Mankind out of one Single Male and Female," but none argued for distinct human species.[95] Indeed, such a claim was untenable so long as "species" denoted a reproductive community, and the interfertility of humans was undeniable. This did not prevent slavery's apologists from asserting that black Africans were animals or the product of interspecific sex, something that Morgan Godwyn used the mixed-race datum to refute in 1680.[96] Although such extreme views were rarely expressed, the related notion that different races constituted separate species within a single human genus had gained a foothold by 1775, thereafter becoming an increasingly prominent strand of polygenesis.[97]

The immediate cause of its rise was confusion regarding the scope of the interbreeding criterion. As outlined in chapter 3, examples of apparently fertile hybrid offspring led Buffon and Linnaeus in the 1760s to redeploy the criterion to govern the boundaries of *genus* rather than *species*. Polygenists seized this opportunity to insist on separate but interfertile human species, justifying this move with a variant of the "nature's parsimony" argument encountered earlier: other animal genera consist of several interfertile species, "Why then," asked White, "should we seek to infringe this apparent law of nature in regard to man, unless to serve an hypothesis?"[98] Dogs were fellow candidates for such recategorization, being both interfertile yet far more diverse in appearance than animals of indisputably distinct species, such as the horse and donkey.[99] For those who lacked pressing religious scruples and who doubted the emergence of present-day human diversity from a unitary origin within the previous 6,000 years, reimagining humanity as a genus containing interfertile species was an attractive prospect. But while authors such as Kames and (arguably) White apparently sought only clarity in their anthropological studies, others capitalized on this terminological flux for less benign purposes.[100]

The most notorious of these was Edward Long, whose *History of Jamaica* (1774) defended slavery by reframing humanity as a genus containing two species, one of which—"Negroes"—he argued was closer to animals than to the rest of humankind. In support, Long singled out the "oran-outang" (chimpanzee) for its advanced cognitive abilities and physical similarities to humans.[101] But his lengthy discussion of this creature did not seek to prove its humanity. Rather, his argument for the vastly inferior status of "Negroes" depended on presenting the "oran-outang" as an animal—perhaps even sui generis and uncomfortably close to humanity—yet the mental and physical equal of the "Negro." His probing, provocative argument that the "oran-outang" *might* be demonstrably human if properly educated to speak aimed not at asserting their humanity but at showing how physical form and intellect can be separated: "The supposition then is well founded, that the brain, and intellectual organs, so far as they are dependent upon meer matter, though similar in texture and modification to those of other men, may in some of the Negroe race be so constituted, as *not to result to the same effects*."[102] Making the "oran-outang" unequivocally human would have considerably blunted Long's argument, not only because "Negroes" would not then be the lowest human species but also because he cannot have been unaware of the ridicule that greeted Lord Monboddo's efforts to assert its humanity.[103] As it was, his extreme views undoubtedly contributed to his decision to publish anonymously, which led to no small confusion over the authorship of the *History* that extended even into the next century.[104]

Long's acknowledgment of "Negro" humanity was begrudging in the extreme, and he spared no effort to insinuate that they were indeed animals. For example, even while he argued that the interbreeding criterion should be applied to a human *genus*, he continued to use it in the traditional manner to distinguish between species. He became notorious for arguing that mulattoes[105] were far less fertile—even infertile—with one another than they were "with a distinct White or Black."[106] He was explicit about the aim of this argument: "It tends, among other evidences, to establish an opinion, which several have entertained, that the White and the Negroe had not one common origin."[107] Long's intellectual dishonesty, evidenced throughout the *History* by extensive plagiarism, is also apparent from this opportunistic use of two incompatible principles—that humanity can be reimagined as a genus because it consists of interfertile species, and that human species can be distinguished by their mutual infertility—in the service of subjugating "Negroes."[108] Compounding this ploy, he insisted that mulatto interfertility could be proven only by fulfilling his unfairly stringent criteria.[109]

Although widely recognized as inconsistent with experience, the idea that mulattoes were inter se unfruitful became an article of polygenist faith in the century following Long.[110] To understand why, it is necessary to distinguish between different scales of experience. At the individual level, of course, countless mulatto couples had perfectly healthy and fertile offspring, and examples poured in from readers who rejected Long's claims. But his gaze was fixed on the broader panorama of Jamaican society, in which the scarcity of such couples during his time on the island (1757–1769) was a prominent feature. Reasons for this include a small mixed population, numbering only *c.* 2,000–3,000; white sexual predation on the enslaved population, exacerbated by an extremely high white sex ratio (more than two males to every female); and the related sociocultural tendency for both sexes to prefer as light-skinned a partner as possible.[111] This tendency is clear from contemporary testimony and later studies testifying to the premium held on whiteness in Caribbean societies, even if it also fed the racist trope that non-white women are inherently attracted to white men.[112] Acknowledging the rarity of mulatto couples, the abolitionist James Ramsay highlighted the obvious truth that this social fact could not support a biological argument for their infecundity.[113] Long himself admitted that his assertion was entirely circumstantial in the notes for an unrealized second edition of the *History*: "The Truth is, the mulatto men can seldom get the women, they are meat for their masters."[114] Ninety years later, the belief in limited mulatto fertility was still being supported by such contingent factors.[115]

Long's biological speculations, while idiosyncratic, reflected the broader discussion of hybrids among naturalists and animal breeders, particularly the widely accepted principle that fertile hybrids are less so with one another than with parent species. This tenet was maintained even by the botanist Joseph Gottlieb Kölreuter, who in the 1760s demonstrated the possibility of transmuting plant species. He produced a hybrid ("bastard") of two species of tobacco, and by continually crossing with one or another parent species, he was able to transmute each into the other. His work profoundly influenced Blumenbach, who thought it "must cure the most partial advocate for the theory of evolution [preformation], of his error," and whose texts on the *Bildungstrieb* widely promoted Kölreuter's ideas.[116] Long and Kölreuter both focused on the pairing of mixed with unmixed individuals, whether hybrids with a parent species or mulattoes with a white or black partner. It was the exclusionary sting of such ideas that the army surgeon John Hunter sought to draw in his 1775 Edinburgh thesis, when he argued that it was precisely because the offspring of mixed children could be completely reabsorbed into one or another parent race that they and their parents are all of the same species.[117]

Most of Long's contemporaries agreed, maintaining that hybridization could only apply to humans as an analogy.[118] This included many apologists for slavery, who were embarrassed by his choice to fight on such uncertain ground.[119] Hybrids provided a decisive example against preformation, but they involved questions of reproductive compatibility that only the most committed polygenists believed relevant to humans.[120] One New York physician, testifying in an 1836 slander suit in which one woman had accused another of committing bestiality with a dog and producing pups, summarized the agreed preconditions for the generation of hybrid offspring, which had hardly changed since Aristotle: both species must be domesticated, they must possess identical reproductive organs and have the same gestation period, and "they must agree in their manner of copulation."[121] He might also have mentioned climate, as Buffon in particular thought that hybrids were fertile only in warm climates—he wrote in 1776 citing the case of a fertile mule in Saint-Domingue (although his Scottish translator undermined the point by describing one from Forfar).[122] Domestication was especially important for overcoming the natural "repugnance" felt by different animal species toward mating with one another, which meant hybrids were almost never found among wild animals.[123] Governed by such factors, hybridity clearly had little direct relevance to human reproduction.

Except in one long-standing debate that was reenergized in the eighteenth century thanks to Edward Tyson's pioneering dissection of a chimpanzee in 1699. Speculation about the possibility of human-ape hybrids was as old as knowledge of anthropoid apes, but while they could not be entirely ruled out, their danger was contained by the comforting assumption that any such creature would be infertile. That is, until Tyson's dissection raised the possibility that the "orang outang" (chimpanzee) might be human after all: he found that its larynx was structured "exactly as 'tis in *Man*," and as nature does nothing in vain, possessing the requisite anatomy implied the ability to speak.[124] Evidenced in part by the continued circulation of stories that apes abducted black women, this single anatomical finding fueled speculation that continued until the Dutch anatomist Petrus Camper (1722–1789) published the results of his dissection in 1779, which showed that apes did not possess human speech organs after all.[125]

In this 80-year window, during which the "orang outang" might have been human, everybody agreed that the principal means to decide the question was reproductive. But the experiment necessary to provide conclusive proof of the "orang outang's" humanity—producing fertile offspring with a human—was highly problematic. In chapter 3 we encountered Rousseau's meditation on its moral and practical difficulties, concerns shared by Monboddo. Yet a rumor that circulated in German-speaking Europe in the late 1770s implied that the British had overcome such scruples and that a London prostitute had been hired as a guinea pig in the experiment to prove the "orang outang's" humanity. Disappointed that this story turned out to be false, the German naturalist Peter Simon Pallas impatiently demanded in the pages of his *Neue Nordische Beyträge* that "corrupt Europeans in the [African or West] Indian plantations" should "in a hot climate procure bastard offspring of an Orang-utang and a female slave for European observers."[126] He stated that the apes brought to Europe were unsuitable for this purpose because they were young, tamed, and placed in a "deleterious and deadly climate."[127] For Eberhard August Wilhelm von Zimmermann, with whom the rumor originated, climate was less important for the potential success of the experiment than the sexual configuration of its subjects: he originally believed that the London attempt had failed because the ape was overexcited at the prospect of mating with a human female; more likely to succeed, he speculated, was the mating of a human male with a female "orang outang" with whom he was familiar.[128]

Racial Embryology at Mid-Century: Maupertuis and Parsons

Eighteenth-century embryological and racial thought shared points of contact beyond the mixed-race datum, which the remainder of this chapter will survey using the work of two polymaths of the mid-eighteenth century—one English, the other French—who were deeply interested in reproduction and in the origins

Fig. 4.2. Benjamin Wilson, *James Parsons* (1705–1770). © National Portrait Gallery, London.

and perpetuation of human difference. These men, James Parsons (fig. 4.2) and Pierre Louis Moreau de Maupertuis (fig. 4.3), were both famous for the breadth of their scientific interests, many of which they shared, and they assumed important positions in the leading scientific societies of their day. Both worked on the life sciences in the decades (*c.* 1740–*c.* 1760) when preformation was in crisis following Trembley's studies of the polyp, experiments that Parsons repeated and illustrated in 1742 as an assistant to Martin Folkes, president of the Royal Society. Widely read in both Britain and France, they differed somewhat in their views, particularly concerning embryology, but shared a common muse in the shape of black Africans with albinism or hermaphroditism who

Fig. 4.3. Robert Levrac-Tournières, *Pierre Louis Moreau de Maupertuis* (1698–1759). 1740 DLn-055-Hz II-83 GK I 10166./Eigentum des Hauses Hohenzollern, Georg Friedrich Prinz von Preußen, SPSG/Wolfgang Pfauder.

were exhibited in London and Paris.[129] They were both fascinated by cases of apparent categorical transgression—black people who were not black, and males and females whose sexual anatomy and physiology did not conform to expectations—and their writings combined race and embryology in a manner sufficiently influential to shape the thought of figures from Buffon to Kant.

Maupertuis: Albinism and the Maternal Imagination

Maupertuis, who headed the Royal Prussian Academy of Sciences from 1746, devoted most of his intellectual career to mathematical and geographical investigations, but in the playful, risqué prose of *Vénus physique* (1745), he clad a highly innovative set of reflections on heredity and reproduction in titillating garb. He was prompted to write by the exhibition in Paris of an albinotic child born to enslaved black African parents. Unlike Voltaire's written response to this spectacle, Maupertuis initially paid little attention to the child and none to its color.[130] Instead, his anonymous *Dissertation physique à l'occasion du nègre blanc* (1744)—which later became the first part of *Vénus physique*—focused entirely on embryological matters. He began by sketching preformationist thought, highlighting its myriad problems, before arguing in favor of epigenesis. Looking back to the previous century, he sought to revive some of Harvey's ideas and rehabilitate the basic elements of Descartes's theory—namely, that generation proceeded from the mix of parental seeds and there was no difference between living and inert matter. Responding to the principal weakness of epigenesis, Maupertuis suggested that the orderly development of the embryo resulted from a process similar to chemical laws of attraction. He soon recognized that this solved very little and postulated that the individual germs, collected from parents and gathered together in their seed, might have a sort of "memory" that controlled organized development. Hereditary phenomena guided his reflections, particularly the questions of familial resemblance and the generation of monsters (a topic of controversy at the French Académie Royale des Sciences), and he was uncompromising in his rejection of the maternal imagination as a possible explanation for either.

In the following year, Maupertuis broadened and deepened his reflections on generation by incorporating a discussion of human diversity in the second part of *Vénus physique*. This began with a breathtakingly short survey of human phenotypical differences (principally pigmentation) before concentrating on the phenomenon of black albinism and the failure of ovism or animalculism to offer a plausible explanation. For Maupertuis, who in common with almost all of his contemporaries assumed that mankind was originally white, the problem of black albinism was the same as that of resemblance to distant family members: how can one account both for the resemblance of children to their parents (which might involve a mixed coloration) and for exceptional cases where children resembled distant ancestors more than their immediate progenitors? For him, the most plausible solution lay in a system of pangenesis, where "parts similar to those of the father and mother, being the most numerous as well as having the greatest affinity, will be the ones to unite most easily and then will form animals like the ones from which they came." But "chance or a shortage of family traits will at times cause other combinations, and then we may see a white child born of Black parents, or even a Black child from white parents, though this is a much rarer phenomenon than the former."[131] Maupertuis couched these powerful, prescient ideas in tentative terms, but while he made a major contribution in exposing some of the links between racial and reproductive phenomena, the brevity of his treatment calls for a deeper consideration, first of albinism and then of the maternal imagination.

ALBINISM

Albinism was inherently important for racial thought because of the light it promised to shed on humanity's original color. Until the 1770s, it was thought to affect only non-Europeans, partly because albinotic peoples were especially conspicuous among non-white populations.[132] Those in the New World were prominent early—hence the widespread adoption of the Iberian term "albino"— but by the eighteenth century significant numbers had been identified in Africa and among its diaspora, particularly in the Caribbean, and were known as *leucæthiops*.[133] All the major racial theorists had something to say on the topic, above all Blumenbach, whose very first paper (1784) to Göttingen's Royal Scientific Society was on albinotic eyes.[134] Typical was Buffon's argument, here recapitulated by Oliver Goldsmith: "We have frequently seen white children produced from black parents, but have never seen a black offspring the production of two whites. From hence we may conclude that whiteness is the colour to which mankind naturally tends; for, as in the tulip, the parent stock is known by all the artificial varieties breaking into it; so in man, that colour must be original which never alters, and to which all the rest are accidentally seen to change."[135]

A remarkable feature of the conversation on albinism is the consistency with which the opposite conclusion was denied: that black was humanity's original color and albinism might show how whiteness originated. Those few authors who argued for humanity's original blackness, such as Prichard and John Hunter (1728–1793), used albinism to support their case.[136] But for reasons that had much to do with intellectual tradition and religious orthodoxy—not to mention racial pride—albinism was generally interpreted as an atavistic phenomenon, a sporadic reversion to the original whiteness of humankind.[137]

This interpretation was contested, especially by those who rejected monogenist ideas. Long's unpublished remarks homed in on its chief weakness. It was a "total contradiction" to claim that white is the original ("primitive") color of humankind and that the proof is that white parents do not produce black children. How then, he asked, could black skin color ever have arisen? The only answer, he triumphantly concluded, was that black pigment originated solely from black ancestors, who must therefore have been created separately from white people.[138] This conclusion notwithstanding, Long's objection was shared by those monogenists who believed that white people did not possess the anatomical or physiological prerequisites for dark skin color (whether the rete mucosum, the fluid *æthiops animal*, or some other entity) and thus were unable to produce a dark-skinned child.[139] Even those who attempted to keep their options open, such as the Rouen anatomist Claude-Nicolas Le Cat, who wrote at length on the *æthiops animal* but was careful not to dismiss the possibility of black children born to white parents, raised important objections to the consensus view. Le Cat pointed out that if albinotic people were a "throwback" to an earlier state of humankind, it is improbable that such a reversion would be limited to color alone. Among other features likely also to be affected is physiognomy, "yet the white moor has all the traits of the black."[140]

Albinism was also a stimulus for thinking about how features might be perpetuated by endogamous reproduction. Most authors recognized that albinism occurred apparently randomly, yet observation of albinotic children born to albinotic parents, coupled with reports of an albino "nation" in Central America or Africa, prompted speculation on the possibility of making this characteristic permanent via in-group reproduction. Such conjecture induced Maupertuis to write about generation, and a host of other authors discussed the question, for and against, with reference not only to albinotic people but also to Jews and other groups who apparently shared "national" features that were originally accidental. Against the backdrop of the precipitous decline in the authority of climatic explanations for physical differences, endogamous reproduction offered a lifeline for committed monogenists, helping them to account for the persistence of features, irrespective of local environmental conditions.[141]

Albinism, however, turned out to be a less than reliable support for this idea. The growing awareness after 1770 that it was a disorder affecting all humans had two linked consequences. The presence of albinotic people within all human societies meant that they were not a distinct race or variety, and some claimed that a feature of the disorder was infertility, which would prevent the emergence of a self-contained "albino" population.[142] Interest instead focused on the origins rather than the consequences of albinism: was it a genuine hereditary disease, and how was it related to other pigmentary phenomena, such as vitiligo ("piebaldism")? Some held that these conditions were the result of mixed black-white parentage, but Buffon and others denied this, pointing to the "mulatto" offspring of such couples.[143]

MATERNAL IMAGINATION

A more common explanation than mixed parentage was the maternal imagination. The belief that a pregnant mother's imagination could exercise a direct influence on her unborn child to cause birthmarks and other congenital phenomena—from minor irregularities to severe malformation—had existed in the West since the pre-Socratics. Despite, or rather because the precise workings of this influence remained unknown, it was a potent source of physical alterations and was often used to explain irregularities associated with human diversity. Albinism was a prime example, and the maternal imagination was consistently mentioned even when authors rejected it: for every Le Cat who used the imagination to account for albinism, there was a Parsons who felt obliged explicitly to discount its influence.[144]

The association of the maternal imagination with physical abnormality has obscured its role in explaining aspects of regular (i.e., nondefective) reproduction. Within epigenetic and pangenetic embryologies, it could explain the consistency of species form and why children were born without the deficiencies of their parents (e.g., sons of circumcised fathers were born with an intact foreskin). The imagination was especially important within mechanistic embryologies, where it could account for family resemblance and, ultimately, the inheritance of acquired characteristics. At least one historian has argued that it received greater attention in the seventeenth century—precisely the years in which mechanism dominated embryological thought—due to the growing inability of climate to explain the external appearance of black Africans.[145]

The close relationship between the maternal imagination and human difference was shaped by its use to explain more than just abnormalities. As François Jacob recognized over fifty years ago, the imagination renders the fetus susceptible to outside influences.[146] It thus plays a role similar to that of nutrition in climatic models: whether via a mother's nutritional intake or her mental processes (which were always believed to result from some external stimulus), the fetus bore traces of its wider environment. As an explanation of skin pigmentation, the most prominent example stemmed from antiquity yet was both apocryphal and somewhat ironic. As retold by Daniel Turner in 1714, "*Hippocrates* did once deliver a Noble Woman, like to suffer as an Adult'ress; for that the Husband and she being white, her Child was born of the Ethiopic Complexion, which the sage old Man imputed readily to a Picture he had observed hanging in her Chamber, exactly resembling the Infant, and which he found she had been often very intently viewing."[147] This story was already recognized as apocryphal in the early modern era (its erroneous attribution to Hippocrates originated with Erasmus), and its irony rested in the fact that Hippocrates, of all the ancient thinkers on generation, was probably the most skeptical about the maternal imagination.[148] Along with the biblical story of Jacob, who produced speckled sheep by placing branches partially stripped of their bark in front of watering troughs so that ewes would gaze on them while mating, the pseudo-Hippocratic anecdote appeared wherever early moderns used the maternal imagination to explain phenotypical features.[149]

These explanations corresponded with wider trends in racial theorizing. In the primarily theological discourse of the seventeenth century, the imagination was used to plug the main gap in the Curse of Ham and explain how enslaved descendants of Canaan came to have black skin. By the early eighteenth century, the tone had become more naturalistic. For some such as the Jesuit missionary Lafitau, the imagination both produced and perpetuated black skin. Others argued that the imagination produced color but that it was perpetuated by other means, such as endogamous reproduction: this was the perspective taken by those who attributed albinism to the maternal imagination. Still others believed that the imagination was responsible only for maintaining the color of a population once it had originated from climatic, cultural, or other factors. This viewpoint was represented by the popular question-and-answer periodicals *The Athenian Oracle* (1691–1697) and *The British Apollo* (1708), as well as some later theorists, such as Stanhope Smith.[150]

By the middle third of the eighteenth century, the scientific respectability of the maternal imagination was in terminal decline. This was principally due to the damage inflicted on the idea by such cases as the Mary Toft fraud (1726), in which a woman claimed to have produced a litter of rabbits, sparking a major controversy.[151] One further reason was the supersession of preformationist embryologies after 1740, and an instructive example is offered by a case that appears in Linnaeus's *Sponsalia plantarum* (1746) but was first mentioned by the Danish anatomist Thomas Bartholin in the 1650s. The case involved a "common Ethiopian" imprisoned in Copenhagen and his white lover, who together produced a son that was as white as his mother except for his penis, which was black. Bartholin attributed this "wholly to the imagination of the mother, which seizing the desired part with a fixed and vigorous mind, impressed its colour on the offspring."[152] Linnaeus, by contrast, said nothing concerning the imagination, using the case solely to discredit preformation: it "evinces, that the beginnings of the coming fetus by no means lie hidden in one sex only."[153]

Reports of the death of the maternal imagination were, however, greatly exaggerated. It continued to feature not only in literature—most famously in the works of Smollett, Sterne, and Goethe—and in popular culture and belief but also in scientific works and the curricula of medical schools, albeit as a "vulgar error" to be combated.[154] Some polygenists used it in the 1770s and 1780s to deny human unity. In 1773, the Edinburgh student Edward Mease employed it to counter Parsons's argument that black albinism proves the original whiteness of mankind.[155] Fifteen years later, John Lindsay (Anglican rector of St. Katharine's, Jamaica) sent a book-length manuscript to Edward Long that claimed it was not necessary to look further than the imagination when seeking to explain differences between parents and their mixed children; any such differences therefore represented no serious challenge to polygenesis.[156] This late bloom of the theory at the hands of polygenists is probably what led Stanhope Smith to attempt to rehabilitate the imagination for monogenism by strictly limiting its purview to the perpetuation of existing features.[157]

The creative power of the imagination was indeed its most dangerous feature, especially when it was consciously manipulated. The biblical and pseudo-Hippocratic anecdotes were part of a long tradition in which parents sought to determine the appearance of their children by directing the mother's attention away from the ugly and deformed and toward the beautiful and well proportioned. This tradition was alive and well in the eighteenth century, featuring in popular sex manuals like Nicolas Venette's *Tableau of Conjugal Love* (1687) or *Aristotle's Master-piece* (1683), as well as such dedicated publications as Claude Quillet's phenomenally popular *Callipædia, or the Art of Getting Pretty Children* (1655). This advised the expectant mother to "Let all her Thoughts on lovely Objects dwell, / Her Fancy in the nicest Charms excel: / No other Phantoms should admittance find, / But what may please the Eye and cheer the Mind."[158] And it was to obviate "Frights and pernicious Impressions" that in 1729 the economic writer Joshua Gee suggested rounding up and placing in a hospital all those who displayed their deformities on London's streets.[159] One Dutch mother apparently used her imagination to counter her imagination: concerned that she had been "surprised with the sight of a *Negro*" and that her child was likely to be black, she washed herself head to foot and focused strongly on her ablutions. "She was at length delivered of a child that was indeed white, yet those parts excepted, where the water . . . had not touched; such as the interstices of the fingers and toes, . . . where the manifest tokens of blackness appear'd."[160]

As this example shows, such a plastic faculty could be easily abused by women themselves. Adulterous women might think of their husbands while pregnant with spurious offspring, a possibility that led to massive anxiety among husbands thereby robbed of one of the few proofs of their paternity (resemblance to their children).[161] Indulgence of the cravings of pregnant women was strongly advised, lest frustration and obsession produce deformity, but tales abounded of this indulgence being abused.[162] Racial counterfeit was also a concern. Lindsay worried that black African parents might have "children, comely in their features, . . . their complexions of the most healthy black, and their whole looks bearing all that sensibility of face, which we commonly ascribe to a Creole Negroe of Creole Parents." He was reassured by the reflection that "this is a phenomenon of rarity—and still the African Race remains distinct."[163] It was the very possibility of abuse that led Alexander Hamilton, professor of midwifery at Edinburgh, to reject the power of the imagination on moral grounds: "Such an effect also would be [a] subversion of all moral society, as by that [means] women might have criminal conversation w[ith] Africans or the like w[ith]out observation."[164] Reproductive satires played on this fear, and one broadside of *c.* 1820 described a mother who had apparently craved charcoal during her pregnancy, leading her husband to be unsuspecting when she produced a black baby. He soon discovered her store of unconsumed charcoal, however, whereupon his neighbors reported the frequent visits of a black sailor to their house.[165] This anecdote indicates quite precisely the extent to which the maternal imagination, or its capacity to cause such pigmentary change, was viewed as credible at the time. The humor derives from the initially successful deception of a credulous husband and a denouement showing the indulgence of his wife to have been foolish. It also depends on the deceit being at least superficially plausible, so while the imagination clearly no longer had the explanatory power it enjoyed a century earlier, it was by no means entirely defunct at the end of our period and indeed survived as a folk belief into the twentieth century.[166]

Parsons: Hermaphroditism

One author who consistently rejected the imagination's influence was James Parsons. Unlike Maupertuis, whose oeuvre in the life sciences is rather compact, Parsons published widely on a range of such topics, from anatomy to botany to zoology. A trained physician, he was educated in Ireland and France before settling into a medical practice in London that focused on obstetrics. He lectured on various reproductive topics, and his first major foray into print was his *Mechanical and Critical Enquiry into the Nature of Hermaphrodites* (1741). He, too, had been prompted to write by the exhibition of an African, in this case an "Angolan" hermaphrodite[167] that from mid-1740 could be seen near Charing Cross.[168]

Parsons clearly set out his motivation for writing and the viewpoint he took: "The Arrival of the *Angolan* Woman encouraged this Undertaking, both from the Belief of the Vulgar concerning her, and the Sentiments of others, who would allow her no Sex but the Masculine."[169] He roundly rejected the existence of genuine hermaphrodites ("the Belief of the Vulgar"), arguing that they were overwhelmingly females with abnormally large or protruding genitals. He began by sketching the various laws about hermaphrodites and the state of early modern knowledge concerning them. In subsequent chapters, he examined a wide range of fields, including ancient medicine, embryology, and ethnology, to enumerate the reasons against a "Hermaphroditical Nature" in humans; provided a historical and scientific account of the causes of hermaphrodites; and gave an overview of what other modern authors had written about the subject. Steadfast in his denial that hermaphrodites exist, Parsons afforded them no role in the determination of their own identity.[170]

Not that he denied the existence of hermaphrodites in other species. His *Philosophical Observations on the Analogy Between the Propagation of Animals and that of Vegetables* (1752) outlined a teleological, purposive model of animal development that was heavily influenced by natural theology. Parsons accepted the existence of hermaphrodites among snails and worms because these animals were slow and vulnerable to predators: such creatures encounter their own species so infrequently that regular sexual dimorphism would excessively limit the chances of successful sexual intercourse and thereby pose an existential threat to the species. For the same reason, polyps regenerate because they are so easy to destroy. This consequentialist model of animal reproduction left no place for the human hermaphrodite because God's purposes for man could be realized without them. Yet Parsons testifies to the currency and importance of the idea of human hermaphrodites by the substantial attention he devoted to them over the years.[171]

HERMAPHRODITES: REAL AND IMAGINED

Hermaphrodites are an important point of contact between racial and reproductive discourses for several reasons. Not only were they thought to be more common among peoples that inhabited the "torrid zone" but they also exemplified the increasingly irresistible urge to read identity from the body, the limited success of such efforts, and the use of reproduction to this end. As a growing body of scholarship has demonstrated, attitudes to hermaphrodites changed dramatically, particularly in the seventeenth and eighteenth centuries, as acceptance of a degree of ambiguity in sexual identity gave way to the demand for absolute certainty. We must be careful not to overstate this development, as the earlier acceptance of ambiguity was never total. For example, the Gnostic belief that Adam may have been a hermaphrodite before the Fall was viewed as heretical from the outset and rejected anew in the seventeenth century.[172]

The very possibility of hermaphroditism also depended heavily on which account of sex determination of the embryo was believed. The view that hermaphrodites were genuinely intermediate between male and female was based on Galenic and Hippocratic thought that respectively emphasized the location of the fetus in the womb or the preponderance of strong/male or weak/female seed as determinants of sex. Many early modern texts addressed the determinative role of location, including the possibility that the womb contains seven cells (in the rightmost three of which a male child could develop, in the leftmost three a female, and in the remaining cell, a hermaphrodite), but by the late seventeenth century, most rejected this explanation in favor of the composition of the seed. Sex thus existed on a continuum, heavily polarized to be sure, but with plenty of intermediate positions. Aristotelian embryology, on the other hand, saw sex as fundamentally binary, and its adherents interpreted hermaphrodites as genuine males or females, albeit with redundant or doubled genitalia. Aristotelians thus denied the reality of hermaphrodites and regarded them as monsters.[173]

Despite serious competition from a resurgent Galenic/Hippocratic tradition that continued to prevail in vernacular literature, the binaristic perspective on sex difference came to dominate learned embryology in the seventeenth and eighteenth centuries. Nicolas Venette's 1687 quinquepartite classification of hermaphrodites was much discussed by authors on the subject in the next century.[174] As is clear from figure 4.4, Venette viewed three specimens as fundamentally male, a fourth as female, and the fifth, or true hermaphrodite, as neither. As he wrote, "These sorts of person are more a species of eunuch than of hermaphrodite, their penis is good for nothing and their periods never flow."[175] In other words, and as figures ranging from Parsons to Blumenbach and Lawrence would argue, hermaphrodites either do not exist or are incapable of reproducing their kind, nullifying the categorical threat they posed.[176] This threat was made clear by Parsons: "That there are certain Limits set to the Things of Generation appears no where better than when Animals of different Species meet and copulate; the Animal that is the Product of such a Congress is no side capable of producing an Off-spring like itself, to this there is an absolute *ne plus ultra*, and why? Because, indeed, if such were capable of Generation, we should, by degrees, have a new set of Heterogeneous Animals upon Earth."[177] The reassuring sterility of hermaphrodites was extended to animals in studies of hermaphrodite cattle, dogs, and poultry, and although

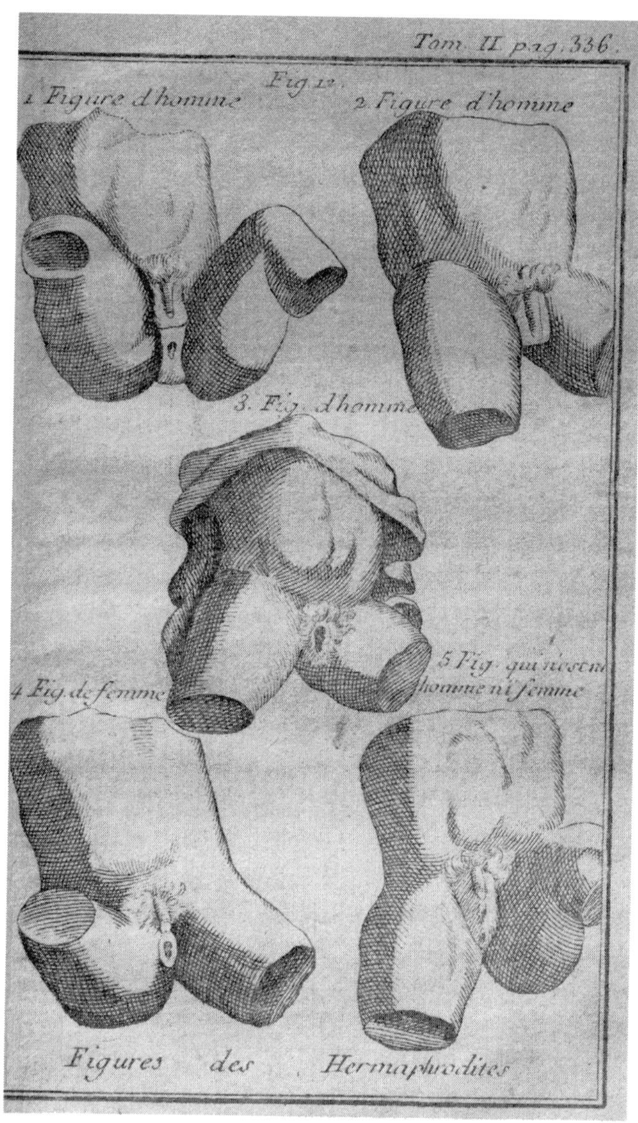

Fig. 4.4. Plate from Nicolas Venette, *La génération de l'homme, ou tableau de l'amour conjugal, considéré dans l'état du mariage* (Londres [Paris?], 1771). Courtesy of the Wellcome Collection, London.

genuine hermaphroditism existed among "lower" animals and plants, it was less threatening there than among creatures closer to humankind.[178]

Hermaphroditism was further neutralized by being remote. It was strongly associated with hot climates: one author estimated in the late 1760s that more hermaphrodites were produced in one year at Surat than in fifty in Sweden.[179] And although they were thought to be "extremely common" in "all those Places especially that are nearest the Equinoctal Line," Africa was especially important.[180] African hermaphrodites had featured heavily in influential earlier texts, and it was the exhibition of an Angolan that had prompted Parsons and others to write.[181] The link to Africa persisted even when the existence of "true" hermaphrodites was denied, and most writers agreed with Parsons (against Venette) that the vast majority of so-called hermaphrodites were females with an unusually large clitoris. These tended to live in hot regions "where the Nonnaturals themselves conduce much to the general Relaxation of the Solids, and, consequently, this unseemly Accretion of that Part."[182] We saw in chapter 3 how European writers were content to dwell on the supposedly prodigious genitals of black Africans, and those discussing hermaphroditism readily followed suit.[183] Everard Home, for example, could not resist talking of the large clitorises of "Mandingo" and "Ibbo" women in his article on a hermaphrodite dog (1799).[184] This genital focus meant in turn that female circumcision featured prominently in discussions of hermaphroditism. Such practices were allegedly taken as presumptive proof of the existence of hermaphrodites in a society: John Ovington reported in 1696 that the "vulgar Opinion" that the native peoples of the Cape of Good Hope were hermaphrodites was "founded only upon Conjecture, . . . the Reason of it, . . . was because the Female Parts were cut in the Fashion of small Teats hanging down."[185]

These aspects of hermaphroditism also featured in accounts of America. De Pauw devoted a chapter of his *Recherches philosophiques sur les américains* (1768–1769) to the hermaphrodites of Florida, whose enslavement apparently meant that young girls with prodigious genitals gladly underwent circumcision to avoid a life of servitude.[186] Eighteenth-century authors reported the confusion of earlier reports or were confused themselves concerning sexual and sartorial customs among the Amerindians of Florida, Mexico, and California, whose practitioners were interpreted as hermaphrodites.[187] As chapter 5 shows, such confusion was heightened by encounters with Pacific sexualities, even though hermaphrodites were imagined in the region long before Europeans voyaged there in the 1760s.[188] In the meantime, Europeans confronted instances of apparent hermaphroditism in their midst, both at home and in the colonies, where

Fig. 4.5. Isaac Cruikshank, *A Man-Mid-Wife* (1793). BM 8376. Courtesy of the Wellcome Collection, London.

such cases as that of the Virginian Thomas/ine Hall (1629) have attracted the interest of modern scholars.[189]

Hall's case exemplifies the strong association of hermaphroditism and individual behavior. Their alleged fornication had first brought them to the attention of the authorities, who then uncovered a long history of Hall flitting between gender identities before and after coming to America. In Europe, sexual

Fig. 4.6. "Mademoiselle de Beaumont, or the Chevalier d'Eon," *London Magazine* (1777). © Trustees of the British Museum.

misbehavior was a substitute for climatic influence as an explanation of corporeal change: female masturbation was viewed as a risk factor for hermaphroditism because it enlarged the clitoris.[190] Anti-masturbation tracts across the eighteenth century explicitly scaremongered that enlarged clitorises were monstrous, that they were the inevitable product of solitary sex, and that their possessors would be viewed as hermaphrodites, "look'd upon as a Creature of vile

Deformity, bringing a Shame upon both Sexes."[191] The linkage between individual behavior and hermaphroditism was strengthened as belief in real hermaphrodites declined in the eighteenth century, replaced by metaphorical or social hermaphrodites. Any possibly ambiguous identity could be framed in terms of hermaphroditism in order to express hostility, for which the representation of men-midwives and the Chevalier d'Eon (1728–1810) in figures 4.5 and 4.6 are cases in point.

Yet the connection between individual behavior and hermaphroditism existed already in the sixteenth century, and this had ramifications for legal thought and practice, an important domain that spearheaded the growing intolerance to hermaphroditic ambiguity. In the early seventeenth century, Edward Coke was able to define an heir as "either a male, or female, or an Hermaphrodite, that is, both male & female. And an Hermaphrodite . . . shal be heir, either as male or female, according to that kinde of the sexe with doth preuaile."[192] This principle remained current into the eighteenth century. When in 1719 a woman stood trial for bigamy at the Old Bailey, her defense was that she had married "a Monster, a Hermaphrodite" named Constantia Boon. Boon appeared in court and acknowledged being a hermaphrodite; once it was clear "by her own Confession as well as other Evidences that the Woman was more predominant in her than the Man," the defendant was acquitted.[193] The vital legal question here was how to establish satisfactorily which sex predominated, and as Lorraine Daston and Katherine Park have observed, an important development of the sixteenth century was the increasing use of outside testimony in preference to the hermaphrodite's own assertions. Hence in addition to "her own Confession," Boon's femininity was proven by "other Evidences."[194]

What these were is unclear, but as hermaphroditism was increasingly denied by jurists in the eighteenth century, the circle of relevant witnesses drastically narrowed. After mid-century—and certainly by Lord Justice Mansfield's ruling concerning the primacy of expert witnesses in *Folkes v. Chadd* (1782)—only "men of science" were acceptable. And they were only too happy to oblige: as one author on medical jurisprudence stated in 1788, "A perfect hermaphrodite or a being partaking of the distinguishing marks of both sexes, with a power of enjoyment from each, is not believed by any one ever to have existed."[195] The trend in legal thought was obvious: in his *Commentaries* (1765–1769), Blackstone followed Coke by distinguishing between monsters in human and nonhuman form yet failed even to mention hermaphrodites, writing simply that heirs were either male or female.[196]

Two final points should be emphasized. The first is that the earlier acceptance of ambiguity in the legal sphere was more apparent than real: the law of inheritance or bigamy demanded that people be regarded as either male or female, and hermaphrodites had to resolve their sex into one or the other. The second is the clear contrast that existed between establishing the sexual identity of alleged hermaphrodites and the legal status of alleged slaves in court. "Men of science" were called in both sorts of cases to testify to sexual and racial identity, but they could only pronounce on the sex of hermaphrodites based on the body they inspected; in freedom cases, the ambiguities of skin coloration and other components of racial identity forced such witnesses to turn to genealogical and even reputational evidence.[197] Had this not been possible, many more plaintiffs (of the pitifully few that successfully sued for their freedom in the antebellum South) would undoubtedly have been condemned to slavery by the lights of a perverted science.

What is clearly exhibited by the intellectual terrain covered by Maupertuis and Parsons, if not by the men themselves, is an intolerance of ambiguity. Maupertuis delighted in the reproductive conundrums he exposed and sought actively to create more by breeding different species in his own private menagerie.[198] But contemporaries saw in albinism, the maternal imagination, and hermaphroditism the possibility to obtain clarity on the mysteries of generation and human diversity, only to be disappointed at every turn. Albinism turned out to be a pathological condition found among all human communities, frustrating the ambition to explore how "racial" characteristics might be established via endogamous reproduction; the maternal imagination was always unstable, being entirely at the mercy of the mother's own fancy before it was struck below the waterline by the Toft fraud and slowly foundered over the course of the eighteenth and nineteenth centuries; and hermaphroditism so offended against the notion of sexual dimorphism that its destructive potential had to be contained, first within the presumption of sterility and later by the denial of their physical existence.

Likewise, Gulliver's madness was arguably a result of being confronted with his own ambiguous status. He grappled with a truth on which most eighteenth-century naturalists were insistent: there was no such thing as a human hybrid or a true human hermaphrodite. Unable to deny his Yahoo nature while aspiring to the dizzying heights of Houyhnhnm perfection, Gulliver's only recourse was to reject himself, his wife, and his children, and to recommend that his own species be exterminated through sterilization. He offers a hybrid's-eye view in which intractable paradoxes of origins and identity were exposed by reproductive means (such as the question of indigeneity raised by the spontaneous generation of the Yahoos, or that of Gulliver's identity provoked by his attempted rape) and could not be resolved without massive violence.

The dangers of the sexual interaction of colonial populations and sexual attraction between colonizers and colonized were there for all to see in the pages of *Gulliver's Travels*. They introduced confusion where clarity was sought and even once found it often proved less than consoling, hence the malevolent efforts of Long and others to isolate white populations from their mixed offspring by erecting barriers of differential fecundity or hereditary immunity. Yet for most people, the simple fact that mixed populations existed served to demonstrate that all humans are one species, as well as that both mother and father contribute equally to the making of their children.

The demand for clarity was inevitably frustrated in a science as opaque as embryology, but its students continued the search in other areas. One of these was language. Both Parsons and Maupertuis wrote works on the history of language, but whereas Maupertuis plowed the same furrow as Condillac, Rousseau, and others writing on the origins of language and its relationship with ideas, Parsons followed Leibniz and foreshadowed Prichard in using language as a tool to analyze the relationships between human groups, past and present. In his long, ponderous *Remains of Japhet* (1767), Parsons sought to show that the Celtic languages of the British Isles were the remains of the Japhetan tongue. He was hardly unique in such an endeavor, but his was the first account written by someone who had a profound interest in both embryology and human racial diversity. As chapter 5 will show, such projects received a huge boost in Parsons's final years, as European encounters with Pacific peoples raised more questions about sex, reproduction, human diversity, and language.[199]

RACE, REPRODUCTION, AND SEX BEYOND THE BRITISH ATLANTIC, 1750–1840

"This Race Benign"

Race and Reproduction in the Pacific, 1760–1820

In April 1768, Louis-Antoine de Bougainville made landfall at Tahiti. On board his two vessels were some 400 weary souls, one of whom was Jeanne Baret, a woman disguised as the valet of her lover, the expedition's naturalist Philibert de Commerson. Remarkably, despite long months at sea in cramped conditions, living cheek by jowl with over a hundred sailors, she had managed to evade detection. This lasted only until the first Tahitian came aboard, a young man named Ahutoru. He immediately identified Baret as female, pointing at her and saying *"Ayenne* which means girl in the local tongue"; on this occasion, she escaped discovery by luckily standing next to an armorer whose "very effeminate features" led the ship's company to believe that he was the object of Ahutoru's interest. Baret was less fortunate the following day when she went ashore with Commerson to botanize, for as soon as they landed "all the Savages tugged at her in one direction or another with a thousand shouts of *Ayenne* and already one determined individual was carrying her off within sight of her master as a starving wolf carries off his prey in the presence of the shepherd. It required an officer whom chance had brought there to drive off this multitude with his sword and frighten the runner who let go. She was put back in the boat right away and sent back to the ship. After this incident, there was no question of going ashore."[1]

Nor could Baret maintain her disguise any longer. Her secret revealed, she was landed with Commerson at the Île de France (modern-day Mauritius) on the home voyage, possibly to avoid awkward questions that would have been posed had the voyagers arrived in France with a woman aboard. They instead brought Ahutoru, whose ability to discover Baret's sex so quickly and with such certainty had amazed the French. It was put to the test in Paris, where Ahutoru was studied by the polymath Charles Marie de La Condamine. In his report, La Condamine stated that the Tahitians' "sense of smell is very acute, such that they can distinguish the odour of a woman from that of a man. A passenger . . . had brought with him a woman disguised as a man; they recognised her sex despite her disguise and indicated this with very energetic signs."[2]

La Condamine's comments, published in a supplement to Bougainville's voyage in 1773, highlight three distinct and interrelated aspects that are at the heart of this chapter.[3] First, he asserts that sexual difference so inheres to the human body that males and females produce different odors. Europeans produce these aromas but cannot detect them.[4] Second, this sensory capacity is attributed to Tahitians in particular and is a specific mark of their exoticism. Third and consequently, this ontological assertion of aromatic sexual difference carries with it an epistemological claim: not merely is the unique sensory capacity of Tahitians a mark of their alterity, but empirical methods are exalted as a (perhaps *the*) crucial means of distinguishing between the sexes.

Astounded by Ahutoru's sensory abilities, the French were less complimentary about his intellect. His lack of French led to suppositions that he was stupid or lazy (or both), which Bougainville did his best to rebut but which circulated widely and made their way across the Channel. When the British themselves brought a Pacific islander back to Europe in 1774—the Raiatean Mai—he received much the same treatment. Condescending assumptions about the simple languages and limited ideas of "savage" societies were applied to peoples of the South Pacific as they had been to Africans and Amerindians for centuries. And while it remained conceivable that the linguistic and other intellectual abilities of such "backward" peoples might improve as they advanced toward civilization, doubts became increasingly widespread in the 1770s as these peoples seemed to be no closer to European standards even after centuries of contact.[5]

This was an ominous development. In the same years as climatic accounts of human difference were under increasing pressure for much the same reason—no discernible change despite centuries of contact—Europeans were devising explanations for just such a combination of enhanced sensory abilities and limited intellectual powers as Ahutoru and Mai seemed to embody.[6] One of the most notorious was proposed by the Mainz anatomist, Samuel Thomas Sömmerring, whose *Über die körperliche Verschiedenheit des Mohren vom Europäer* [On the Bodily Differences of the Moor from the European] (1784)[7] claimed that the brains of black Africans differed from those of Europeans, not in overall volume but in containing more nerve fibers. Sömmerring interpreted this as proof that a greater proportion of the African brain was dedicated to processing raw sensations, making it less capable of contemplation and higher reasoning.

Examining European discourses on the Pacific and its peoples from the mid-eighteenth to the early nineteenth centuries, this chapter emphasizes that such a close linkage of language, sense, and sex was at the heart of newer forms of somatic identity (i.e., identities based primarily on the physical body). Language, specifically speech, served to distinguish humans from other animals, and its relative complexity acted as an empirical marker of intellectual ability. This had long been recognized, but what Pacific encounters demonstrate with particular clarity is the increasing importance of the *aesthetics* of the physical body—not merely its beauty or ugliness but also its own sensory capacities (an important but often overlooked eighteenth-century meaning of "aesthetics")—in the development of new racial ideas.

On the other hand, the Pacific was nothing if not confusing for the first generations of Europeans to engage with the region. Its complex societies and sexualities confounded European efforts to comprehend in terms of preexisting categories of (noble) "savagery" and "civilization." And while it serves to highlight how Europeans were rethinking identity in terms of the body and intellect, it also underscores the limitations of that process. The Pacific had long been a site of heteronormative sexual fantasy, but its cultures featured same-sex activities and third-sex identities that were hard for Europeans to fathom. For Europeans, the region encapsulated a reproductive paradigm that not only shaped perceptions of sex but also ensured the continuing vitality of the older meaning of "race" as shared lineage. Language, too, contributed to this effort, being an empirical marker of humanity, intellectual ability, *and* common descent. The different appearances of people later labeled "Melanesian" and "Polynesian" prompted the suspicion that they belonged to different "races," yet what cemented this view was less physical anthropology than comparative linguistics.

These issues are explored in this chapter's three sections. The first examines the sexualized depiction of the Pacific in Europe, as contained in the travel narratives produced by British and French explorers (above all Cook and Bougainville) or texts deriving from them, such as Diderot's *Supplément au voyage de Bougainville* (1772) and the cycle of satirical pamphlets produced in response to the Cook voyages in Britain. Stress is laid in this section on the heavily reproductive focus both of European fantasies and of early imperial projects in the region, which reveal complex attitudes to sexual mixture between Europeans and Pacific islanders held by both groups.

The second section concentrates more particularly on the place of the Pacific in European racial thought of the second half of the eighteenth century. Pacific encounters had a transformative effect on racial ideas, as the aesthetic judgments leveled at encountered peoples dovetailed with the more fundamental role of the senses in concepts of somatic identity, both in terms of discernible characteristics sensed by the observer and the apparently enhanced sensory capacities of the person or group observed. Ahutoru's ability to smell sexual difference was only the most noteworthy of such capacities, which themselves served as markers of a difference between Pacific and European bodies that was being essentialized as it was described.

The final section addresses the key empirically sensible characteristic that indicated intellectual ability: language, particularly speech. This, too, was liable to subjective interpretation according to European standards of simplicity or sophistication and served as a helpful adjunct, especially for those who sought to consign peoples with simpler languages to the lower ranks of the human species. Language was, however, always a more precise marker of specific rather than racial difference, and in the case of the Pacific it ultimately served to undermine the effort to fix identity entirely in the human body. Studies of Pacific etymologies and language history, while helping to cement the notion of two "races" in the region divided by linguistic heritage, sustained the older concept of race-as-lineage and thwarted the possibility of unequivocal knowledge of racial identity without recourse to extracorporeal data.

The South Seas: A Reproductive Paradise?

Occurring while Europeans were busily creating modern sexual difference by embodying their gender ideals in human physiology and anatomy, voyages to the South Seas were fortuitously timed. The visibility of Pacific bodies—and especially Pacific sex—marked an ideal opportunity for Europeans to confirm that their own gender ideology was "natural." Basing European ideals on a natural entity (the human body) would be all the more convincing if traces of those ideals could be found among a people that "came unimproved out of the hands of nature."[8] In other words, the credibility of the project whereby European *cultural* norms received *natural* justification in the fabric of the body would be enhanced by finding the rudiments of those norms in a putatively "natural" society. Consequently, serious effort was put into "discovering" elements of a bourgeois gender order in the "primitive" island societies encountered.[9]

But, as with race, things were never that simple. It was relatively straightforward to imagine Pacific societies as "natural": the broad commitment to stadial models of progress led many to believe that exotic societies were so far behind their own that they opened up a perspective on their own ancient past. Unencumbered by civilization, the customs of these societies were deemed to be fundamentally dictated by nature. It was more difficult, however, to digest the range and strangeness of the many sexual customs that voyagers encountered. As with race, the effort to tidily assimilate exotic customs into existing models of thought foundered on the rocks of a messy empirical reality.

These difficulties were visible even before Europeans set foot in the region, by which time it had already enjoyed a lengthy career as a surface onto which they projected their fantasies.[10] Relatively unknown and of uncertain extent, the "South Seas" was an ideal setting for utopias of all kinds. Here, sex was conspicuous, a result as much of its value as a means of social critique as of the conventions of the utopian genre, in which the regulation of reproduction for the good of the state had featured since Plato's *Republic*.[11] Most South Seas utopias followed this pattern. Bensalem, the utopian city in Bacon's *New Atlantis* (1627), possessed both an institution where fertile hybrids were produced (Salomon's House) and a strict marital code that organized unions on rational foundations, privileging reproduction. *The History of the Sevarites* (1675) by the Huguenot refugee Denis Veiras (Vairasse) was set in part of the Terra Australis Incognita[12] and featured a society in which marriage was compulsory, but temporary sexual partners were granted to travelers. Its leading city, Sevarinde, was situated in such a rarefied climate that the slightest moral delinquency (sexual or otherwise) left a visible mark or deformity on the body. This widely read text therefore contained not only a typical utopian fantasy of optimal sexual regulation combined with the availability of (exotic) females but also the guiding conceit of later models of racial and sexual difference: that outward signs could be a reliable guide to inward constitution. Tellingly, such transparency was made possible by the local climate.[13]

Another seventeenth-century South Seas utopia featured a complex, obscure sexual order in which the "hermaphrodite" body reinforced the correspondence between interior constitution and outward shape. The narrator of Gabriel de Foigny's *La terre australe, connue* (1676, English translation 1693), Jacques Sadeur, had been born a "hermaphrodite" and was shipwrecked after a storm, landing on a coast populated by xenophobes who were themselves "hermaphrodite" and usually killed strangers on sight. They happily made an exception because of Sadeur's physical similarity and apparent bravery. Foigny's text is unusually tight-lipped about sex itself: an extreme privacy surrounded the social organization of reproduction among the Australians, who rejected sexual dimorphism and destroyed male and female infants at birth as "monsters." An "Old Man," who sought both to protect and enlighten Sadeur, explained that sexual dimorphism is for the animals and that humanity consisted in the perfect union of body and mind. Without saying as much, this emphasis on the correspondence between the body and an implicitly unsexed mind reinforced François Poulain de la Barre's claim that "the mind has no sex," made only three years earlier (1673).[14]

Eighteenth-century fantasies introduced more explicitly "racial" themes. For example, the eponymous protagonist of Simon Tyssot de Patot's *Voyages et aventures de Jacques Massé* (1714 × 1717, English translation 1733) describes the postmortem examination of an enslaved black man who had committed suicide. Set in 1643, Massé anticipates Malpighi's 1665 discovery of the "rete mucosum" as the seat of skin color, which serves as a pretext for an endorsement of polygenesis. Other philosophical asides, clearly intended to showcase the learning of its author, included a critique of animalculist preformationism and a rejection of spontaneous generation. Polygamy among the social elites also predictably appears in Massé's description of the paradisiacal society he encountered.[15]

Later imaginative fictions in English also combined a Pacific setting with discussions of marriage, sexual reproduction, and mixed offspring. In Robert Paltock's *Life and Adventures of Peter Wilkins* (1750)—described by a contemporary reviewer as "the illegitimate offspring of no very natural conjunction betwixt *Gulliver*'s travels and *Robinson Crusoe*"—the protagonist marries a winged woman named Youwarkee and they produce seven children, some winged like their mother, others not. Flight also featured in *The Life and Astonishing Adventures of John Daniel* (1751), in which the protagonist discovered that his shipmate, Thomas, was a woman named Ruth whom he then married. Daniel later persuaded Ruth to allow their children to couple incestuously with one another after the example of Adam and Eve's children, an argument predicated on a commitment to monogenesis that also featured in Henry Neville's miscegenist fantasy, *The Isle of Pines* (1668).[16]

Finally, *The Travels of Hildebrand Bowman* (1778), written in the wake of the first British and French voyages to the Pacific, imagined five nations in the South Seas that corresponded exactly to the different stages of stadial theory and which featured various marital customs, including polygamy. In "Olfactaria,"

Bowman overcame his reluctance to marry on being told that marriage was casual and that children would be cared for by the nation, "as we think it every man's duty to raise children for the state." Abandoning his pregnant wife to continue his travels, Bowman visits an idealization of Elizabethan England ("Bonhommica") and of present-day Britain ("Luxo-Volupto"), where flying people were also to be found. Unlike Youwarkee's people, only the dissipated developed wings in Luxo-Volupto; they therefore served, like the Sevarite climate, as an unconcealable, indelible marker of moral turpitude. The recurrence of this device—an obvious and unmistakable physical indicator of moral condition—in South Seas fiction points to one of the driving forces behind the development of somatic forms of identity: anxiety concerning the deceptiveness of appearances, which made the discovery of such an indicator highly desirable. This became a matter of urgency when actual encounters with Pacific sexualities left Europeans highly confused.[17]

Most bewildering was the apparent sexual abandon of several cultures encountered in the early voyages, which was radically different from anything deemed respectable in Europe. Worse still, societies that were otherwise praised (such as the Maohi of Tahiti) often had the least acceptable sexual customs, while those that were generally denigrated (e.g., New Caledonia or Vanuatu) might feature a praiseworthy sexual morality.[18] The pains taken to locate a naturalized bourgeois morality in Tahiti were considerable. To do so, Europeans needed to resituate Pacific societies in relation to the European past, to marginalize unacceptable sexualities, or to employ a combination of these approaches. In seeking to make the European gender order seem natural, the importance of "discovering" it in the Pacific was so great that efforts were made to defend even highly objectionable native customs.

Initial contact, however, was suffused with perplexity. John Hawkesworth, whose *Account of the Voyages Undertaken [. . .] For Making Discoveries in the Southern Hemisphere* (1773) was the official but heavily edited account of Cook's first voyage (1768–71, in HMS *Endeavour*), confessed an inability to understand Tahitian sexuality in any terms: "There is a scale in dissolute sensuality, which these people have ascended, wholly unknown to every other nation whose manners have been recorded from the beginning of the world to the present hour, and which no imagination could possibly conceive."[19] Particularly difficult to comprehend were instances of public sex, such as the coupling of a young man with a barely pubescent girl at "Point Venus" in May 1769.[20] Despite the enormous popularity of Hawkesworth's *Account*, his description of such events produced a fierce backlash that may have contributed to his early death. Divines such as John Wesley exhibited the greatest outrage: "'A nation . . . without any sense of shame! Men and women coupling together in the face of the sun, and in the sight of scores of people!' . . . *Hume* or *Voltaire* might believe this: But I cannot."[21] More moderate was the Bluestocking Elizabeth Montagu, who wrote to her sister in July 1773 expressing disapproval of Hawkesworth but unable to "enter into the prudery of the Ladies, who are afraid to own they have read the Voyages, and less still into the moral delicacy of those who suppose the effronterie of the Demoiselles of Ottaheité will corrupt our Misses; if the girls had invented a surer way to keep intrigues secret, it might have been dangerous, but their publick amours will not be imitated."[22]

Such homegrown efforts to dismiss the threat posed by Tahitian morals bolstered attempts to understand Tahitian sexuality in European terms that were already underway during the *Endeavour* voyage. Cook was more circumspect than Wesley, stating that the "Point Venus" ritual "appear'd to be done more from Custom than Lewdness," a comment Hawkesworth later rendered as a question whether shame is "implanted by nature or superinduced by custom."[23] This conundrum is a noteworthy example of the provisionality of identity in the Pacific discussed by Jonathan Lamb: if shame is instinctive, then Tahitian sexuality rests on an unknown custom sufficiently powerful to overcome instinct; if, however, it is customary, then European morality rests on an equally powerful and obscure custom.[24] Hence there was a fundamental barrier to reading Tahitian society in European terms. The literal reading of Tahiti as an arcadian New Cythera by Commerson, Bougainville, and others was unpersuasive, but the alternative view of Maohi society as a "schoolroom of natural law" that exempted exotic females from European regulations threatened to overturn European order.[25] Either Tahiti was unconvincingly identified with the classical past or it proved that European morality was not natural/universal.[26]

Two possible resolutions existed. A more convincing model relating Tahiti to the European past could be applied, or it might be possible to argue that the Tahitian sexual order was fundamentally equivalent to the European and that any differences were superficial. Both featured in the textual responses to Cook's second voyage (1772–75), the latter being Cook's own approach. His ghostwriter, John Douglas, sought to expunge the reputation for promiscuity imposed on Tahitian women by Hawkesworth's *Account*, writing that those who engaged in sexual commerce with European sailors were unmarried and prostitutes, and that married women (or unmarried of the better sort) were innocent of this behavior.[27] In marked contrast to his first voyage, where Cook baldly stated that "chastity indeed is but little Valued especialy among the middle people," he now confined aberrant sexual behavior to the same margins it occupied in Europe, asserting the congruence of the European and Tahitian sexual orders.[28]

This was not universally popular. For the teacher, writer, and minister Vicesimus Knox, the taint of Pacific eroticism remained too strong. He complained that "several celebrated writers have inferred the absurdity of many, not only innocent, but laudable and beneficial notions and practices, from their being unknown, or different from those established in savage nations. In order to imbibe ideas of decency and moral fitness, they have obliquely referred us to the groves of Otaheite."[29] Other critics of Hawkesworth heaped opprobrium on his account of the *arioi* (a religious order dedicated to the war god 'Oro), which was described as an infanticidal society devoted to free love.[30] For many the *arioi* was a source of deep anxiety: it indulged in "sterilizing excesses" amid an otherwise idyllic island society at a time when British (and other European) feminine ideals were oriented around maternalism and pronatalism, and were being more fundamentally based on sexual biology.[31]

Efforts were therefore made, particularly after the second voyage, to rehabilitate the *arioi*. This process exemplifies the way in which the European (gender) order was projected onto and then read back from Pacific societies. Ever since the *Endeavour* voyage, for instance, Cook had reported that membership of the *arioi* was limited to the elite of Tahitian society.[32] As such it was either likened to a select society ("one may almost compare these men to Free-Masons")[33] or freighted with connotations of aristocratic libertinism and sexual excess.[34] A fuller explanation was provided by Georg Forster, who, while clearly and strongly opposed to infanticide, remarkably conjectured that the *arioi* was originally devoted to celibacy.[35] Incompatible with nature, celibacy was doomed to failure, but by destroying the offspring resulting from lapses, so Forster reasoned, the society was able to fulfill the goal of sexual abstinence.[36] He offered two possible reasons for the existence of the *arioi*. The first was that it was originally an elite group of warriors, who refrained from enervating sexual activity. The second claimed that the *arioi* was a population control measure, intended "to prevent the too rapid propagation of the race of chiefs."[37] The danger posed by the *arioi* to the project to locate a Pacific prototype of European morality was thus contained: either its membership was limited to a potentially corrupt elite or its purpose was identified as the promotion of military strength or the prevention of oppression.

But Pacific islanders did not stand idly by as Europeans used them for their own preoccupations. They responded directly to Europeans in several ways, including satire. In May 1774, after several days of attempting to persuade *arioi* members on Raiatea not to kill their children, Cook's crew witnessed a dramatic performance featuring a mock childbirth. Both mother and child were fully grown men, and the latter ran around the stage, dragging an ersatz umbilical cord and placenta while being chased by midwives. This play may have been traditional and "staged mainly for amusement," but the timing of its performance suggests an element of satire. Unaware that they were perhaps the butt of a Raiatean joke, Cook, Johann Forster, and others obliviously wondered in their journals about the immodesty of a people that would permit such performances.[38]

Back in Europe, their fellow islander Mai also stubbornly resisted the moralizing of those, such as the early abolitionist Granville Sharp, who encouraged him to abjure polygamy.[39] Mai soon featured prominently in satirical literature of the 1770s that responded to Pacific voyaging by exhibiting a prurient interest in racial mixture. These texts, which began by lampooning Joseph Banks's purported relationship with "Queen" Oberea during the *Endeavour* voyage, swiftly incorporated Mai to provide a sexual focus as the topicality of Banks's behavior wore off.[40] One pamphlet called explicitly for sexual mixture of Europeans and Pacific islanders as a means of countering both prostitution and effeminacy, advocating the mixture of "southern passions" with "northern art."[41]

French texts took up this theme. Voltaire's short story "The Ears of Lord Chesterfield" (1775) paid homage to the sexual religion of Otaheite, in which worshippers "labour in the formation of a reasoning and thinking creature, . . . the most noble and pious action, that can possibly be performed by man."[42] The most significant text was undoubtedly Diderot's *Supplément au voyage de Bougainville* (1772), which circulated widely in manuscript in the early 1770s and was revised by Diderot later in the decade, but was not published until 1796.[43] As with his reproductive thought experiments in the *Rêve de d'Alembert* (1769), Diderot took an interest in improving the quality of people by proto-eugenic means. In the words of his Tahitian intermediary, Orou, speaking to a French chaplain, "we saw at once that you surpassed us in intelligence, and we immediately marked out for you some of our most beautiful women and girls to receive the seed of a race superior to ours."[44] Diderot thus reversed the "fatal impact" perspective of European-native contact that formed a strong (counter)current of Enlightenment thought—examples include his own contributions to Raynal's *Histoire des deux Indes*—by imagining the Tahitians exacting a "tribute" from Europeans: "from your person, from your very flesh."[45] European sexual exploitation is transformed into native agency, but the radicalism of this agency is blunted by the characterization of Tahitians as desirous of sex with

Fig. 5.1. John Webber, *Poedua* (1777). © National Maritime Museum, Greenwich, London.

Europeans, in fulfillment of colonialist fantasy.[46] Indeed, a measure of the complexity and polyphony of the *Supplément* can be seen in an earlier section, in which an "old man's farewell" manifests the "fatal impact" viewpoint by thundering forth a devastating indictment of European colonialism and exploration.[47]

British texts that were not part of the boom in occasional, satirical literature tended to be more interested in the absolute increase of bodies.[48] Thomas Malthus's use of data on the Pacific demonstrates its importance for questions of population, and visual representations from the voyages could promote maternal ideals as well as erotic fantasies. John Webber's portrait of Poetua (1777; fig. 5.1), in which a youthful, alluring, and bare-breasted Polynesian woman is depicted, is a case in point.[49] French texts also exhibited interest in the absolute size of the populace. In his revisions to the *Supplément*, Diderot inserted a story that originated in the Anglo-American press in the 1740s. This was the fictional account of Polly Baker, encountered in chapter 1, whose courtroom plea was a paean to pronatalism.[50]

Nor was interest in population growth limited to humans. The second and third (1776–1780) Cook voyages witnessed a concerted effort to establish European livestock on Pacific islands. Cook and his officers went to enormous lengths to ensure the survival of livestock on the voyage and their optimal placement in the region. Subterfuge was sometimes necessary, and Cook was always careful to impart to islanders the importance of caring for these animals. Recipients of this genetic capital were carefully chosen.[51] Animals were always deposited or given as a pair, and the British were incredulous at the "stupidity" both of islanders, who failed to recognize the bounty they had received, and of the Spanish, who reportedly left just one sex of animals with Pacific islanders.[52] This activity was unmistakably motivated by British "benevolence" and an improving ethos, but it was thematically linked to the increase of native populations, especially as racial, species, and gender difference all acted as analogies of each other in Pacific narratives.[53]

A focus on absolute population growth rather than racial mixture in British texts was also a result of the dangers of cross-cultural sex, which became apparent from the first visit to Tahiti, where the sexual commerce of islanders and sailors endangered the existence of both. For the first British crew on the island, simple ironware, particularly nails, soon became the commodity of exchange for sex, and inflation rapidly set in. Their captain, Samuel Wallis, had to intervene to prevent this commerce when he found that nails were being pulled out of the ship.[54] Cook took the precaution of issuing orders that no metal goods were to be exchanged for anything but provisions.[55] Such sexual activity was no less dangerous for Pacific islanders. For one thing, South Seas women who engaged in this activity were perceived by Europeans as prostitutes, facilitating their social and racial marginalization. This was to have devastating consequences with the onset of missionary activity and the racial policies of colonial powers in the nineteenth century.[56] Second, European responsibility for the introduction of venereal disease to the region was recognized early on and became a staple of "fatal impact" interpretations of Pacific exploration. When they were not blaming the French or Spanish for its introduction, it was a source of British national guilt, but this moral discomfort itself produced a degree of callousness toward native peoples. This was most evident in Johann Forster's remarkable unpublished suggestion that it would have been better if the first Māori woman to have been infected had been murdered immediately thereafter by her partner: "Howsoever detestable the murder of such a poor wretch must be, it would be a real benefit to the whole community & preserve a harmless brave & numerous Nation from all the horrors of being poisoned from their very infancy."[57]

In addition, sex posed danger to the wider social order. The powerful desire for sex and the ensuing relationships between sailors and native women were the motive force behind several, mostly failed, desertions during the Cook voyages. They were also widely blamed for pushing the crew of the *Bounty* to mutiny in April 1789.[58] By this time several Pacific peoples (Hawaiians and New Hebrideans, for example) had come to be viewed more negatively because of their lack of civilization or for having killed Cook. With the *Bounty* came a powerful blow to the exalted status of Tahiti, as its famed hospitality was now perceived to be deleterious to naval discipline and a factor that enhanced the degeneration consequent on distance from home.[59] This change in perception finally dispelled any lingering commitment to the noble savagery of Tahitians, still further undermining models of human universality and reinforcing existing processes of embodying difference.

Racial and Sexual Aesthetics in the South Seas

Inscribing identities on the body only made sense if they could be reliably read back. This section explores how the body was read in discourses of Pacific exploration and in wider racial thought, focusing on the role of aesthetics in both. Almost inevitably given the sexualized perception of the region, aesthetic categories such as "beauty" and "ugliness" recurred frequently. In aesthetic works such as Edmund Burke's *Philosophical Enquiry into the Origin of Our Ideas of the Sublime and Beautiful* (1757), the concept of beauty was abstracted from lust but remained strongly based on it, which created a powerful association between (female) beauty and (male) sexual desire.[60] Sexual desire entailed a presumption of beauty, and vice versa; so, *mutatis mutandis*, with ugliness. In Pacific and racial texts, "beauty" and "ugliness" were based on color and body/facial shape, although it was generally the case that color was foregrounded before other aspects were considered.[61]

Color itself was interpreted in light of the optical advances of the late seventeenth century, which had a lasting impact in casting blackness as privation. But while interpreting color was comparatively simple, establishing exact rules by which to judge facial peculiarities proved difficult. In the 1750s, William

Hogarth attempted to map beauty and ugliness to a single line, a technique that remained influential among prominent racial theorists. Furthermore, early encounters with the Pacific coincided with the high point (followed by the slow demise) of physiognomy, a popular set of ideas that sought, especially in the face, to substantiate Montaigne's belief that "there is nothing more likely than the conformity and relation of the body to the spirit."[62] Perhaps the most enduring contribution of physiognomy's eccentric late-eighteenth-century spokesman, Johann Caspar Lavater, was highlighting the interpretive act at the heart of racial thought.[63] Such an act is exemplified by the dual meaning of "complexion" as both exterior skin color and internal (moral) temperament, an ambiguity that was mirrored in imaginary travels to the South Seas, which emphasized physical indicators of moral status.

Travels

The essential prerequisite for reading the body is reliable data, and travel accounts had long been the principal source of information on foreign bodies. Cook's voyages marked the arrival of a new standard of reliability in travel literature, and with it came a fresh confidence in the data on which social, cultural, and scientific theories were based, and thus a new resilience in the face of skeptical readers.[64] Many voyages were now being undertaken for supposedly no other reason than the expansion of knowledge; they carried the official sanction of the state, and their accounts its imprimatur; they were manned by resourceful seamen and professional scientists, who painstakingly measured and collected, making available these technical data within their writings and providing specimens to museums.[65] Reliability did not necessarily connote objectivity. Older myths, such as the existence of Patagonian giants, might be debunked, but others took their place. Empiricism was not value-free, and "seeing" was an activity strongly related to, if not entirely determined by, context.[66] For one thing, artistic representations of the voyages moved from a neoclassical to Romantic mode that produced greater verisimilitude but owed much to changing European tastes.[67]

One example that neatly illustrates this new standard in action is anthropophagy in New Zealand. Europeans had not expected to find cannibalism in the Pacific, and it initially met with a measure of public disbelief.[68] One indication that it was not treated seriously was its popular attribution to all Pacific peoples rather than solely to the Māori, a distinction that Hawkesworth and his sources had been careful to make. A facetious anecdote about Mai joked that he was afraid to bow before George III, fearing that he would be eaten.[69] Indeed, it was widely assumed that South Seas islanders believed Europeans to be cannibals: in 1775, one newspaper hoped for the success of Mai's repatriation, if only to disabuse Pacific islanders of this suspicion.[70] Such ease with the notion of cannibalism, to the point of self-attribution, suggests that the initial reports of it were not taken seriously.

Cook was sufficiently stung by this skepticism to seek incontrovertible proof. When a severed head was purchased from the local Māori by one of his officers aboard the *Resolution* in November 1773, Cook, "being desirous of becoming an eye-witness of a fact which may be doubted," ordered a piece of flesh from this head to be cut, grilled, and eaten by one of the natives.[71] With cannibalism established as a "fact," the naturalists on board the *Resolution*, the father-and-son team of Johann Reinhold and Georg Forster, could incorporate it within their models of human progress, and Cook was able to respond conclusively to skeptics.[72] As he told James Boswell at a Royal Society dinner in April 1776, "He candidly confessed to me that he and his companions who visited the South Sea Islands could not be certain of any information they got, or supposed they got, except as to objects falling under the observations of the senses; . . . He gave me a distinct account of a New Zealander eating human flesh in his presence and in that of many more aboard, so that the fact of cannibals is now certainly known."[73]

This episode reveals much about both the part played by native actions—"indigenous countersigns," in the words of one scholar—in the formation of European knowledge and the paramount authority of eyewitness evidence. Cook demanded irrefutable proof, but proof of what? Surely not what he thought: "That the New Zealanders are Canibals."[74] Rather, that one individual New Zealander had, when required by Europeans, eaten human flesh. This is not to deny the contemporaneous existence of anthropophagy in New Zealand. But it neatly shows how European preoccupations distorted the ethnographic evidence they collected, how they readily generalized from particular incidents, and, most importantly, how native action was essential for European knowledge. (The Raiatean play discussed in the previous section also illustrates this point.) At no stage did Cook or the other writers who commented on this incident stop to consider that what they had seen was a self-conscious (and possibly ironic) performance by an autonomous actor.[75] Instead, they readily assumed that this individual was a mere cipher of his culture, a viewpoint that typified the travelers' gaze in the Pacific.[76] The role of Europeans in influencing the events they witnessed, and their obliviousness to this, demonstrates their lack of objectivity, as defined by Lorraine Daston and Peter Galison ("knowledge that bears no trace of the knower").[77] Nevertheless, the fetishization of empirical proof overrode such concerns to produce a potent, if perhaps ill-founded, confidence in the data streaming back to Europe.

The transformation in European racial thought after the 1760s discussed in chapter 3 is arguably due to this confidence, as indicated by the prominence of the Pacific in this discourse. Kames used the peoples of the Pacific to argue in favor of separately created, autochthonous populations, and he asked his friend James Lind, who was due to (but ultimately did not) participate in Cook's second voyage, to obtain evidence in support.[78] The effect of the Pacific was more transformative on Blumenbach's ideas. Like many who discussed human variation, Blumenbach was particularly remembered for "his familiarity with voyages and travels," on which he based his division of the world's peoples into four varieties (European, South and Southeast Asian, African, and American) in the first edition of the *De Generis Humani Varietate Nativa* (1775).[79] The second (1781) edition revised this schema in light of the Pacific voyages, now identifying five groups (Ethiopian, Malay, European, American, Mongolian). As Blumenbach had mentioned the Pacific islands in his first edition, this revision seems to have been directly caused by the accounts published following Cook's second voyage, particularly Johann Reinhold Forster's *Observations Made during a Voyage round the World* (1778).[80] Notwithstanding the claim that this text was somehow non- or preracial, Forster's bipartite division of the Pacific's peoples became an orthodoxy by the early nineteenth century, enshrined in Dumont d'Urville's 1832 division of the Pacific into three groups constituting two races (Micronesian and Polynesian on the one hand, Melanesian on the other).[81]

In Britain, the role played by Joseph Banks in popularizing Pacific voyages and nurturing racial thought was crucial.[82] His popular association with the Pacific was widespread and lasting: the *Endeavour* voyage was commonly named "Banks's voyage" in early reportage, and his ties to the region form the visual lexicon of James Gillray's cartoon, *The Great South-Sea Caterpillar* (1795; fig. 5.2).[83] Satirizing his investiture with the Order of the Bath owing to his association with the king, Gillray presents Banks as a creature of the South Seas, juxtaposing his fine clothing, prominent sash, and star to the flora and fauna in which he is draped, in an echo of his earlier reputation as a "botanizing macaroni."[84] Banks was the focal point of a wide network of correspondents, many of whom were naturalists and among whom information and specimens frequently passed. He corresponded with Blumenbach, Bougainville, Buffon, Camper, the Forsters, Benjamin Franklin, and the Manchester Literary and Philosophical Society (of which Charles White was a prominent member), and visited Monboddo.[85]

Fig. 5.2. James Gillray, *The Great South Sea Caterpillar, Transform'd into a Bath Butterfly* (1795). Courtesy of the Lewis Walpole Library, Yale University.

But it was his close relationship with Blumenbach that constituted Banks's most significant link with racial thought.[86] He supplied Blumenbach with skulls from the Pacific and elsewhere, despite stiff competition for crania from the fractious rivals Petrus Camper and John Hunter.[87] Although at least one attempt to procure skulls to order was thwarted by the *Bounty* mutiny, Banks continued to send Caribbean and Pacific crania, and he received Blumenbach on a visit to London in early 1792.[88] All this earned Banks a lengthy dedicatory letter in the preface to the third (1795) edition of Blumenbach's *De Generis*.[89] Nor was Blumenbach his only protégé: Banks had hosted Camper in 1785 and introduced him to Sir Joshua Reynolds, all three apparently discussing the complexities of the facial angle (see below).[90] A testament to British maritime dominance, Banks continued to act as a clearing house for skeletal remains from across the British Empire until the early nineteenth century, particularly for Aboriginal remains from Australia.[91]

Banks also testified to the role of feminine beauty in assessing the Pacific's peoples. In 1773, he wrote a short set of "Thoughts on the manners of Otaheite [. . .] for the amusement of the Prince of Orange," which exemplify how beauty and ugliness were discussed in relation to other cultures. His remarks begin by expressing the view encountered in chapter 3: that women in the lowest/earliest stages of society were oppressed by their menfolk, and that European women are more beautiful than those in other, apparently more favorable climates because European men pay them so much attention. In a similar fashion, the bodies and souls of Tahitian women are molded to their "chief occupation" ("love") because they are not forced to endure the drudgery of their less fortunate sisters elsewhere. Banks states categorically that he had "no where seen such Elegant women as those of Otaheite," but he offered a vital caveat—namely, "the article of Complexion[,] in which our European ladies certainly excell all inhabitants of the Torrid Zone." Nevertheless, whiteness was apparently prized in this society, as Banks suggested when offering his hypothesis on the origins of tattooing: "I am inclined to think that as whiteness of skin is Esteem'd an Essential beauty these marks were originaly intended to make that whiteness appear to greater advantage by the Contrast[,] Evidently in the same manner as the patches us'd by our European beauties."[92]

Colors

By testifying to the importance of color, Banks reinforced the role of sight as the most immediate and important sense for perceiving difference, and color is its most pertinent characteristic. Simple color contrasts were a staple of racial commentary in this period. Accounts of black Africans often mentioned their teeth, which apparently appeared whiter by contrast with their skin, making their teeth "their only beauty."[93] As a commentary on the Song of Songs put it in 1776, "A person that is not of a fair complexion, being in the company of one that is, looks abundantly worse than if viewed alone."[94] The aesthetic judgment here is unmistakable: the person not of fair complexion does not look *darker* by the comparison, but *worse*. In addition, the very variability of colors in white populations, as opposed to the "monotony" of "gloomy" black-skinned peoples, was used as an argument for their superior beauty.[95] As Virey asked in 1824 (echoing White a generation earlier), "Are not the fine mixtures of red and white, the expressions of every passion, by greater or less suffusions of color in the one, preferable to the eternal monotony which reigns in the countenances, that immovable veil of black, which covers all the emotions of the other race?"[96]

Three points concerning color are vitally important. First, many commentators recognized that perceptions of color (and of beauty more generally) were culturally relative. In 1646, the physician Thomas Browne set about exploding the "vulgar error" that blackness was a curse by emphasizing relativistic aesthetic standards: black people "esteeme deformity by other colours, describing the Devill, and terrible objects White."[97] This idea appeared frequently across the long eighteenth century, but the recognition of different beauty standards was not as open-minded as it might appear. For one thing, it failed to prevent chauvinistic aesthetic judgments based on an absolute standard, the beholder's own.[98] For another, not all standards were valued equally: different standards could be—and were—used to prove the backwardness of other peoples. And aesthetic relativism reinforced the notion, popular among polygenists, that each community of humans either does or rather should prefer its own kind.[99]

Second, color was a signifier of such clarity that it required no sophisticated hermeneutic apparatus to interpret it. It is therefore significant that skin color itself was described in the eighteenth century most often as "complexion," a multi-faceted term. The synonymity of "color" and "complexion" was commonplace, as manifested in Samuel Johnson's definition of "complexion": "The colour of the external parts of any body."[100] But an older sense of complexion as "the temperature [temperament] of the body according to the various proportions of the four medical humours," remained current well into the nineteenth century.[101] Color, via "complexion," thus signified both the external color and the internal constitution of an individual, and the elision of "color" and "complexion" lulled interpreters of the one into assuming that the other was equally easy to grasp.[102]

Finally, the important seventeenth-century work on color undertaken by Boyle, Hooke, Newton, and other members of the Royal Society shaped discussions of human difference across the next century.[103] Boyle's *Experiments and Considerations Touching Colours* (1664), for example, dealt explicitly with blackness in Africans: this monogenist account repeated Browne's comments on aesthetic relativity, rejected the influence of heat on color, and waxed skeptical on the Curse of Ham but crucially recognized that the color *white* reflects light while *black* absorbs it.[104] The conception of black as a privation of light directly influenced the Virginia physician John Mitchell, who in 1744 argued that blackness resulted from the skin's absorption of light due to the thickness of the epidermis and not, as was generally believed, to a pigmented layer of skin.[105] In an argument that sought to profit from the huge cachet of "Newtonianism," he recognized the demonstrable transparency of the epidermis yet claimed it became increasingly opaque the thicker it was; eventually, no color from beneath could be transmitted, making the individual black.[106] Directly echoing Newton, he wrote, "Where-ever there is a Privation of Light or Colour, there, of course, ensues Darkness or Blackness. But as most solid Bodies . . . do generally reflect some Colour, which we know no black Body does, we shall next inquire into the particular Make of their Skins, *by which they are rendered incapable to reflect, as well as to transmit, the Rays of Light.*"[107]

This argument constituted a reckless misuse of scientific data. The skin pigmentation of black people manifestly does not absorb all light wavelengths, but Mitchell argued for precisely this point.[108] His tendentious use of Newton and color science exemplifies the important role played by seventeenth-century science in changing conceptions of color—above all blackness, which became powerfully associated with privation.

Yet by the time Europeans began large-scale voyages to the Pacific, the use of color—like climate—in learned discussions of human variation was in crisis. Human skin pigmentation was simply too varied, and what it signified was increasingly unclear. The seemingly infinite variety of skin color, as well as its radical alteration within one generation in the case of mixed offspring or albinism, meant that it failed to provide a consistent bodily marker on which identities—above all, "slave" or "free"—could be based. Early efforts to provide the desired clarity focused on the more "authentic" color of skin hidden from elements, hence the attention to the coloration of neonates or the genitals.[109] But these efforts also failed, not least because it was impracticable to require the existence of infants or exposure of the genitalia to establish an individual's identity.

What skin was supposed to signify was also increasingly uncertain, as Pacific voyages demonstrate. On the one hand, well-established patterns in the discussion of skin color were followed in the voyage literature, where lighter skin was associated with civilization and darker with savagery. In Tahiti, the Society Islands, and New Zealand, lighter-skinned populations were viewed as civilized—hence the sustained effort to comprehend their sexualities in European terms—while populations elsewhere, particularly in Australia, New Guinea, New Caledonia, and Vanuatu, were darker and thought to be less civilized. These associations were furthermore applied to groups within Pacific societies, as darker Tahitians were thought to be of lower social status.[110]

On the other hand, this comfortable set of preconceptions ran into trouble almost immediately. Cook himself muddied the waters by undermining the abjectness of "savagery": in a set of reflections the orientalist Hawkesworth failed to incorporate in the *Account*, Cook speculated on the "Tranquillity" and happiness of Aboriginal Australians, whose basic needs were met in their environment and who were not plagued by the "superfluities" of civilization.[111] Mai's color was variously interpreted, depending on whether he featured as an object of reportage (in which case he was lighter skinned) or satire (where he was explicitly negritized; see below). And the ambiguities of color in the Pacific were at the heart of the debate between Georg Forster and Immanuel Kant, whose entire racial theory was based on the consistent inheritance of color.

Smells

Doubts about color encouraged other ways of perceiving difference. One of relatively long standing was smell. Buffon echoed François Pyrard de Laval (1570–1621) in writing "that the sweat of the Indians, whether male or female, has no unsavoury odour, while the stench of the African negroes, when they are over-heated, is perfectly insupportable."[112] Edward Long linked this "extreme fetid Smell" to intellectual ability: "The most stupid of the Negroe race, are the most offensive; and those of Senegal (who are distinguished . . . by greater acuteness of understanding . . .) have the least of this noxious odour."[113] A German medic who spent time in Senegal disagreed, stating that the "foul and nasty vapours" he encountered there proved that African bodies were better adapted to that climate, as they excreted "foul and noxious matter" that accumulated dangerously in the bodies of Europeans, causing disease.[114] Jefferson focused on the racial difference implicit in these ideas, blazing a trail followed by other polygenists.[115] By contrast, Samuel Stanhope Smith ascribed the odor of black peoples and the curl of their hair to the same cause: heat and sweat, with the smell attributed to the presence of volatile salts.[116] Like Smith, Lawrence attributed this characteristic to climatic and other environmentalist factors, even if he came close to making it a racial feature: the odor "is said by those, who are well acquainted with this race, to be very characteristic, and to be transmitted to the offspring, as well as their other peculiarities, in the mixed breeds."[117] Odor was such a key characteristic by 1799 that White—following Buffon—used it to establish a gradation in living creatures.[118] Indeed, the description of Aboriginal Australians in such terms, as well as William Wells's denial that it is universal to all black peoples, testified to the importance of odor as a racial characteristic.[119]

Olfaction was not merely something racial theorists did; it was also a faculty that they analyzed in others, and Tahitians were not alone in apparently possessing unusually sensitive noses. Albrecht von Haller and Oliver Goldsmith both claimed that "the Negroes of the Antilles, by the smell alone, can distinguish between the footsteps of a Frenchman and a Negro," and James Cowles Prichard believed it was "universally remarked" that "Africans have . . . a very perfect perception of odours."[120] Nor was smell the only extraordinary sensory ability that attracted comment. Writing about Amerindians, Raynal stated that "their sight, smell, and hearing, and all their senses, were remarkably quick, and gave them early notice of their dangers and wants."[121] Unusual abilities beyond the senses were also frequently noticed: individuals with prodigious memories or arithmetic skills were mentioned, particularly by those such as the abbé Grégoire, who used these cases to argue for a purely social and cultural basis for the backwardness of other peoples.[122]

The efforts of Grégoire and others could, however, be risky. It was one thing to point to intellectually or artistically gifted black people, such as Phyllis Wheatley or Ignatius Sancho. But too strong an emphasis on prodigious feats of memory or arithmetic might rather reinforce difference than undermine it.[123] The fact that the principal writers to emphasize the unusual sensory capacities of non-European peoples were polygenists goes some way to illustrate this point. The fullest discussion came in the works of Long, Sömmerring, Virey, and White, and they met with powerful opposition from (among others) Lawrence, Prichard, and Smith.[124]

Bones

The variability and subjectivity of beauty, color, and odor led racial thinkers after 1770 to seek refuge in a more solid foundation for explaining and classifying difference.[125] The skeleton had long been used to compare species, but newer studies argued that the skull in particular was a predictable and legible signifier of racial identity.[126] The most significant new contribution was Sömmerring's *Körperliche Verschiedenheit*, a landmark in the history of comparative anatomy that strongly influenced his friend, Georg Forster.[127] Although never fully translated into English, extracts appeared as appendices to polygenist tracts in Britain and America.[128] Sömmerring's observations followed the contours of the skeleton, starting at the head and moving to the extremities before finishing with conclusions on the brain.[129] White was engaged in a similar project, although more concerned with the similarities of black humans to apes, and embarked on a long mathematical examination of the ratio of forearm lengths in order to demonstrate that black people's forearms are longer than those of white people, and that they therefore approach closer to apes.[130]

Thanks to this "skeletal turn," the skull became one of the most important parts of human anatomy for the delineation of races.[131] Its size was viewed as an indicator and a limit on brain size and hence mental ability. In addition to social custom, by which certain societies deliberately modified the head shape of neonates, Blumenbach and Sömmerring argued that an involuntary process of constraint shaped the skull, where head muscles made an impression on the bones beneath. This mechanism, by which the environment could penetrate the skin and shape the skeleton, could also work the other way round: Blumenbach noted that the cranial interior was molded to the brain it surrounded.[132] Indeed, White's interest in two craniofacial features in particular are dictated by this principle. As he wrote, "All the nations of Africa, and the inhabitants of the southern [i.e., Pacific] isles have either very narrow skulls . . . or they have a flat receding forehead, and hind-head: and the bony sockets which contain the eyes, are more capacious than those of Europeans."[133] White here emphasizes inferior cognitive power (from the limited space available for the brain) and superior sensory ability (from the greater space available for the eyes), both on the basis of the anatomy of the skull. He was not alone: a generation later, Virey used the development of the skull to argue that black people are less intellectually gifted than white, "undoubtedly, because the ossification is completed too soon, and prevents the perfect development of the brain."[134]

Lines

Cranium size was but one metric employed in the "skeletal turn." Other reliable measures of the face and body had long been sought, and William Hogarth's *Analysis of Beauty* (1753), a seminal work in British aesthetics, did much to promote the reduction of facial peculiarities to simple lines. Hogarth believed

Fig. 5.3. Detail from title page, William Hogarth, *The Analysis of Beauty* (1753). Courtesy of Metropolitan Museum of Art (Watson Library copy: Gift of Mary C. Schlosser).

there was a deducible principle behind the human ability to distinguish between faces, and he argued that faces conforming to a serpentine (or elongated "S" shaped) line (fig. 5.3) were the most beautiful. Those in which the curves of the "S" were more pronounced were uglier, exemplified by the figure of Silenus, about whom Hogarth wrote, "Human nature can hardly be represented more debased . . . where the bulging-line figure . . . runs through all the features of the face, as well as other parts of his swinish body."[135]

Hogarth's reduction of the concept of beauty to a line was profoundly significant. He believed he was redressing an imbalance, as "nature hath afforded us so many lines and shapes to indicate the deficiencies and blemishes of the mind, whilst there are none at all that point out the perfections of it beyond the appearance of common sense and placidity."[136] Such "deficiencies and blemishes" managed to work their way to the skin's surface in a way analogous to Blumenbach and Sömmerring's conception of the muscular shaping of the skull: "It is by natural and unaffected movements of the muscles, caused by the passions of the mind, that every man's character would in some measure be written in his face, by [the] . . . time he arrives at forty years of age, were it not for accidents which often, tho' not always prevent it."[137] The subtext here is clear: beauty and ugliness map directly onto virtues and vices. This was hardly a new idea, but Hogarth's application of it in his linear model was an innovation that had significant ramifications.

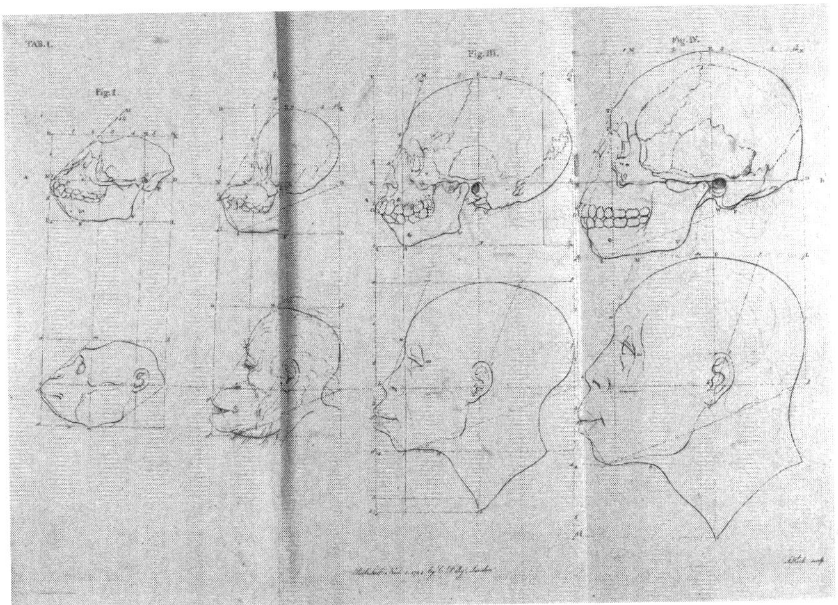

Fig. 5.4. Comparison of ape, "orang-outang," "Negro," and "Calmuck" heads from [Petrus Camper], *The Works of the Late Professor Camper* (1794). Courtesy of the Metropolitan Museum of Art (Elisha Whittelsey Collection, Elisha Whittelsey Fund, 1960).

Perhaps the clearest of these was its influence on Camper, whose work was also influenced by Pacific exploration. Camper sought a rule to standardize graphical representations of "the inhabitants of these principal parts of the earth, including the islanders of the South Seas, inhabitants of New Holland and New Zealand."[138] His English translator explicitly linked Camper's work to that of other artists who "have attempted to introduce certain principles, and to propose invariable rules, which they deemed conducive to the improvement of taste in the higher branches of painting, and precision in all."[139] The subtitle of Hogarth's work—"Written with a view of fixing the fluctuating Ideas of Taste"— suggests it was an important influence.

Camper's innovation was the facial angle, formed by the intersection of two lines, one horizontal and another connecting the foremost teeth to the forehead (fig. 5.4). In humans, this angle ranged between seventy and a hundred degrees, "from the negro to the Grecian antique; make it under 70, and you describe an ourang[-outang] or an ape."[140] The relation of "negroes" to apes in Camper's work was not just a case of mathematical or diagrammatical contiguity. He

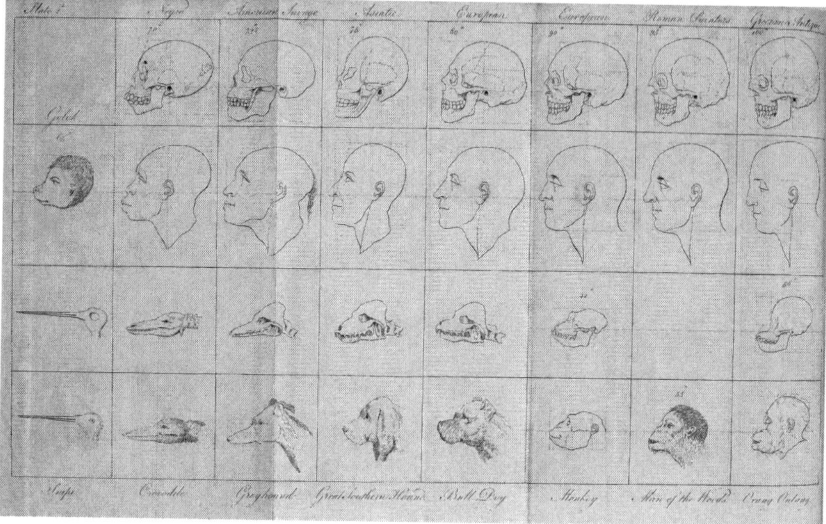

Fig. 5.5. Table from Charles White, *An Account of the Regular Gradation in Man* (1799). Courtesy of the Wellcome Collection.

asserted aesthetic similarities too: "The lower jaw, as well as the upper, is also much forwarder in the negro, Caffre, and Calmuck; and therefore it is that these people approach nearer to the figure of an ape than either the European or the antique."[141] But Camper was no polygenist, and his animalizing comments about black people were, somewhat paradoxically, motivated by *anti-racist* sentiments. The goal of his facial angle theory was twofold: to provide an aid for anatomical draftsmanship and, more importantly, "to prove that non-European characteristics were as natural as European ones. Camper was trying to prove empirically that all races are equal."[142] By placing Europeans and black people on the same scale and demonstrating human intervention was not responsible for racial traits, Camper was attempting to combat racism.[143] But he was arguably—and sadly not uniquely—naive if he believed that slavery's defenders would refrain from using a scale that appealed directly to the empirical methodology of the new science at precisely the moment when they sought new and convincing justifications for the institution.[144]

Not all who misused the facial angle to support racial taxonomies defended slavery. White's *Account* disavowed slavery while supporting polygenesis, and his elaborate table of profiles (fig. 5.5) is a direct but expanded copy of Camper's.[145] In addition to adding labels to each head-skull dyad, White has included the value of the Camperian facial angle as well as a visualization of what had

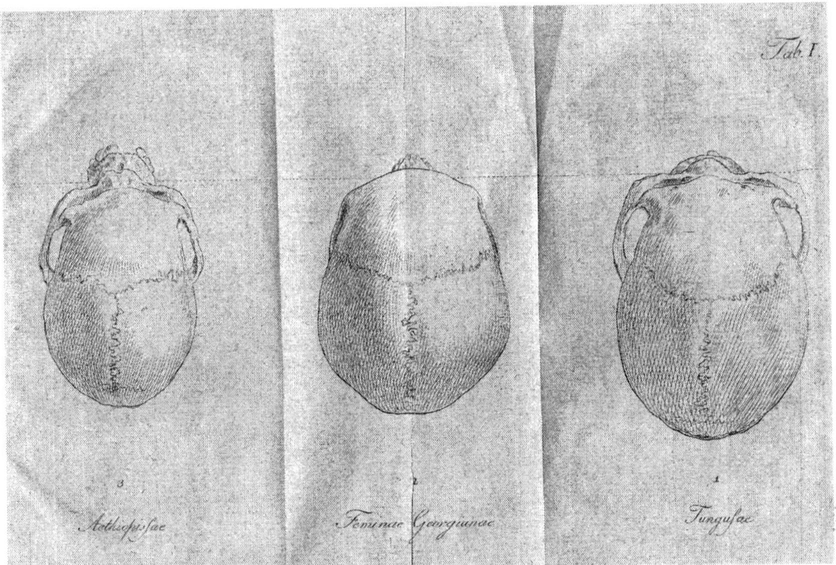

Fig. 5.6. Blumenbach's *norma verticalis*, showing his three "primary" races (*l-r* Ethiopian, Caucasian, Mongolian). Plate 3 from Bendyshe. Courtesy of the Max-Planck-Institut für Wissenschaftsgeschichte, Berlin.

been for Camper merely a throwaway remark: that by making the facial angle more acute, it would be easy to draw a dog or a snipe.[146] It is a measure of the derivative nature of White's text that he followed these instructions to the letter, the sequence running (from top right to bottom left) in his table from humans to apes to dogs and finally a bird, the snipe. It is also very easy to tell which diagrams were interpolated by White, as the "Orang-Outang" and "Monkey" on the bottom row of White's table are exactly as they are in the two leftmost positions in figure 5.4.

Somewhat more original was Blumenbach's ranking. As we have seen, European voyaging to the Pacific led Blumenbach to argue that humanity was divided into five varieties formed of three primary (Caucasians, Ethiopians, and Mongolians) and two intermediary (Americans, between Caucasians and Mongols, and Malays, between Caucasians and Ethiopians) groups. He did not arrange these groups in a visual hierarchy, although contemporaries (and later editors) did.[147] Blumenbach was dissatisfied with Camper's method, which he thought was unevenly applied and too subject to variation, preferring his own vertical scale (*norma verticalis*). Blumenbach provided a visual example of this for only the three primary groups in his categorization. In figure 5.6, the three skulls typical of the primary varieties (from left to right, Ethiopian, Caucasian, and Mongolian) are viewed from above, with a line drawn at the level of the cheekbones. From this perspective, "all that most conduces to the racial [*gentilitium*] character of skulls, whether it be the direction of the jaws, or the cheekbones, the breadth or narrowness of the skull, the advancing or receding outline of the forehead, &c. strike the eye . . . distinctly at one glance."[148] Part of the significance of this project for Blumenbach is its creation of an instantly legible system of marking human variation.

Attempts to reduce beauty, race, or variety to a line were attempts to produce a simple metric of human difference. In the case of facial features, the moving spirit for this project was the Swiss cleric Johann Caspar Lavater, whose works, *Von der Physiognomik* (*On Physiognomy*, 1772) and *Physiognomische Fragmente* (*Essays on Physiognomy*, 1775–1778), were phenomenally popular. Like Blumenbach and Sömmerring, he held that the bones of the skull were shaped by the influence of both muscles and the brain.[149] Comparing skulls, Lavater concluded that the "Calmuc" was the only concave skull, which to him signified "weak powers of mind, which endeavour, as all natural weaknesses do, to supply and conceal their deficiencies by the strength of cunning."[150] But Lavater was no racial theorist, and while he did discuss certain human types (principally, the "negro"), he was not up-to-date with the latest literature (he failed to mention Pacific peoples) and concentrated on discussing European physiognomies. Furthermore, the publication of *Physiognomische Fragmente* in the late 1770s marked the peak of Lavater's popularity among the learned. By the time it was translated into English in 1789, his program was already falling into disrepute, principally because it failed to produce a reliable system, and several prominent figures distanced themselves from Lavater, none more eagerly than Goethe.[151]

Yet Lavater's insights remained popular among the wider public, and he did make important contributions to racial theorizing in Britain. On the one hand, he introduced some of Kant's racial writings to an English readership: parts of Kant's essay, "Von den Verschiedenen Racen der Menschen", were translated for the first time in Lavater's *Essays*.[152] Kant's direct impact on racial thinking in Britain was limited at least until De Quincey, but Lavater hereby indicated Kant's influence on his work, marking it as more racially informed than many critics have allowed.[153] The second, and more important, of Lavater's contributions was to make explicit the semiotic act at the heart of all attempts to make the body signify identity. And of all the sensible body parts, his concentration on the face highlights just how essential the head and skull had become by this time. As he asked rhetorically about the human face, "Shall not the correspondence between the interiour and exteriour [*sic*], the visible and the invisible, the cause and the effect, be there apparent?"[154]

Language: Genius and Genealogy

Many felt that a more direct means of judging the interior life of an individual was already at hand. Speech, and language more generally, had long served as evidence of the existence of reason, hence of a rational soul, and had a lengthy career policing the animal-human boundary.[155] The venerable use of language as a means of distinguishing between peoples assumed a new importance as the project to classify human diversity took shape. As a tool of racial thought, language was used in at least two ways: native languages themselves were analyzed and compared with other (European/ancient) tongues in order to situate a foreign society on the civilized-savage spectrum; and language genealogies were reconstructed and relationships with known languages identified, which allowed theorists to make decisive pronouncements on membership of racial categories in default of an unambiguous body. Both uses of language appear in the literature of Pacific exploration, and this section examines the key influences on linguistic thought in eighteenth-century Britain and France, before proceeding to explore the Pacific case study.

Language and Difference in Eighteenth-Century Thought

The foremost influence on eighteenth-century thinking about language was undoubtedly John Locke, for whom spoken language is empirical proof of the existence of thought.[156] As he put it, "Those who cannot distinguish, compare, and abstract, would hardly be able to understand, and make use of language, or judge, or reason to any tolerable degree: but only a little, and imperfectly, about things present, and very familiar to the senses."[157] By implication, those whose intellectual faculties are stronger would possess a more sophisticated language. Buffon's comments on Amerindians echo such views: "As they have but few ideas, their expressions are limited to the most common objects; and, though every mode of expression should differ from another; yet the smallness of their number necessarily renders them of easy acquisition. It is not, therefore, so difficult for a savage to learn the language of all other savages, as for a polished man to learn the language of another people equally advanced in civilization."[158]

This correlation between mental simplicity and primitive language was ubiquitous (Vico was a rare dissenter) and present even in sympathetic accounts of non-Europeans.[159] The Jesuit Lafitau, for example, attempted to defend

Amerindians from the criticism that their language showed they possessed no notions of religion, virtue, politics, or science by conceding precisely this point: "Lack of words on the one hand and ignorance of many things on the other, must make their languages less rich than ours."[160] Those who criticized Amerindians for their ignorance were Eurocentric, he said, and exaggerated native intellectual poverty. Unable to escape the critics' logic, which he shared, Lafitau contented himself with quibbling about the scale of the problem.

The most significant linguistic model for writers on human diversity was that pioneered by Locke and further developed by others such as Condillac, Adam Smith, and Monboddo. The key elements of this model were its refusal to accept language as divinely revealed and its consequent account of linguistic development. Demonstrating the Lockean commitment to the nonexistence of innate ideas and to the mind of a newborn as a tabula rasa, the most influential eighteenth-century account of the origin of language, that of Condillac (1746), centered on a thought experiment involving two infants, abandoned and without language. Initially alone, cohabitation necessitated communication that was at first entirely gestural. For spoken language to develop, nature simply had to take its course: this couple started a family, and the serial production of generations of offspring, in whose infancy the speech organs were more flexible and readily molded, facilitated the emergence and increasing sophistication of speech. Implicit in this simple philosophical parable are many of the aspects encountered in previous chapters, above all the role of biological reproduction in ensuring the survival and further development of language. Also essential for the wider acceptability of such a model was its compatibility with the biblical creation narrative. Condillac first dealt with this by supposing that the two infants were abandoned after the Flood.[161] Situating such a Lockean account in the postdiluvian era proved to be a convenient way of reconciling sacred history with an account of the development of language that claimed scientific validity. Lord Kames recognized this when he invoked the story of Babel as an orthodox—and likely disingenuous—disclaimer to the polygenism he was outlining.[162]

Condillac's conjectural scenario became a popular topos of Lockean argument, appearing frequently in later works on language. Adam Smith's *Considerations Concerning the First Formation of Languages* (1767) is a key example. In his version, the protagonists are two savages rather than two infants. This small change sidestepped theological problems with the timing of this event, freeing Smith of the need to specify at which point in Genesis his fable would have taken place.[163] It also engaged with stadial theory and demonstrated a commitment to a universal human nature.[164] The use of savages in these philosophical fables worked both to highlight the process of language acquisition (using the Other as a conjectural guinea pig) and to emphasize a common humanity, without which such conjectural histories would have been meaningless. There were exceptions to this commitment to a universal human nature: Kames's *Sketches*, for instance, remarkably featured literary criticism in the service of polygenism by advocating for the authenticity of the Ossian poems. In arguing that Ossianic language was too refined for the first stage in society, he viewed Ossianic Scots as a unique—"miraculous"—exception to the general picture of human development. Kames's engagement with the Ossian controversy was thus an intrinsic part of his polygenism, notwithstanding the pious lip service he paid to the story of Babel. More broadly, however, not only was a shared human nature emphasized by all serious conjectural historians but the use of the earliest humans—whether biblical or modern-day savages—in a fable concerning the origins of language promoted the view that language was natural to man.[165]

Yet here, too, there were exceptions. The most notable, James Burnett, Lord Monboddo, served alongside Kames on the bench of the Scottish Court of Session.[166] His *Origin and Progress of Language* (1773–92) is for several reasons a fascinating and important text. Swimming against the enlightened tide, this book was a neo-Aristotelian attack on the Lockean/Newtonian consensus, in which Monboddo aimed to demonstrate that language was not natural to man.[167] Despite their distinct motivations, Monboddo and Rousseau were the most famous proponents of this idea in the period, and they both sought to replace language as the chief signifier of mental sophistication. In part, this was inevitable given their respective arguments. Monboddo claimed that speech was acquired by man, not revealed to him, so society must therefore have existed for some time.[168] As an invention of man, language is a sign not merely of thought but of advanced thought. Rousseau made explicit the implications of this view in his *Discours sur l'origine de l'inégalité* (1755): "General ideas can enter the Mind only with the help of words, and the understanding grasps them only by means of propositions."[169]

Language was thus essential for the formation of complex ideas. Monboddo argued that only civilized men were in fact capable of ideas of reflection, stating categorically that "savages have no such ideas."[170] Ironically, since his arguments were prefaced with an appeal to examine savages in order properly to understand human nature, Monboddo and Rousseau were both arguing that it was not humans but only mentally advanced humans that were capable of the full use of language. Indeed, Monboddo's definition of language (*"the expression of the conceptions of the mind by articulate sounds*")[171] completely ignored gestural languages, thereby excluding the hearing- and speaking-impaired. In addition, feral children, to whom Monboddo and Rousseau devoted a substantial amount of attention, had profound linguistic difficulties, yet few ever seriously doubted their humanity. For these writers, language could not function as an unambiguous border separating animals and humans because there existed some humans who had not fully realized their linguistic potential.[172]

Provocatively, Monboddo and Rousseau both placed the "orang outang" among such humans.[173] They continued to believe that language was a quintessentially human characteristic, but if it was possible for some clearly human individuals to lack language, then why, they asked, should an anthropoid creature that possessed the anatomical prerequisites for speech (as was believed between 1699 and 1779) not be accepted as human? This proposition was, perhaps paradoxically, both weak and dangerous. Its weakness is apparent from the tactics used by Monboddo and Rousseau to bolster their case. Apes themselves proved poor allies, as they stubbornly refused to speak despite the offer of immediate baptism (at least according to Diderot), so Rousseau and Monboddo based their claims on physical and behavioral characteristics, among which was the suggestive fact that "they carry off negroe [*sic*] girls, whom they make slaves of, and use both for work and pleasure."[174] It was in such evidence that this proposition was at its most dangerous, as it lent support to Edward Long's views (outlined in chapter 4), which would have likely horrified both Monboddo and Rousseau. The sting was thankfully drawn within a few years by Camper's precise dissection that found no humanoid larynx in the orangutan (1779), although this naturally did little to convince hardened polygenists. But by undermining language as a discriminatory tool, Rousseau and Monboddo contributed to making external form or behavior the only means remaining for distinguishing humans from animals.[175]

The ideas suggested by those who continued to associate language and reason could be just as damaging. Five years before Camper laid the question of "orang outang" speech to rest, Kames had written that, despite having organs of speech "in perfection," the "orang outang" lacked the requisite understanding to make meaning from words. After all, "a parrot can pronounce articulate sounds, and it has frequently an inclination to speak; but, for want of understanding, none of the kind can form a single sentence."[176] The parrot was used throughout the early modern era as the example par excellence of speech without knowledge or reason. White's assertion to the contrary proves its ubiquity: "It should not be understood that they are destitute of thought and reflection about what they speak, since many authentic instances might be adduced of having discovered much reflection and discriminative accuracy in the application of their speech to particular occasions."[177] By far the most notorious reference to parrots, however, was David Hume's "racist footnote" in his essay "On National Characters."[178] In this he compares Francis Williams, a black polymath who was educated in Britain and almost elected to the Royal Society, to a parrot that repeats what it has learned but has no understanding of its own.[179] The implication is clear: a learned "negro" is as a parrot, a simulacrum of intellect.[180] Language here works as a fiction, apparently indicating mental sophistication, but where language is encountered that is sophisticated enough to identify the speaker as European, it is trumped by race. Williams's racial identity as a "negro" tells a more fundamental truth than his language; his racial identity had come to signify his reason far more than could his language.

Language and Difference in the Pacific

The application of these ideas in the Pacific took two forms, synchronic and diachronic. The synchronic approach characterized the earliest European reactions to the region and had two guiding conceits. The first shared with the diachronic form the view that native languages reflected mental simplicity. The other was the climatic explanation voiced by the founder of modern aesthetics, Johann Winckelmann: accents, dialects, and languages can be attributed to the local environment, which makes the tongue less flexible in cold climates.[181] The French were in no doubt that Tahiti was the home of the noble savage, whose language was described as one of "noble simplicity," more immediate and with fewer words.[182] British voyagers were less enraptured, but they agreed with the overall assessment of the French.[183] The superiority of European languages was a constant refrain in the literature of Pacific discovery, and this viewpoint was reinforced by comparison with South Seas visitors to Europe, who were frequently criticized or ridiculed for being unable to grasp European languages.[184]

Other thinkers used language diachronically as a more certain guide to the genealogy of human groups. The centerpiece of efforts to account for European ethnic diversity, this was a project of remarkable duration, running throughout the early modern era and into the twentieth century.[185] In the Pacific case, it began with the assertion of a genealogical relationship between New Zealanders and Tahitians given the similarity of their languages, and entered the public discussion even before Hawkesworth's *Account* was published.[186] The *Account* then endorsed this relationship, and the texts from Cook's second voyage went even further by distinguishing linguistically between the two principal groups into which the Pacific's peoples were being divided.[187] By the early nineteenth century, the use of language to fix categories of human difference by establishing diachronic, genealogical relationships reached its fullest flower. Prichard was a major practitioner: as he wrote, "The most important of . . . aids [to tracing affinities of peoples] is the comparison of language."[188] Prichard still gravitated toward the convention that a savage state entailed few ideas and hence required few words, but the spread and prevalence of savage states produced distinct languages, meaning that "discrepancy of language is no proof whatever of diversity of origin."[189] This conception of savagery ensured human unity, but at the cost of marking savages as inferior: the association of intellectual shortcomings and linguistic poverty only made sense by assuming a common psychology and hence species identity.

Prichard's approach was anticipated in many respects by Johann Reinhold Forster. Both men were polyglot (Prichard knew at least nine languages, Forster seventeen) and consumed prodigious quantities of travel narratives. Forster translated Bougainville's voyage into English and shared with the Bristol doctor a profound interest in natural history. He was also influenced by others who emphasized the importance of acquiring linguistic knowledge when exploring: this was the culminating item in the *"Hints* offered to the consideration of Captain Cooke, Mr Bankes, [and] Doctor Solander" by the President of the Royal Society on the eve of the *Endeavour* voyage, as well as part of J. D. Michaelis's *Fragen an eine Gesellschaft gelehrter Männer* (1762), a volume containing over a hundred detailed instructions for travelers, a copy of which Forster took with him on the *Resolution*.[190]

Forster's *Observations* described the Pacific's peoples in typical fashion. In the lengthy chapter containing "remarks on the human species in the South-Sea isles," Forster began with a brief estimate of the total population of the islands. He then introduced his bipartite division of the Pacific's peoples and described each group in relation, first to their color and beauty and then to other physical characteristics such as size, form, and culturally modified features (like the skull, genitals, and breasts). In addition, Forster used feminine beauty as his strongest argument against the "orang outang's" humanity.[191] Crucially, though, these factors served at most to "make it appear *highly probable* that [Pacific islanders] may be descended from two different races of men"; it was through analyzing languages that he hoped to *"prove* [this], by an historical argument."[192] His ultimate contention, that similar vocabulary suggested that "Malays" and "Polynesians" had a common ancestor, was less conclusive. But despite leaving the final decision "to better instructed, and more enlightened ages," his is a clear example of how language, and therefore genealogical evidence, served to reinforce racial categories that had been provisionally established by somatic means.[193]

Notwithstanding the growing purchase of the diachronic model of language in the making of racial categories, at a popular level the synchronic form prevailed. The clearest example of this is offered by the popular response to Mai during his time in Britain (1774–1776).[194] Mai's visit was significant for several reasons. First, other ethnographic "specimens" that had been brought to Europe were either animals or dead (Ahutoru died on the return journey).[195] Second, the 1770s was a turning point in the history of celebrity culture, as the press began to fabricate celebrities and make them more accessible to a wider audience.[196] Mai was a consummate celebrity and was lionized by Fanny Burney, played chess with Samuel Johnson's friends, and spent a great deal of time with Banks, Solander, and Sandwich, First Lord of the Admiralty.[197] He was even portrayed on the stage. Garrick first had the idea to use him as Harlequin, but it was the Covent Garden Theatre that produced a pantomime, *Omai*, almost a decade after its eponymous hero had returned to the Pacific. Phenomenally popular, *Omai* was viewed by tens of thousands, it apparently brought George III to tears, and was repeated several times by royal command.[198] In the pantomime, Mai was rendered the quintessential South Seas islander who functioned as a metonym for British imperial and maritime power.[199] Finally, thanks to his visit, Mai became the archetype of Pacific islanders within racial theories of the later eighteenth century, theories that manifested the growing confidence of the British intelligentsia to classify and comprehend all human societies. Banks supplied Blumenbach with a picture of Mai, and another was included as a plate in William Lawrence's *Lectures* (1819), acting as the exemplar for the "Malay" race in a work which followed Blumenbach's racial schema almost to the letter.[200]

Mai's celebrity and prominence, his metonymic relationship with both South Seas islanders and British imperial power, and the increasingly prominent activities of racial theorists all contributed to democratizing anthropological speculation. Mai's press coverage enabled the reading public to formulate opinions concerning Pacific peoples and to communicate these in print. But Mai was an ambivalent celebrity. Used in traditional fashion as a mouthpiece of satire and social criticism, there was nevertheless a strong current of opinion that viewed him negatively.[201] This facilitated a popular discussion of him in racial or ethnographic terms. Much of what appeared on Mai in the press between 1774 and 1776 was saturated with stereotypical imagery of non-Europeans, such as their lack of foresight, fickleness, stupidity, or hypersexuality.[202] Like Ahutoru, Mai was criticized for being intellectually backward, unable to speak English or to understand Britain in anything other than his own terms.[203]

As a result, Mai was explicitly negritized in popular culture.[204] This influenced both his image and speech, and his very presence in Britain was explained by some as a consequence of his features: "A flatness in his nose, which indicated a mixture of the negro breed and his family, and made him less respectable in those islands, where blood is considered in the highest degree, contributed to make him more ready to undertake this voyage."[205] Great attention was paid to his metaphorical coinages, such as "stone water" (ice) and "soldier bird" (wasp). Perhaps because he was a potent representative of British imperial might, it was not Mai but another "Otaheitean traveller" whose speech in *Omai* was a crudely caricaturized African patois:

> In de big canoe
> I o'er ocean swim me,
> Jack and merry crew
> Give good liquor to me.
> Over sand and rocks
> Teach me sail, no paddle;
> Teach me den to box,
> So to use my daddle [fist].[206]

In cheaper or printed pictorial representations of Mai, he was portrayed as black-skinned, as in figures 5.7 and 5.8. His hair in both of these images is long and has a tendency to curl and could in no way be mistaken for the "wool" typical of representations of Africans in the period.[207] Yet the notion that he was black had taken so thoroughly that he was being referred to as such while still in Britain, and at least one horse owner named a black stallion "Omiah."[208] Attitudes to Pacific peoples more broadly became less indulgent or friendly following Cook's death, but the example of Mai shows that there was already in the mid-1770s a significant current of opinion which regarded Pacific peoples negatively. Ever the ambivalent figure, Mai was still being referred to as "a black man" by Granville Sharp in 1800, long after he had become the avatar of British imperial pride in *Omai*.[209]

European voyages to the Pacific in the second half of the eighteenth century represent a crucial chapter in the history of racial thought. They proved a testing

Fig. 5.7. Francesco Bartolozzi after Nathaniel Dance, *Omai a Native of Ulaietea* (1774). Courtesy of the Alexander Turnbull Library, Wellington, New Zealand.

Fig. 5.8. J. Page after unknown artist, *Omiah: A Native of Otaheite Brought to England by Capt. Fourneaux* (from *London Magazine*, August 1774). Courtesy of the National Library of Australia.

ground for theories developed on the basis of other peoples and played a vital role in reshaping models of human difference. The emergence of a new empirical standard meant that what the voyagers said was taken more seriously than had been the case in the past, and what they said made an enormous contribution to racial ideas. Voyagers such as the Forsters themselves theorized, where others distributed materials to investigate. The texts addressed crucial topics of interest to racial thinkers, and the transportation of living human specimens back to Europe facilitated a widening of the discourse, as every newspaper reader could form anthropological opinions. At the same time, the metropolitan gender order received a boost by being ultimately "discovered" in "nature," a centrifugal process that marginalized nonbourgeois, nondomestic sexualities in the Pacific through assertions of class or racial identity. Miscegenetic fantasies were at once a manifestation of and a causal factor behind the embodiment of identities, being invoked to solve thorny anthropological problems (such as the *arioi*), reflecting pervasive pronatalism, and contributing to the erosion of universalistic models of human progress at the hands of racial concepts and vocabularies.

But perhaps more importantly, the Pacific demonstrates the ultimately inconclusive nature of models of identity that were based on the physical body. Subjective aesthetic judgments, based on unreliable sensory data, drove theorists to look ever deeper inside the human body for a characteristic that might reliably determine racial identity. The ultimate futility of this project was not to become undeniable until the twentieth century, but one final example shows how Pacific bodies were only too obviously ambivalent, even in the European imagination.

This concerns the flexibility of gender in the discourses of Pacific discovery.[210] The discussion in this chapter began by illustrating, through the example of Baret, how even the earliest voyages manifested this flexibility. In the popular press, Banks and Mai were both lampooned and attacked as macaronis, effete individuals whose foppish dress and penchant for peppering their speech with foreign terms earned them a reputation as ungendered.[211] While this may not connote homosexuality, it nevertheless played a role in the voyage accounts.[212] Sodomy was a nautical stereotype difficult to dislodge, but certain Pacific sexual identities, such as the *aikāne* of Hawai'i, were particularly hard for European voyagers to comprehend with a binary (homo/hetero) conception of sexuality.[213] Despite such difficulties, early contact was characterized by relative ease and understanding, and even self-reflexivity. One example of this appears in Cook's second voyage, where he describes how the Europeans came to realize that some of their men were being sexually harassed by native men on Tanna not because of any same-sex desire but because the European men in question were performing a typically female duty (carrying items).[214] This soon gave way to a binaristic and judgmental code. By the 1820s at the very latest, what little relativism had been shown regarding same-sex activity had been obliterated as such activities were once again described as "unnatural practices."[215] Banks was criticized for his gaudiness into the 1790s and beyond, but never again suffered parody as a macaroni. This stereotype itself was on the wane by the end of the century, perhaps because it had become ontologically incompatible with a new gender order intolerant of fluidity or liminality.[216]

Relating this cultural shift to the body, it is salutary to return to Diderot's *Supplément*. In the Diderotian sexual order, an immediately scrutable marker of sexual identity was offered that told whether individuals could engage in legitimate utilitarian intercourse. An elaborate system of veils of different colors demonstrated whether a woman was fertile, sterile, or menstruating. In keeping with the fluid, falsifiable, and ultimately inscrutable body, it bears stating that these veils were items of clothing. Worn and shed at will (albeit at risk of punishment), their use as a solid marker of sexual identity within Tahitian society was thus of severely limited use. Remarkably, more permanent and corporeal markers of identity were *already* available. Bougainville had commented on tattooing in his *Voyage autour du monde* (1769), and although popularized as a result of the Pacific voyages ("tattoo" is itself a word originating in the Pacific), tattooing had never wholly disappeared from Europe after its heyday in antiquity.[217] It would have been quite possible for Diderot to mark the sterile with a tattoo rather than preserve their autonomy in the form of a veil. That he chose not to do so indicates that the embodiment of identity was not yet hegemonic.

CHAPTER SIX

Colonial Ethnogenesis and the Sexual Making of Race

The abolitionist James Stephen hated it, and for good reason. The anonymous novel *Marly; or, the Life of a Planter in Jamaica* (1828) claimed to be above the fray but came down squarely on the side of the West India lobby, advocating a long, drawn-out process of gradual emancipation, and what it lacked in plot and character development, it made up for in vivid and unduly positive descriptions of life in the Caribbean. Stephen described his experience of reading this novel as a "painful sacrifice" that he felt bound to make for the cause of abolition, "as the Surgeons pore over a putrid corpse when they dissect it, but I have not a surgeons apathy."[1]

One section of the text that may have provoked his ire was, ironically, a plan for the abolition of slavery proposed by Wogan, an overseer. Wogan's plan called for interracial sex on a massive scale: mixed-race children born to black enslaved mothers would be free from birth, and mothers who had borne four such children were to be manumitted. The scheme was both genocidal and ethnogenetic; it envisaged the eradication of "the black colour," but it also entailed the creation of mixed-race children over the course of "probably, hundreds of years . . . before the population is generally whitened."[2] Wogan enthused about the possibility of extending this plan to Africa through large-scale European colonization, but he acknowledged that such white-black concubinage—for marriage between black and white partners was out of the question—might get stuck on the "nice distinctions between virtue and vice" that were made at home but "very little understood in this western archipelago."[3]

One obvious question raised by this plan is why the eradication of blackness would necessarily entail that of slavery, and vice versa. This question, and several other issues raised by Wogan's plan, such as the production and classification of people of color or the distinct moral orders of Britain and the Caribbean, are the topic of this chapter. It contains a comparative analysis of ethnogenesis—the creation of mixed-race individuals through the process of racial sexual mixture—in the Caribbean and British India that demonstrates the pivotal importance of chattel slavery for this complex of racial and sexual ideas. Sexual activity not merely gave rise to mixed-race people but also exposed a colonial moral and gender order that differed profoundly from metropolitan moral codes. The stark contrast between these orders contributed strongly to what Nancy Stepan has termed the "ontologizing via embodiment" of metropolitan notions of masculinity and femininity in the shape of biological sex.[4] At the same time, elaborate systems of racial classification, which widely differed in complexity but were equally arbitrary, point to the staggering diversity of racial thought and practice, and show that the key to race's longevity lay in its usefulness. Ideas of race were vulnerable to attack from all sides, but they nonetheless survived because race was simply too useful a notion to do without.

Ethnogenesis and Racial Classification

Wherever Europeans settled, they engaged in sexual relations with the local population. In the early phases of European colonization in the New World, *mestizaje* (also known as *métissage*, or "mixture") took place with local Amerindians. By the middle of the sixteenth century, sexual intercourse also regularly took place between Amerindians, Europeans, or their mixed offspring and populations of enslaved or free Africans. The complicated mixture of populations of black, white, and red was a prominent and visible component of European colonialism. The term "ethnogenesis" is used here to encompass the whole range of such mixtures (*mestizaje/métissage* relates primarily to European-Amerindian mixing) and to highlight the creation of new peoples through sexual coupling. As the British Caribbean and India are the foci of what follows, mixture between European and Amerindian populations will not feature strongly. Indeed, by the end of the period under consideration (roughly 1760–1840), the indigenous population of Britain's Caribbean islands had all but disappeared.

Interracial sexual activity in the British Caribbean thus took the shape of white-black coupling. These color-based terms are used in preference to "European" or "African," which are strictly valid only for those who were not "creoles" (i.e., born in the colonies). Nevertheless, an overriding concern of this chapter is to demonstrate that color-based terminology was never watertight and actually goes much of the way to proving the fictiveness of race. It does so, in part, by associating particular characteristics with certain colors. In the Caribbean and many parts of North America, "black" (or "negro") was, by the eighteenth century, a synonym for "slave," while "white" came to replace "Christian" as a foil for the growing range of undesirable characteristics heaped on "black." "White" and "negro" were not constructed solely with reference to slavery or to each other, however: they often encoded particular gendered and sexual meanings as well, and insofar as these became naturalized within particular color identities, they also went some way to fleshing out understandings of sexual difference.

In addition to these two foundational terms, an elaborate vocabulary of racial mixture emerged, much of which was imported from the Spanish and Portuguese. The terms "mulatto," "sambo," "quarteron," and so on were made to stand for more or less precise mixtures of black and white in a system of classification that, on paper, appeared sophisticated and powerful.[5] It varied between (and even within) colonies, but thanks to proslavery authors such as Bryan Edwards and Edward Long, the Jamaican variant is one of the best documented of these systems.[6] Long had spent twelve years in Jamaica and became a scientific correspondent and widely acknowledged expert on the colony.[7] Edwards spent longer in Jamaica than Long and, like him, was a wealthy planter and a historian of the West Indies, who produced a *History, Civil and Commercial, of the British Colonies in the West Indies* in 1793.[8]

The system as they presented it was based on the elaborate typologies of the Spanish and had between five and six ranks: "Negro," "White," and mixtures of these two, such as "Mulatto," "Terceron" (or "Quadroon"), "Quarteron" (or "Mustee"), "Quinteron" (or "Musteephino"/"Mestizo") and "Sambo."[9] No two versions of this system were exactly alike, but the overall picture (including key variations) is shown in figure 6.1. Apart from "Sambo" (the offspring of a "Mulatto" and "Negro"), each category was the product of a union of a "White" and another category. The child of a "Sambo" and "Negro" was a "Negro," demonstrating both the ease with which one could be reabsorbed into the "Negro" community and the difficulty of becoming "White."

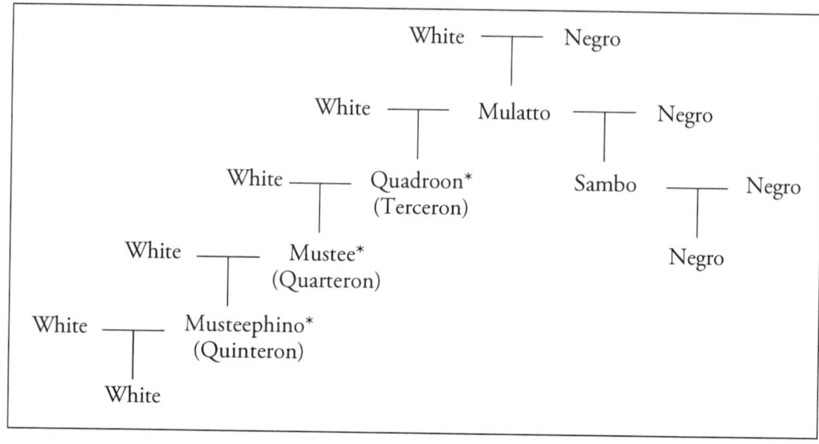

Fig. 6.1. Jamaican racial gradations. Sources: Long, *HJ*, 2:260; Edwards, *History*, 2:15–17; Moreton, *West India Customs*, 123; Higman, *Population and Economy*, 140. *Quadroon* and *Mustee* were reversed in the classification described by J. B. Moreton. Parenthesized terms were used by Long. Edwards omitted *Musteephino / Quinteron*.

Two features of this system are significant. The first is that each category owes its existence to sexual intercourse with either a white or black person. The production of mixed-race people solely from relationships with "Whites" or "Negroes" led to the claim that it was impossible for mixed-race people to reproduce together.[10] As shown in chapter 4, such a claim had a basis in colonial social life as well as in clear racism. It is, however, noteworthy that taxonomies of mixed race graphically supported this claim of mixed-race sterility. In describing people of color as the product of a union that must involve either a "White" or "Negro," authors reified the notion that people of color were unprolific with one another: their ability to reproduce with other people of color was taxonomically prohibited.

The second feature is the fact that the white partner is always male.[11] Indeed, Edwards's summary of the system is a concise statement of the operation of Caribbean ethnogenesis:

A Sambo is the offspring of a Black Woman by a Mulatto Man, or vice versa.
A Mulatto is the offspring of a Black Woman by a White Man.
A Quadroon is the offspring of a Mulatto Woman by a White Man.
A Mestize or Mustee is the offspring of a Quadroon Woman by a White Man.
The offspring of a Mestize by a White Man are white by law.[12]

The only racial category that is the product of a coupling in which the mother can hold a "higher" racial rank than the father is one in which neither partner is "White" and where the result is the rank closest to "Negro" (i.e., "Sambo"). In all other ranks, the "White" partner is male. This points to several features of Caribbean society highlighted throughout this chapter—namely, the inability to contemplate a racial mix in which a "White" woman pairs with a colored or "Black" partner,[13] the male-heavy white sex ratio in the Caribbean, the matrilineality of slavery, and the ability of white people to sexually assault slaves at will.[14]

The following sections explore ethnogenesis in the British Caribbean and the Company Raj in India. They have a parallel structure, first exploring and tracing changes in attitudes to mixed-race people in each context. As perceptions of mixed-race people depended largely on existing attitudes toward their non-white parent(s), these sections proceed to explore attitudes to black and Indian populations. Climate understandably plays a large part in these views, which nonetheless encompass a wide range of stereotypes, including many of a sexual nature. These sections then explore the contexts of interracial sex, highlighting their similarities in both the East and West Indies. The final part of the chapter builds on the preceding analysis by asking why no complicated system of racial classification, such as that outlined above, emerged in India. It is here that chattel slavery emerges as the crucial *differentia* between the two contexts, one that is central to the making of racial taxonomy but that also highlights race's contradictions.

Ethnogenesis in the Caribbean

Our exploration of ethnogenesis begins in the British Caribbean, specifically Jamaica. The size and dominance of this island makes its importance undeniable. In 1807, there were approximately 775,000 slaves in the British Caribbean. Jamaica's share of this population was approximately 347,000 or 45 percent; its closest rival (Demerara, Essequibo, and Berbice, later united as British Guiana) had just over 10 percent of the total.[15] Jamaica's white population (28,000) was also the largest and 80 percent higher than its nearest rival, Barbados.[16] Yet being outnumbered by over twelve to one could not but be a source of profound fear for whites in Jamaica.

As a result, white attitudes to people of color in the Caribbean became increasingly negative over the eighteenth century. Although most people of color had a white parent, they represented a threat because perceptions of mixed-race people depended overwhelmingly on attitudes to their non-white parent(s). And blackness came to be synonymous with slavery, a trend that was already evident by 1700 in the British Caribbean (arguably much earlier in the Spanish and Portuguese Empires) and which accelerated in the following decades.[17] The characteristics of the "slave" soon became indistinguishable from those of the "black" (more accurately, the "negro") and vice versa. These features owed much to existing ideas about climate and the suitability of Africans for labor in the torrid zone, but they also derived in part from the contexts of interracial sex on the tropical plantation. This sexual activity had a marked impact on ideals of masculinity and femininity, and contributed to render the Caribbean a zone of debased gender norms, against which European standards could be defined and then reified in the form of opposite sexes.

As the living, breathing consequences of this sexual activity, mixed-race people were often treated in a manner that reflected moral condemnation of the circumstances of their birth. While their population was small, they did not symbolize a systematic moral depravity but were rather the result of individual lapses and were generally tolerated. Mixed-race people could inherit property, and in the very early days in Jamaica they were entitled to vote, to work in public offices, and to count as white for the purposes of deficiency legislation (laws that specified a minimum number of white employees on a plantation). Even once these rights had been restricted, individuals could petition the assembly for exemption.[18]

These rights were swiftly and drastically curtailed once the mixed-race population began to grow. The principal reason was not public morality but white security. As their numbers increased, it became clear that people of color were inheriting significant amounts of property from their white ancestor(s), which threatened the social, political, and economic dominance of the white population. In 1761, immediately following Tacky's Revolt, the most severe slave rebellion to hit Jamaica before the 1830s, a cap was placed on the amount of property an individual of color could inherit, throttling the economic development of the mixed-race community. This and other restrictions were repealed only in the 1820s, when the white population sought to make common cause with free people of color against the greater twin dangers of metropolitan abolitionism and an increasingly restive enslaved population. Even then, as the bell was tolling for chattel slavery, concessions were granted only slowly and with great reluctance.[19]

Alongside these official restrictions there existed a wide range of cultural stereotypes that were no less socially debilitating than legal measures. People of color were deemed improvident, indolent, or weak and effeminate.[20] They were thought to be particularly taken with gaudy clothes and engaged in conspicuous consumption, depleting the financial resources of their white partner, over whom they were thought to exercise an unnatural degree of control.[21] Women of color were held to be particularly dangerous, partly because of their reputed sexual incontinence but also because they held their white partners in thrall, distracting men from relationships with white women.[22] Their hyper- or hypofecundity was a particular mark of their abnormality.[23] In general, people of color were believed to lack the paternalistic noblesse oblige of whites and to seek all the advantage they could from the pigmentocratic social order; they were thought to treat their own slaves with condescension and cruelty.[24]

Belief in these characteristics was not uniform: some believed people of color to be active, charitable, hardy, industrious, intelligent, lively, and resistant to disease, and several wrote about the beauty of mixed-race women, especially those with lighter skin.[25] The perspective an author took on people of color usually (but not always) depended on their attitude to slavery.[26] Indeed, the fight to preserve slavery in the early nineteenth century produced radical changes in attitudes and policy toward mixed-race people throughout the Caribbean. But their treatment remained inextricably bound to the security concerns of the day.[27]

The Blackness of Slaves

At the most fundamental level, people of color posed a threat to white security because blackness was practically synonymous with slavery, and their non-white heritage meant they were unable to escape this association, irrespective of their legal status. The ties between blackness and slavery in Caribbean thought were profound, and one of the strongest was climatic. The belief that black people were better suited than white to laboring in the "torrid zone" was near universal: even some abolitionists stated openly that "no one is senseless enough to propose that the colonies should be cultivated by Europeans."[28] Modern scholars have cautiously acknowledged that immunity to tropical diseases may have conferred a selective advantage on those of African descent.[29] But debates of the period focused more on the heat of the climate than on disease. "Let us hear no more then of white men working," wrote the medic and planter David Collins in 1803, "where they have so much difficulty to exist, even without work." Indeed, "heat not only extinguishes the power, but the will, for exertion."[30] Even as they turned decisively toward emancipation, anti-slavery campaigners explicitly claimed that those of African descent were uniquely suited to the Caribbean climate, to emphasize the horrific decline of this population in a supposedly natural habitat.[31]

At the heart of the connection between blackness and slavery was the legerdemain whereby the body was made to signify a social status. As described by one author in the early 1770s, "The distinction of black and white . . . we have *so unreasonably* made the marks of freedom and of slavery."[32] Fifty years later, James Stephen directly identified the problem, commenting that "the odium and contempt, excited by the civil and moral degradation, is actually *transferred to the bodily peculiarity of the* slave. . . . The term slave is not in the West Indies, as in other countries where private bondage has prevailed, a term of obloquy, or colloquial reproach; but the bodily designation is substituted for such purposes in its stead. Amidst all the reviling epithets, used in anger towards these poor bondmen, '*you slave*,' or any allusion to the condition, is never heard; but '*negro!*' pronounced with an angry or contemptuous emphasis, is a word of superlative reproach."[33]

Stephen's recognition of the process whereby "negro" and "slave" became synonymous did not, however, have any appreciable effect even on the rhetoric of his fellow abolitionists: Elizabeth Heyrick, for example, treated "negro," "slave," and "African" synonymously in her *Immediate, not Gradual Abolition* (1824), and she was far from unique.[34] As long as they unwittingly repeated the terms of this equation, abolitionists contributed to the elision whereby the characteristics of the slave were naturalized within the category "negro."

This conflation was more common among those who did not oppose slavery. Lady Nugent, wife of Jamaica's lieutenant governor (1801–1806), referred to slaves uniformly as "blackies" despite clearly recognizing differences of color.[35] But the conflation was most frequent—and of much longer standing—among those who traded and owned slaves.[36] The extent to which this was an unconscious act is clear from the estate records of Waltham in Grenada. Its overseers recorded many of the mundane details of plantation life in its day books: entries such as "Negroes in their Gardens" (July 26, 1834) or "Gave the Negroes a General allowance of Fish" (July 28, 1834) were commonplace. But on July 31, 1834, slavery officially came to an end in Grenada, replaced by four years of "apprenticeship" before full emancipation. Within 48 hours of the end of slavery, "negroes" had suddenly become "labourers": the same hand recorded on August 2 that "apprenticed labourers [were] in their Gardens," and two days later the overseer "Gave . . . Labourers a General all[owan]ce [of] Fish."[37] Thus, by the 1830s (but undoubtedly much earlier), the putatively racially descriptive label "negro" had become fused with the category "slave," such that it was no longer appropriate to refer to ex-slaves as "negroes." With such a strong association in place, the characteristics of the slave undoubtedly influenced wider perceptions of the "negro," whether enslaved or not.

Several characteristics of the slave shaped attitudes toward black people, all of which stem from the dehumanization of the enslaved and which in turn contributed to ideas that black people were somewhat less than human. This dehumanization took several forms. One was the insistence that the slave was a chattel or piece of merchandise, sometimes distinguished only by a number: in 1750, for example, a New York merchant wrote to his brother-in-law to inform him that "I have Sold one y^r She Negroes N°. 9 at £40 . . . and the remaining boy N°. 5 at £42. for Cash."[38] Another was a condescending paternalism that imagined the slave as a child. For example, George Canning stated openly in the House of Commons in 1824 that "we must deal with the negro as with a person possessing sense, but the sense only of an infant." (He went on to liken the consequences of immediate emancipation to those of Frankenstein's creation.)[39] The infantilization of slaves was common and often implicit, as when they were described as "children of nature" or "children of Africa."[40] It conveyed an understanding of the slave as fundamentally unable to govern him- or herself, a belief at the heart of much planter discussion of the materialistic whimsicality of slaves, endlessly talking nonsense and governed by their passions and spite.[41] At least when trying to convince metropolitan audiences of the legitimacy of slavery, however, planters held that the slaves were indeed incompetent to govern themselves but not irredeemably so.[42]

Less common, but more spectacularly offensive to modern eyes, was the literal brutalization of slaves: likening them to animals. This clearly appalled abolitionists, who by continually referring to it, kept the issue at the forefront of debate.[43] Nevertheless, there is disagreement about the extent to which slaves and animals were linked in planters' minds.[44] Even if naming patterns did not deliberately set out to humiliate the enslaved by giving them names in common with livestock, there were certainly parallels between slaves and animals in planter thought. In 1816, for example, when Matthew Lewis first visited his estates in Jamaica, he commented on the widely held opinion that, like slaves, the only cattle that could labor in the West Indies are those "which have a great deal of black in them."[45] It is possible to overstate the interrelation of domestic animals and slaves, yet "dehumanization" is not merely a shorthand for cruel treatment: it accurately describes the place allotted to slaves in planter thought.[46]

Such broad stereotypes were maintained despite awareness of African ethnic diversity. Planters frequently associated particular African groups, such as Ibos and Coromantis, with certain traits and purchased slaves on this basis.[47] Slaves were, however, usually discussed in the aggregate.[48] Both pro- and anti-slavery writers were guilty of homogenizing the black population: sometimes implicitly, like the abolitionist Thomas Cooper, who wrote that it was partly due "to the universal profligacy of manners prevailing among Blacks and Whites, that the Negroes in Jamaica are a very unprolific race"; sometimes explicitly, as with the proslavery writer Reverend George Bridges, who stated that "the African hordes resemble each other in their barbarous lives . . . while all that is monstrous, vile, and contemptible, is universally found amongst those savages, whose native cowardice can be stimulated only by the consciousness of numbers."[49] Metropolitan authors habitually spoke in general terms of "the negro" or "the slave," but Cooper and Bridges both spent considerable time in the Caribbean, and others well acquainted with colonial life generalized in equally broad terms.[50]

This homogenized "negro" was therefore perceived largely in terms derived from the equally homogeneous "slave." This was a reciprocal process: as the characteristics of the slave permeated creole perceptions of black people, so too did the stereotypical features of black people penetrate the image of the typical slave. The hybrid entity "negro slave(ry)," which was so called throughout early nineteenth-century discourses of slavery and abolition, was the result of this cross-fertilization.[51]

Interracial Sex in the British Caribbean

The racial identities imposed on and assumed by people of color were also deeply influenced by the context of their birth, as a result of interracial sex within a slave society. This affected both the construction of blackness and whiteness, and highlighted the place of sex and gender in these racial categories. For not only were these racial groups produced by a mixture of white and black but the white progenitor was almost without exception male. These features are directly based on the demographic and social context in which mixed-race individuals were born.

Perhaps the most significant feature of white Caribbean society was its unbalanced sex ratio. In 1844, at the time of the first census in Jamaica, this stood at 143.1 males per 100 females (double this in some parishes): an imbalance that was probably greater under slavery.[52] Many white men therefore looked to the enslaved population for sexual gratification, encouraged by several features of plantation life.[53] First, estates were often geographically isolated and their white staff were (as with domestic servants in Britain) expected to be single.[54] Second, this staff was often transient, moving from one plantation to another and seeking to amass wealth before returning to Britain.[55] Third, the composition of the slave population was also unbalanced, having either too many or too few females: in the last decades of the slave trade, 100 were imported for every 150–180 males, but the numbers of males fell after 1807 to fewer than 95 per 100 females by the early 1830s.[56] Women were therefore either in abundance or in demand. Both possibilities were discussed at length in pro- and anti-slavery literature, especially regarding the resentment of slaves at white sexual predation.[57]

There are several indicators of the extent of this sexual rapacity. The strongest comes from the size of the colored population. This trebled between 1790 and 1820 while the white population fell, suggesting that white settlement in the Caribbean, for which the presence of white women was necessary, had only limited appeal.[58] Furthermore, the rapidly expanding colored population stemmed predominantly from white fathers. Of course, people of color could be produced from the mixture of black, white, or Amerindian parents, of one of these with a person of color, or of two colored parents. But the possibility of widespread Amerindian or white female parentage of people of color may be immediately dismissed: there were simply not enough of either category. Furthermore, the prospect of sexual relations between white women and non-white men filled the white community with horror: the rare occasions on which it occurred (for instance, with white prostitutes) produced savage punishment.[59] In addition, social and cultural constraints limited the number of children of color with two colored parents.[60] Both white men and women of color sought relationships with a partner whose skin was as light as possible.[61] Light skin was deemed more beautiful by the white population, and the power and privilege of whiteness meant that women of color often hoped for a better life for themselves and their children through a relationship with a male whose skin was lighter than their own.[62] This pigmentocratic tendency hindered men of color from finding women of their own racial stamp, and the dearth of such couples led, as we saw in chapter 4, to the presumption of their infertility. It is therefore likely that the majority of the colored population was fathered by white men, varying according to the size of the enslaved population and how long it had existed in any given location.[63]

However frequent it was, interracial sexual activity was nonetheless trenchantly criticized, even by those ultimately responsible. Responding to abolitionist exposés of white sexual abuse, the upper ranks of white society in the Caribbean sought to defend slavery by blaming their staff for debauching slaves.[64] In this, planters found unlikely allies: where once they blamed planters for offering the sexual services of their slaves to guests, abolitionists now joined them in condemning lower-ranking whites, in the interest of obtaining a consensus in favor of gradual emancipation.[65] The making of such common cause must, however, not be exaggerated: several fissures in white Caribbean (especially religious) identity were apparent by the 1820s, but the overwhelming fear of slave rebellion in the generation following the Haitian Revolution prevented it from fracturing completely.[66]

Indeed, enough was common to white male behavior in the Caribbean to highlight how remote it was from metropolitan norms of masculinity. The ideals of compassion, restraint, sensitivity, and the sense of paternal duty attached to men in Europe were conspicuously absent in portrayals of their Caribbean counterparts.[67] To representations of white male sexual abandon, female abolitionists added the maltreatment of pregnant or nursing women, successfully making the flogging of female slaves a cause célèbre of the 1820s. Consequently, discourses of slavery and abolition helped to naturalize and reinforce metropolitan gender norms by pointing to their negation in Caribbean contexts. The very use of gender norms to combat slavery indicates their cultural power.[68]

Although many argued that the system of slavery was responsible for immorality, both pro- and anti-slavery writers blamed slaves themselves.[69] As we saw in chapter 2, slaveowners attributed the failure of their efforts to encourage reproduction to the incorrigible promiscuity and polygamy of their enslaved workers. Proslavery texts made the most direct accusation that this behavior was intrinsic to black people; the most hard-line examples came before the emancipation campaign was launched in 1823 and slavery's apologists had to change tactics.[70] For one member of the Lords Select Committee on the West Indies, the Caribbean climate was directly responsible for promoting licentious behavior, as it forced people to go about in a state of seminudity.[71] Earlier, however, immorality was made part of the fabric of the black body: Collins wrote in 1803 that "negro women have ardent constitutions, which dispose them to be liberal of their favours; and it had been found by experience, that they who resign themselves to the indiscriminate caresses of men, are seldom very prolific."[72] By the later 1820s, apologists for slavery began to talk in similar terms to anti-slavery writers: slaves were morally educable, but only over an extended period. Naturally, they differed on the precise duration, but on this topic, as on many others, emancipationists continued to fall into the trap of collapsing the "negro" and the "slave" into one another. As one anti-slavery society stated in 1824, ignoring the possibility of non-enslaved black people, "This debased state of character is not to be attributed to any peculiarities of constitution in the Negro, but that it is solely and entirely to be attributed to the system under which he has been placed."[73]

Enslaved women bore the brunt of the attack on sexual misconduct, in a discussion characterized by the widespread use of stereotypes. Though crude, these stereotypes changed across time and were sufficiently vacuous to be usable by all parties: planters, abolitionists, and even the enslaved themselves.[74] Several—including the "Sable Venus," "She-Devil," and "Drudge"—have been analyzed by historians who have almost entirely neglected women of color.[75] This both is and is not surprising: on the one hand women of color were particularly prominent as domestic slaves, but on the other all slaves of color were largely invisible in a discourse that concentrated on the most oppressed sectors of the enslaved population.[76] In addition, these stereotypes also contributed to the conflation of "black" and "slave" visible elsewhere. For example, the "Quasheba" stereotype (the female version of "Quashee") was oriented around the assertion of African roots.[77] The licentious, treacherous, and scheming nature of "Quasheba" was related directly to stereotypical African traits and stamped an African identity on any slave so described, even if she was one of the very few "Quinteron" enslaved.[78]

Partly as a result of blaming immorality on the female slave, the Caribbean was perceived as a site of deviant femininity. And just as descriptions of corrupt masculinity in the region included white as well as black people, so too did those of debased femininity. White women were routinely described as idle, spendthrift, vain, and vindictive, and were from their earliest infancy supposedly extremely harsh to their slaves. Ironically, given the improbability of their having a white female parent, many accounts of mixed-race women portrayed them in identical terms. Thus, with men as with women, debased gender norms suggest racial mixture or contamination. This may perhaps help to explain the efforts taken by many women in the Caribbean to whiten their skin artificially, as described by Janet Schaw, a Scot who visited North America and the Caribbean on the eve of the American Revolution.[79]

Schaw's journal also indicates that the crucial division between European and Caribbean femininity was reproductive. While contemporary ideals emphasized chastity and maternalism, little could be less feminine than sexual promiscuity and infanticide, of which Schaw directly accused Black women: "The young black wenches lay themselves out for white lovers, in which they are but too successful . . . and as even a mulattoe child interrupts their pleasures and is troublesome, they have certain herbs and medicines, that free them from such an incumbrance, but which seldom fails to cut short their own lives, as well as that of their offspring."[80]

Perceptions of mixed-race people were therefore clearly based both on attitudes to their non-white parent(s) and on the contexts of interracial sex. Their treatment always depended on security concerns: the genealogical association with the enslaved population meant that repression intensified or slackened according to the perceived vulnerability of white power. This association also fed the notion that they were especially unsuited to owning slaves themselves—either from their reputed cruelty or the hatred slaves bore to non-white masters—and fueled the perception of mixed-race people as conceited, greedy, spoiled children or as animals. This latter perspective is evidenced in the very name for the largest group of colored people: "mulatto," a term that efficiently communicates animality, hybridity, and sterility. The Caribbean climate and its attendant seminudity reputedly provoked sexual incontinence among all groups of the population, slave or free (with the possible exception of white women). The unbalanced white sex ratio and a planter order characterized by proprietor absenteeism, high staff turnover, absolute control over the enslaved workforce, remote plantations, and disincentives to marriage all led to white sexual predation on an appalling scale. This did not stop white people, whether metropolitan or Caribbean, proslavery or abolitionist, from blaming the enslaved for their own sexual exploitation. But the moral cancer of the slaveholding Caribbean could not be restricted to slaves themselves; it encompassed all sectors of the population to make the West Indies a byword for debased masculinity and femininity. This ultimately had the dual effect of reinforcing by negation the bourgeois gender order in metropolitan Britain and undermining the racial order in the Caribbean, as outrage at the immoral and un(wo)manly behavior in the region strengthened calls for the abolition of slavery itself.

Racial Mixing in British India

Ethnogenesis in India was remarkably similar to its Caribbean counterpart. The demographic situation of the white population was identical: it was greatly outnumbered by the non-white populace, and few white women were available as partners. The climate was believed to exercise a similar effect, and many of the assumptions about black slaves were also held about South Asians in general (that they were lazy, superstitious, and immoral) and South Asian women in particular (that they, too, had an early sexual peak and were oppressed by their menfolk). Of the few differences between the Caribbean and India, by far the most significant was that in the subcontinent there was no elaborate taxonomy of racial mixtures. The complicated hierarchies encapsulated by the term "people of color" in the Caribbean had no South Asian equivalent. There was less agreement in the East than in the West Indies about the name for mixed-race people—"Indo-Britons," "East Asians," and "Eurasians" were all debated—but these terms were not shorthand for the labyrinth of distinctions so common in the Caribbean.

Nevertheless, as some contemporaries recognized, attitudes to mixed-race people were broadly the same.[81] At first, while their numbers were few, they appear to have been tolerated. Many European fathers, especially if wealthy and influential, sent their Eurasian children back to Britain to be educated, and most provided for them in their wills. When the Bengal Military Orphan Society was established in 1782, it also sought to educate orphaned Eurasian children in Britain. Although disallowed by the Court of Directors in London, such children were nonetheless able to obtain positions in the East India Company.

Attitudes gradually became more negative, however. Certain stereotypes were of long standing—for example, that Eurasians had poor English and poor fashion sense.[82] They were believed to be cowardly, timid, and unintelligent, a viewpoint sufficiently common by the 1830s to be parodied.[83] They also suffered from the imposition of more concrete restrictions after 1790. Eurasians were excluded from covenanted service in the Company, severely limiting their employment opportunities, and many Eurasian and native women were denied pensions or benefits arising from their husband's or partner's military or Company service.

As in the Caribbean, this change in attitudes to mixed-race people was largely due to security concerns, and like there, the British felt sufficiently secure to lessen restrictions against this group only in the 1820s. Eurasians were barred from covenanted service in part because their very existence threatened British power and prestige. As adherence to metropolitan codes of Christian morality steadily became a sign of superiority (and thus justification for rule) over Indians, the presence of Eurasian children (many of whom were illegitimate) was increasingly embarrassing. Worse still, their mothers were often of a low caste: such children, it was argued, ought not to be in a position of responsibility over local natives.[84] Obviating the danger to British power and prestige was therefore central to these increasing restrictions, even if they were also imposed for financial and other reasons.[85]

Indeed, biological and economic motives combined in the financial exclusions meted out to non-European wives. The amounts disbursed to "natives of India" for support were always less than those given to European wives, even when official restrictions were lifted. This was partly due to the family support networks that local women could supposedly draw on, but Company officials also felt that European women needed more support because they were less accustomed to the climate. As one memorandum from *c.* 1820 put it, "Born in India and habituated to live chiefly on rice the wants and wishes of the Half-Caste are much more confined than those of European women."[86]

The similarities of the Indian and Caribbean contexts of ethnogenesis highlight the need to explain the key difference: that no system of racial gradations developed in South Asia. It will become clear that the respective place of slavery in the East and West Indies accounts for the dissimilar nature of racial thinking in both locations.

The Indianness of Eurasians

As in the Caribbean, climate was crucial to understanding human difference in India. In both locations, an earlier confidence in the ability of Europeans to acclimatize broke down, albeit at different times and for reasons beyond the scientific. As Company rule in India was consolidated in the early nineteenth century, the belief in British superiority over Indians grew. This idea was part cultural (manifested by adherence to Christian codes of conduct) and part biological, and admitting the transformative potential of climate threatened British power by suggesting that Europeans could degenerate. As a result, the possibility of acclimatization was increasingly denied, and Europeans came to believe that through clean living and temperance they might combat the vitiating effects of the climate. Any apparent change was associated with those (principally the lower orders) who did not possess these virtues. Likewise, the slave economies of the Caribbean were predicated on the inability of white people to labor in the tropical climate. Thus, for economic and political reasons, white adaptability to local climates was increasingly denied in the decades around 1800.[87]

Not that the Indian climate was thought uniformly benign prior to this date. Paradoxically, Africans, who were not indigenous to the Caribbean, were deemed particularly suited for labor in the West Indies, whereas Indians were thought to have degenerated because of their *own* climate. A broad swathe of opinion held that the perceived "effeminacy" of Indians was climatic in origin.[88] Many recognized that India's climate varied significantly, leaving open the possibility of combating its effects by relocating to a more salubrious region, such as the north of India. But movement could entail change for the worse: one writer of 1757 stated that Kashmiris were "convinced, that the climate and soil of those [more southerly] provinces only serve to rob them of that hardiness they contract in the hills."[89]

Ideas about the climate fed into other, more general stereotypes of Indians. "Effeminacy" was the most frequently used term to describe South Asians, and it encompassed several distinct traits.[90] The frequent conquest of India by outsiders was explained by reference to general weakness and luxuriance; South

Asians were supposedly profoundly untrustworthy, a result of both mendacity and incompetence; they were deemed to be improvident and stupid, with little concept of the future.[91] Untrustworthiness was occasionally explained by greed and ambition: as one author wrote in 1784, "The striking characteristic of a Gentoo [Hindu] is avarice, and that of a Mahometan is ambition."[92] Indians were viewed as cowardly, lazy, and vindictive, although this was often ascribed to the effects of tyrannical (Mughal) government.[93] Occasional voices could be heard to claim that effeminacy could be positive, helping to refine politeness, but they were rare.[94]

"Effeminacy" also had biological aspects. For one thing, it led to problems of gender recognition. One British officer wrote that "all those natives have such a genteel and delicate mien, that, together with their dress, a stranger is apt to take them for women." Apparently, a group of Scottish soldiers did exactly that, "and it was only in attempting to use a few familiarities with them as such that the Highlanders discovered their mistake."[95] Far less sympathetic were later discussions of *hijras*, an Indian third-sex community of biological males, many of whom undergo a total castration (testes and penis). Generally labeled "eunuchs" or "hermaphrodites," by the early nineteenth century they were commonly described as "disgusting" or "degraded beings."[96] Around 1810, the Flemish artist Balthazar Solvyns went even further: "Could it be believed, that there is a country in the world which tolerates a set of men, whose whole life is an outrage to morality and common decency, by the ostentatious display which they continually make of the privation of the marks of their sex? This is the case in Hindoostan."[97]

Such a linkage of effeminacy and sexual immorality was common and largely blamed on despotic government, pagan religion, and the climate. The traditional European association of tyranny and Islamic government was used to explain the confinement of women in harems (*zenanas*), which were principally associated with Muslim Indians.[98] It was thought that Hindus did not practice polygamy, that Hindu women were exceptionally chaste, and that most of the women found in a *zenana*—whether kept by a European or Indian—were Muslim.[99] But Hindus still came in for criticism. For some, Hindu customs and institutions seemed designed for the gratification of lust, and early marriage was explained "with the view of guarding in some measure against this dreadful depravity."[100] It was claimed that Hindu men were functionally polygamous through widespread concubinage, and as Thomas Williamson wrote, in his *East India Vade-Mecum* (1810), one of the reasons for early marriage was to prevent bored sons from impregnating the women of their father's *zenana*.[101]

The two incompatibles, polygamy and pregnancy, could thus coexist in India, where the phenomenally fecund climate nullified the deleterious effects of polygamy and tyranny and promoted population growth.[102] But the prospect of consequence-free immorality was disturbing, leading to efforts to restore the linkage of polygamy and population decline. One tactic was to argue that *zenanas* were dangerously productive of homosexuality. The confinement of women, it was warned, resulted in lesbian relationships, while men who had their own harems became so sated that they turned to same-sex relations. But most simply asserted the old truth that polygamy was detrimental to population, even in India. Indeed, one author claimed that it was due to the engrossing of available women within the *zenana* that poor men took to their own sex.[103] And the existence of *hijras* seemed to confirm both the existence and extent of this "unnatural vice."[104]

Even the chastity of Hindu women came under attack.[105] Some were concubines or in *zenanas*—particularly if they had lost caste—while others filled the ranks of dancing (*nautch*) girls.[106] European men took little care to distinguish *nautch* girls from common prostitutes, and one author stated in the 1750s that they were "bound to refuse no one for their price," but were well trained in "the art of pleasing" and had "none of that nauseous boldness which characterises the European prostitutes."[107] Fifty years later, Williamson, quoting a friend, was more nuanced and distinguished between different groups of *nautch* girls in an effort to minimize their association with prostitution (a gambit that resembled the attempt to rehabilitate Tahitian women's morality, encountered in chapter 5). As he argued, "'It is said these women always consider their first lover as their real husband,'" in a manner suggestive of the range of "marriages" available in India, to be discussed shortly.[108]

Beyond immorality, native women in India were thought to share other features with enslaved women in the Caribbean, which in turn shaped attitudes to their mixed-race offspring. It was widely believed that Indian women attained sexual maturity very early, were especially fertile, and had exhausted their reproductive capacity years before their European counterparts.[109] Their fecundity was attributed to the climate and was used by commentators such as Robert Orme and Alexander Dow to explain how India's population remained high despite the twin evils of polygamy and despotism.[110] Dow even argued that polygamy was necessary given an Indian woman's brief reproductive life: "One man retains his vigour beyond the common succession of three women through their prime; and the law for a multiplicity of wives is necessary for the support of the human race."[111] To their accelerated and concentrated development was added the stereotype, common to women throughout the "torrid zone," that childbearing was uncommonly easy and straightforward.[112]

However easy it may have been to become a mother, Indian women were deemed little better suited to the task than enslaved women in the Caribbean. Stories about the carelessness or even murderousness of Indian mothers were common, despite the wealth of evidence to the contrary. For example, Robert Grant, in a paper written in the 1790s but first published by Parliament in 1813, acknowledged "that examples are not very uncommon of parents who show much tenderness to their children, especially during their infancy; but instances on the other side are so general, as clearly to mark the dispositions of the people."[113] He told of a mother who sold her child during a time of famine, whose own survival was ensured by the "liberal contribution of the inhabitants of Calcutta" but who nonetheless failed to reclaim her child once the crisis had passed. Some mothers were implicitly accused of carelessness in exposing their children to sexual vice, while others adhered too strictly to their religion: the Company servant John Henry Grose mentioned a Hindu mother who killed her children in order not to violate her beliefs.[114]

Indian mothers were also accused of abortion or infanticide in order to avoid interrupting their sexual pleasure, in much the same manner as enslaved Caribbean women. Missionaries such as the abbé Dubois paid particular attention to this. As he wrote, "modesty and virtue place no restrictions on them; their only fear is that their misconduct may be found out. Consequently, abortion is their invariable resource to prevent such a contingency, and they practice it without the slightest scruple or remorse. There is not a woman among them who does not know how to bring it about."[115] James Mill, in his magisterial *History of British India* (1817), quoted a Baptist missionary who wrote of parents that murdered their children "by burying them alive, throwing them to the alligators, or hanging them up alive in trees for the ants and crows before their own doors, or by sacrificing them to the Ganges."[116] The deliberate targeting of female children, or those born "under an unlucky star," was deplored by these commentators.[117]

Mill's *History* also exemplifies the treatment of female oppression, something linking Hindu with other women throughout the "torrid zone," and discussed by practically every writer on India in the period. Earlier comment pointed to harmonious and affectionate marriage, but by 1800 much ink had been spilled on the supposedly unusual degree to which women were oppressed in Hindu society. Although commentators differed on questions such as the practicability of divorce for Hindu women, many stated that women were unable to remarry if widowed, had marriages characterized by conflict and unhappiness, and were treated with severe contempt.[118] The state of Hindu wives was, for many, encapsulated in the question of *sati*, the ritual immolation of widows, banned throughout British India in 1829.[119] European assumptions about the unhappiness of Hindu marriage and the oppression of Indian wives are apparent in the supposition that *sati* originated in an effort to prevent wives from poisoning their husbands.[120] This explanation was treated with skepticism almost as soon as it was posited (by Strabo) and in British accounts as early as the 1750s.[121] Nevertheless, the belief that women were deeply subjugated persisted and was used by Mill—following in the footsteps of Scottish Enlightenment thinkers—as a measure of the relative backwardness of Indian society.[122]

European representations of Indians, Hindu and Muslim, thus share several key features with those of Caribbean slaves. The importance of the climate as a causal factor behind the "effeminacy" of Indians and the ability of Africans to labor in the Caribbean is clear. So, too, is the association of sexual immorality with slaves and Indians, and the particular blame laid on the shoulders of women, for being both immoral and unusually fecund but unusually negligent, if not murderous mothers. As the next subsection shows, these similarities also extended to the contexts of interracial sex.

Interracial Sex in the Company Raj

Many of the factors characterizing interracial sexual activity in the Caribbean also existed in India. First and most importantly, the white population was small and overwhelmingly male. According to one contemporary estimate, only 6 percent of Europeans in Bengal were women.[123] This sexual imbalance was believed at the time—and since—to have been the primary cause of interracial sexual activity. Marriages where both parties were white, it was claimed, were simply "unfeasible." Several eligible women returned from India unwed, however, so this unfeasibility was only partly due to numbers. Also blamed was the expense of keeping a European wife, a claim that strongly reinforced existing stereotypes of female prodigality.[124] Thomas Williamson in 1810 estimated that the expense of a European wife was £300–600 per annum.[125] Complaints about the cost of supporting marriages were not unique to India: in the West Indies, too, overseers were discouraged from marrying partly because of the expense that a married household would entail for a plantation owner.[126]

Second, mixed-race relationships were deprecated in both India and the Caribbean. This attitude apparently became more entrenched in India after 1800 because an influx of European women made partnerships with native women less necessary and therefore less excusable.[127] They nevertheless continued to occur, especially outside of the presidency towns (Madras, Bombay, Calcutta). Nor was this change radical: such relationships were discouraged already in the mid-eighteenth century. Williamson, for example, points out that marriage between Europeans and Indians was "wisely" discouraged by the Company, which also forbade the transport of orphans ("i.e. those born of native mothers") to Europe. Even married women of Indian birth were unwelcome at public functions, and "whatever may be their merits, come under the censure of public characters [i.e., are treated as prostitutes], and, of course, are in a manner proscribed."[128] These examples demonstrate both the real constraints that existed in the eighteenth century and the flexibility of such labels as "orphan" and "prostitute" (of which more later).

These restrictions may have been real enough in the presidency towns, but they had limited reach into the *mofussil* (countryside). Rank-and-file soldiers were often exempted, in part because the domestic labor provided by native wives benefited the "service family" of a regiment.[129] Yet travel restrictions remained, and the inability to pay the deposit of 500 rupees for a native wife to travel to Britain hindered the maintenance of family relationships.[130] (This is not to say that they were lacking in affection nor that their duration was uniformly short.) Such policies were however rarely official. When, in 1817, the military authorities in Madras inquired whether European soldiers could send their Eurasian or native families to Britain, the advocate-general replied that although there were no legal restrictions, it was discouraged in practice.[131] While some interracial relationships arose from the perceived absence or expense of European wives, they were all subject to a range of restrictions that meant they were viewed as cohabitation or concubinage. This led to doubts about the legitimacy of all mixed-race people, even those born within wedlock.[132]

Therefore, a third shared feature of interracial sex in India and the Caribbean was the effort by the white male population to distance itself from such deprecated relationships. Only in India, however, was the attempt made to recast such liaisons as legitimate.[133] Several authors claimed that concubinage was functionally equivalent to matrimony, at least from the perspective of the Indian partner. They discussed the various forms of marriage under Hindu law: in one remarkable example, even the seduction of *nautch* girls was deemed a form of wedlock.[134] More typical was one author of 1784, who wrote that 90 percent of the Company's employees were "unable to enter into matrimonial connections of a suitable kind," making it "fruitless and unjust" to restrict sexual activity "which, though it may not be entirely defensible, is obviously rendered unavoidable by the very nature of our establishments in India."[135] Of course, missionaries in India opposed these suggestions and viewed marriage in starkly Christian monogamous terms.[136] Military officers were also deeply critical of European sexual behavior; Captain Innes Munro even judged a preference for Indian women "unnatural."[137] But clear efforts were made to counter the structural and cultural factors that delegitimized relationships with non-European women, in a manner not replicated in the West Indies.

The principal motive for this endeavor was to provide for children. The legal status of a marriage and the legitimacy of offspring were crucial in determining eligibility for benefits. European policy dictated that illegitimate children were "to be considered of the same country as their mother," legitimate of their father.[138] Yet even this ethnocentric principle was undermined by the strength of ethnic prejudice. A dispute of 1820 between the British Army and the East India Company makes this clear. It concerned the negligent behavior of John Wardell, erstwhile commanding officer of the 1st Battalion, 66th (Berkshire) Regiment, which had decamped from Ceylon to Bengal in 1814, bringing with it the native partners and children of many of the soldiers. It later embarked for St. Helena, abandoning the women and children and leaving the Company to pay for their repatriation to Ceylon; a handful of orphaned children were accommodated in the Orphan School in Calcutta.[139] The army refused to reimburse the Company for the repatriation or pay for the maintenance of orphans, as the adjutant-general made clear: "Unless any of them shall be the children of <u>Soldiers born of European Women in Wedlock</u> there is no possible mode of providing for them . . . any of this latter description who are orphans will be received into the Military Asylum at Chelsea. But [the commander-in-chief] is totally at a loss for the means of arranging the disposal of the natural Children of Soldiers by Native Women, nor could [he] entertain the least expectation that His Majesty's Government could be induced to sanction any expense for such objects."[140] Unaccommodated for in this response were the legitimate children of European soldiers and native wives: to qualify for benefits, children were required to be not merely legitimate but also of European parentage on both sides. Otherwise, they were classed as "natural"—that is, illegitimate. As with "orphans" and "prostitutes," such a combination of "racial" thinking and categorical sleight of hand was common in both India and the Caribbean.

Another aspect of interracial sex shared by the East and West Indies was the perversion of masculine and feminine social norms. In the case of India, a central figure in this process was the "nabob," the stereotypically wealthy British East India Company servant who by the 1770s had come to personify corruption, despotism, and sexual luxuriance. By using vast and potentially ill-gotten wealth irresponsibly, the nabob threatened to introduce oriental despotism with his oriental riches. The number of Company servants who returned to Britain with both wealth and political ambitions was tiny, but the stereotype was reinforced by the Company's financial problems, by its perceived mismanagement of famine-stricken Bengal (blamed by many on the most famous nabob of them all, Robert Clive), and by the campaign against Governor-General Warren Hastings orchestrated by Edmund Burke.[141]

The stereotype of the nabob owed much to the play of that name by Samuel Foote (1772).[142] Through the nefarious acts of its protagonist, Sir Matthew Mite, the play repeatedly emphasizes that nabobs illicitly acquired enormous wealth (by exploiting helpless Indians) and that they failed to use it responsibly (through conspicuous consumption, corruption, and gambling). Even sympathetic representations of nabobs centered on these features: one pamphlet of 1783 that sketched an idiosyncratic typology of nabobs did not deny that such behavior occurred but limited it to just one type: "mushroom nabobs," sprung "from the rank hotbed of a Court."[143]

Nabobs were primarily attacked because of the dubious sources and uses of their wealth, but other offenses against metropolitan moral norms were also highlighted. For example, they were apparently out of step with a developing humanitarian sensibility. Foote's play premiered a week after Lord Justice Mansfield's judgment in the Somerset Case, and it portrays Mite postponing the import of castrated Bengalis to staff a proposed seraglio in Britain, "till the point is determined whether those creatures are to be considered as mere chattels, or men."[144] The sexual depravity of the nabob was reinforced by titillating accounts of liaisons with married women in the *Town and Country Magazine*, or more severely by poems such as *The Nabob: or, Asiatic Plunderers* (1773), which pulled no punches: the nabob was "in blood a tyrant, and in lust a beast."[145] Although sexual irregularity was never at the heart of attacks on nabobs, it was nonetheless regularly adduced to reinforce a general picture of immorality and of an exotic and sexualized Orient.[146]

Not that sexual impropriety moved unidirectionally from India to Britain. On the contrary, many writers blamed immorality in India on its existence *within* and consequent exportation *from* Britain. Elizabeth Griffith's play, *A Wife in the Right* (1772), for example, features a nabob lamenting that the everyday sexual infidelity of Britain "would taint the very air of India, and breed a pestilence there."[147] Within India, many "nabobs" indulged in sexual relations with native and Eurasian women. Sir David Ochterlony, a high-ranking officer in

Fig. 6.2. James Gillray, *A Sale of English Beauties in the East Indies* (1786). BM 7014.
© Trustees of the British Museum.

the Company's army, reputedly had his own *zenana*, and William Hickey, a lawyer in Calcutta, had children by more than one native woman.[148] One partner, Kiraun, produced a son, "whom I certainly imagined to be of my own begetting, though somewhat surprized at the darkness of my son and heir's complexion; still, that surprize did not amount to any suspicion of the fidelity of my companion."[149] It was only once Hickey caught Kiraun in flagrante delicto with one of his own servants that he was satisfied he was not the boy's father.

The behavior of men was not the only cause for concern. The small but significant number of British women who traveled to India in search of a husband (unflatteringly termed "the fishing fleet") was also a source of anxiety.[150] This is best illustrated in Gillray's *A Sale of English Beauties in the East Indies* (1786; fig. 6.2), which shows scandalously exposed British women being examined as merchandise by Indians and Europeans in "Indian" attire, while an auctioneer conducts their sale. In the background, one woman is weighed against a barrel full of rupees, while unsold brides make their way, sobbing, into a "warehouse for unsaleable goods," due to be returned to Europe.

Gillray's target is the degeneration of European moral norms of masculinity *and* femininity in India. Men lack masculinity and women femininity because they are both prepared to undergo the indignity of this transactional scene. European women are depicted in the same grotesque manner as became commonplace in representations of their enslaved counterparts: with the accoutrements of femininity (fine dresses and hairstyles) but none of its genuine attributes, such as modesty.[151] Above all, men were supposed to protect women, not exploit them. Yet less than a month before *A Sale* was published, one of the charges laid before the Commons against Hastings attacked the exposure of "women naked in the marketplace."[152] The women in question were not British but Indian *begums* (princesses) who had apparently been forced to expose themselves to public view due to poverty, for which the greed of Hastings was supposedly to blame.[153] Avarice had apparently overpowered the masculine duty of protection, owed especially to high-ranking women. Scholars have interpreted *A Sale* as a satire either on British morals in general or on the "fishing fleet" in particular, but it ought also to be read in conjunction with the specific charges raised against Hastings in the month prior to its publication.[154]

Nabobs were unmistakably associated with this traffic in women. Part of Mite's scheme in *The Nabob* was to transport the sisters of his intended, Sophy Oldham, "to Madrass or Calcutta, and there to procure them suitable husbands."[155] In 1784, one newspaper commented that Pitt's India Bill was missing a provision that would act as "a check on those wretches who bring up their daughters, and *let them out* to debauched Nabobs, on condition to furnish them to India, and to put them in a way to procure husbands."[156] For their part, women who went to India for this purpose became nabobs themselves, as Griffith stated in her epilogue: "British features bear a premium" in India, such that "even this homely face would charm, they say, / Amongst the copper beauties of Bombay; / . . . / Pantheon, Opera, Playhouse, Fantoccini, / Farewel–I'll go, and be a Nabobini."[157] Able thus to transcend gender, the nabob was a representative figure for the subversion of normative gendered behavior consequent on growing imperial entanglements and profoundly undermined metropolitan ideals of both masculinity and femininity.

The Instrumentality of Race
Slavery and the Making of Racial Categories

Given the extensive common ground shared by ethnogenesis in the Caribbean and India, why did an elaborate system of racial gradations only emerge in one of these contexts? The answer to this question lies fundamentally in the uses to which race was put, uses suggested by the varied nature of the colonial enterprise in the East and West Indies. In the Caribbean, following the demographic catastrophe inflicted on the native Caribs that was already largely complete by the time the English colonized Barbados in the 1620s, there were few indigenous inhabitants in the way of colonization.[158] European activity in the region mostly consisted of establishing agricultural units to grow and export exotic produce, necessitating the ownership of land, a legal system designed to uphold this, as well as a large workforce for labor-intensive farming. After early use of white indentured laborers, the planters turned toward an enslaved workforce, and slavery soon became the primary economic relation within Caribbean colonial society. By the early eighteenth century, very few enslaved laborers were of Amerindian origin, and blackness came to be identified with enslavement. This united with the dominance of slavery over all aspects of life in the British Caribbean to mean that skin color was a status marker of the utmost importance.

Paradoxically, as a chiaroscuro viewpoint made blackness the crucial marker of the slave, it was a more particular focus on the degree of blackness that determined an individual's place within slave society. Mixed-race slaves were given the comparative privilege of working in a domestic, workshop, or urban setting, while other enslaved people toiled in the fields. A sophisticated taxonomy of mixed race was therefore a useful tool in dividing the slave population against itself and consolidating white rule. This succeeded so well that arguably one cause of the 1831–1832 Jamaican slave revolt was the subversion of the system by white colonists themselves. Under economic pressure caused by a drop in sugar prices during the 1820s, planters had taken to working slaves of color in the fields, thereby robbing them of their comparatively privileged status.[159]

In British India, things were very different. Political instability in the Mughal Empire from the 1660s had allowed European powers to gain toeholds on the subcontinent, but there was no demographic catastrophe on the scale of the Columbian Exchange that facilitated colonization, the establishment of commercialized plantation agriculture, and the wholesale imposition of metropolitan legal and political systems. English and later British interests were confined to trading and the financial exploitation of existing administrative structures, such as the *Diwani* of Bengal (granted in 1765). Notwithstanding the development of indigo plantations in the *mofussil* following the loss of America, planter or settler colonization remained unpalatable prior to the demise of the East India Company after 1857. As the territory under the Company's control expanded rapidly from the mid-eighteenth century, elements of British legal tradition were introduced, but there remained a commitment to separate legal systems for British, Hindu, and

Muslim people. The simple fact of a large indigenous population in India meant that British imperial rule here differed from that within the Caribbean.

Most crucially, of course, slavery was not introduced by European powers to India, even if they did import, use, profit from, and trade in slaves until the mid-nineteenth century.[160] Slavery in India was inflected by a range of culturally specific contexts, further complicated by the impermanence of one's status as a slave: slavery could be a transitional process or condition through which people moved toward "freedom" or closer integration in the household and kinship networks of their ostensible "owners." It was perceived to be far more domestic than agricultural and, where agricultural, to be distinct from capitalist production in the Caribbean mold. Agricultural slavery, which was often not termed "slavery" at all due to the complexity of Indian forms of unfree labor, was also linked with discussions of caste. These perceptions were directly informed by efforts to distance East from West Indian slavery, in the interest of promoting abolitionist sentiment and providing a "free labor" alternative to West Indian sugar. While there existed some dissenting voices that sought to describe slavery as slavery—particularly the West Indian lobby, which was keen to demolish the idea that India provided a more ethical alternative to their produce—British discourses on Indian slavery in the era of abolitionism promoted the idea that it was largely benign and inoffensive, indeed hardly "slavery" at all.[161]

This tendency flew in the face of the statistical evidence. According to one nineteenth-century estimate, Indian slavery oppressed between 8 and 9 million people in 1841, dwarfing the 775,000 enslaved people in the British Caribbean in 1807.[162] The proportion of the Indian population subject to slavery was much smaller: approximately 4 percent, compared with over 80 in the West Indies. However large and appalling the figure of 9 million unfree laborers, its proportion of the total population (*c.* 225 million in 1840) makes it clear that slavery was not the primary economic relation in British India.[163]

Much as slavery was less central to India than to the Caribbean, so too was the importance of skin color. This is not to say that pigmentation was entirely unimportant in India. On the contrary, it had been used by the Mughals as a means of standardizing descriptions of criminals, rebels, and other troublemakers.[164] To European eyes, skin color distinguished Indians from Europeans and Africans, and it was used as an index of greater or lesser "effeminacy" among the Indian population, as more temperate climates were associated both with a greater hardiness and a lighter pigment.[165] It also served as a supposedly reliable marker of caste identity (which, as Susan Bayly and others have demonstrated, was neither a timeless feature of Indian society nor a wholly invented relic of colonialism), with "higher" castes viewed as being generally lighter skinned, although there did exist disagreement about this.[166] Yet skin color went unused as a marker of slave status.

An important reason is the geographical origins of slaves in India, which were so wide-ranging as to prevent any one particular pigment from being indicative of slave status. The East African slave trade remains far less studied than its Atlantic counterpart, but there was a small but steady importation of Africans to India until the 1830s. The sheer size and diversity of the population of South Asia, however, with distinctions of religion, caste, and local culture, effectively prevented there being one single and instantly recognizable class of slaves. In the last analysis, it was neither necessary nor feasible to impose a complicated system of racial classification on an already diverse population to support a variegated institution neither wholly nor even mainly controlled by Europeans.

There were, of course, more uses for complicated taxonomies of mixed race than simply the maintenance of chattel slavery. Such classifications arose in the New World wherever two or more populations mixed and there was a need to assert the supremacy of one, which was usually white-skinned and in the minority. This might have little to do with slavery. In Spanish America, for example, the *sistema de castas* emerged in the sixteenth and seventeenth centuries and divided the population into an ever-expanding array of categories. Among the reasons for its development were the increasing political and economic exclusion of non-white peoples, anxieties about both religious orthodoxy and the success of conversion efforts, questions about the (im)purity of Amerindians and black Africans, and the growing desire to establish an authentic creole culture in contradistinction to peninsular Spain. The role of slavery in establishing and maintaining the *sistema* was limited, as indeed was slavery's own importance in New Spain. By the end of the eighteenth century no more than 10,000 (or 0.167 percent) of a population of 6 million were enslaved.[167]

But New Spain was, in many respects, an exceptional case. Systems of mixed-race classification usually emerged in colonial contexts in which slavery played a major part. Even if such systems were intended primarily to assert European supremacy than explicitly to uphold slavery, maintaining slavery where it existed was an intrinsic part of sustaining supremacy. In Brazil, for example, slavery was more important than anywhere in Spanish America.[168] Brazilian systems of racial classification contained a bewildering array of categories (at least 150, according to some estimates), the profusion of which was partly due to the perceived need to isolate and restrict the free descendants of slaves. Although such systems could be extremely fluid, inconsistently applied, and their categories often deployed as a basis for collective action by those they were designed to isolate, it remained the case that full social privileges were reserved for those with non-African (and hence nonslave) ancestry.[169]

What makes the absence of taxonomies of mixed race in India more remarkable, and which clearly demonstrates the paramount importance of chattel slavery in explaining this absence, is that the processes of racial mixture were well understood. In 1784, for example, one author argued in favor of educating Eurasian children in Britain by denying that their later marriage to Britons "would give rise to a race, in which, after the third or fourth generations, there would, in any respect, be a perceptible degeneracy from the male [European] stock."[170] In other words, there would be no permanent change wrought by intermarriage between Eurasians and Britons. Nevertheless, the elliptical suggestion of differences *within* the three or four generations after such a marriage shows that the author was keenly aware of the processes of racial mixture and their likely duration.

Such a perspective was still influential in the early nineteenth century. The 1816 proposals for the Bengal Military Fund stipulated that the parents of any potential beneficiary, marrying after the fund was established, must both be "European, or of unmixed European Blood, though born in other quarters of the World, four removes from an Asiatic or African being considered as European blood."[171] Once again, the absence of specific terminology to describe the gradations of these "four removes" (generations) did not entail ignorance of the processes of racial mixture and ultimate dilution.

The similarity of attitudes to mixed-race people in India and the Caribbean, the analogous circumstances of their origins, and a shared understanding of the processes of racial mixture all point to the importance of the sole remaining major difference between the two contexts in explaining the presence or absence of taxonomies of mixed race: the place of chattel slavery.

Confounding Race

But why does the existence or nonexistence of detailed taxonomies of mixed-race people—an apparently trivial element of racial thinking—matter? There are at least three reasons. First, it highlights the diversity and complexity of racial thought and practice. Typologies and classifications of mixed-race people expose the ego ideal of "scientific" race: an advanced, rational, and sophisticated taxonomy that brings the full weight of modern science and the natural world to bear in an exposition of the innate physical and moral differences between peoples.[172] This self-image of race had, however, little to do with its actual manifestations as an idea or in racist practice.[173] Race and racism could flourish just as well in contexts where no elaborate typology of mixed race existed. This should give us pause, because it shows that indiscriminately labeling such a diverse range of ideas and practices as "racial" or "racist" ultimately lumps together phenomena that may be better understood when treated individually. Furthermore, taxonomies of mixed race came to influence directly the shape of later racial theories. They acted for the French anatomist and anthropologist Paul Broca, for example, as decisive proof of one of the most important issues to be debated by racial theorists: that mixed-race people could reproduce with their "parent races."[174] For Broca, the absence of specific terminology for such mixtures in Australia hinted at the potential sterility of Australian Aboriginal peoples. On the other hand, such terminology was also manifestly lacking in racial studies of British India, such as Herbert Risley's *The People of India* (1912), which sought to fashion a complex ethnic tapestry out of the twin threads of race and caste.[175] The presence or absence of a taxonomy of mixed-race people therefore demonstrates the complexity of race and also determined—at least in part—the shape of racial theories.

The second reason is even more important because it highlights what is perhaps the most fundamental characteristic of race: its utility. Mixed-race taxonomies foreground the fact that race was a category invented, deployed, reinvented, and redeployed (and so on) because it was *useful*. This is clear from the example of both India and the Caribbean, where the (non-)incorporation of a taxonomy of mixed-race indicates *how* race was used in each individual context. In India, race was used as a means of asserting British prestige by socially and physically distancing Britons from native Indians.[176] The existence of Eurasians was a source of embarrassment, as it highlighted the fragility of a British prestige based on adherence to a superior moral code. Hence, they were largely overlooked in racial thought pertaining to the subcontinent. In the Caribbean, on the other hand, race was directed to the maintenance of chattel slavery until 1834 and of white supremacy thereafter. An extensive taxonomy of mixed race was used to divide and rule over both the enslaved and free populations of color: slaves of color could expect comparatively advantageous work (such as domestic or skilled labor), and the free population of color had a clearly marked path to social advancement ("whiteness").

Of course, once racial ideas were expressed in an increasingly powerful idiom (science) or were linked with other popular notions (that identity could be read from the physical body, for example), they became more mobile and persistent, existing independently of their original context and applicable across a range of others. But it was the overall usefulness of race that accounts for its longevity and pliability as well as its remarkably stubborn resistance to its own contradictions.

The third reason why ethnogenesis and taxonomies of mixed race matter is because they highlight these inconsistencies with particular clarity. The contradictions and paradoxes of race and the ways in which they are exposed by ethnogenesis can be exemplified by examining the range of terms used to categorize people of mixed race more closely. Such terms varied not merely between the European empires but also between (and even within) individual colonies. In the British colonies, a range of racial terms were used in slave registration returns (introduced throughout the British Caribbean from 1813). In addition to those we have already met, these include "Yellow," "Red," "Cabre," "Mongrel," "Brown," "Castee," and "Indian." Of these, the most significant was "Yellow," which appears frequently in literature emanating from the colonies throughout the period.[177] In addition, the terms "Mulatto" and "Coloured" were used widely to denote all people of color, even including those without European heritage.

At Waltham, the range of terms used to describe the slave population expanded over the course of the 1820s. Registration in Grenada began in 1817, and, in addition to a full return every four years, there were annual increase and decrease returns, listing slaves born, died, purchased, and sold. In 1817, slaves were listed as "Negro," "Black," or "Mulatto"/"Mulatress." Furthermore, it was noted (under "remarks") that several slaves in the first two of these categories had "yellow skin." This was also a separate category in 1818 and 1833–1834 (where it appeared as a category, it was not listed in the "remarks"). From 1824, "Mustee" appears, as does "Cabre" in 1828. This expanding vocabulary could simply be the result of ever more mixed offspring, which might explain the disappearance of "Mustee" in the returns after 1830 (all three children so listed from 1824 were dead by that date). But categorical confusion appears to have taken hold at Waltham, where "Cabre" children were listed as such at birth but in subsequent returns always as "Black."[178]

Outside the British colonies, terminology was just as confused and arbitrary. The *sistema de castas* in New Spain even contained a category entitled "I don't understand you" (*no te entiendo*), in addition to others equally obscure or related to animals (such as *tente en el aire*, *lobo*, or *coyote*).[179] In the French Caribbean, attempts were made to fix terminology with an implausibly precise degree of mathematical accuracy, yet there is no evidence that it was any more applicable to a real-world situation. Moreau de Saint-Méry, a jurist who wrote extensively on Saint-Domingue in particular, described no fewer than 13 distinct classes of racial intermixture, arising from supposedly "pure" black or white people, and which consisted of a spectrum running from the extremely (but not entirely) black "sacatra" at one end to the extremely (but not entirely) white "sang-mêlé" at the other. He extended his calculus of racial mixture to an absurd extent, distinguishing an individual who was 8,191 parts white to 1 part black from a "pure" white. To this extreme degree, skin color could be of no use in any effort to apply this calculus to real-world individuals, leading to a reliance on genealogical or less objective knowledge for making judgments as to racial identity. Indeed, color categories were themselves highly subjective, varying across time and space, and were applied differently by people in different regions. The published figures for the 1789 census of Saint-Domingue, for example, did not even include a total for free people of color (*gens de couleur libres*), let alone a breakdown of this population by pretended racial group.[180]

Categorical confusion was not confined to the Caribbean. As we have seen, in India the meanings of "prostitute" and "orphan" could be staggeringly broad. Just as wide was disagreement on the term by which mixed-race individuals were known, and some suggestions—"East Asians" in particular—were especially confusing. But perhaps the least well-defined term of all was "Portuguese."[181] This was often used to designate mixed-race individuals (the process of racial mixture had largely begun with the Portuguese in Goa) or Indians who had converted to Christianity; only rarely did it refer to descendants of Portuguese settlers. Indeed, the "Portuguese" were often not held to be European at all, being segregated in the "Black Town" of Calcutta, for example. This confusion was so extreme that it became its own moral exemplar: their degraded state was held by some as an example of the dangers of racial intermixture. But the "Portuguese" did demonstrate the ongoing importance of religion (only those who were Protestant were accepted in mixed-race communities of British origin) and the performativity of race. Indians who spoke a pidgin Portuguese and wore European clothing (especially a hat) were able to pass as Eurasian, but as they were termed "Portuguese," this latter term became yet further confused and circular.[182]

The unworkability of race is here made manifest. Mixed-race taxonomies, so elaborately detailed by Long, Edwards, and others, depended on the visibility of racial status being maintained from "Negro" to "Quinteron" (and beyond). Without one or more empirical markers, foremost among which was skin color, it became impossible to tell whether an individual who appeared white was truly so or was a member of one of the categories of the potentially enslaved. It became necessary to know individuals' genealogy to determine their status as slave or free, which completely undermined the empirical value of skin color.[183] Not merely was this an unsatisfactory racial marker, as metropolitan racial theorists had already come to realize, but the rapid growth of the free colored population from the middle of the eighteenth century also problematized the association of dark skin with slavery. Such a situation could result in the promotion of alternative empirical markers (such as odor or the skeleton) or in the destabilization of the racial system. Ultimately, however, the system of racial gradations in the Caribbean was too intricate for any bodily marker to act as a reliable signifier. The inevitable result of this situation was a confusion of terms beyond the most basic and immediately apparent categories ("White," "Negro," and "Mulatto"), hence the frequent use of "yellow" to refer in general to the colored population.[184] The extent to which this confusion was due to planter apathy with the system is now impossible to determine, but this apathy was itself arguably a consequence of the inherent limitations of the system of racial gradations. Whatever the motive for terminological confusion, the system was no less undermined. Yet, simply because it was too useful to do without, race persisted.

This comparative analysis of ideas about mixed-race people in India and the British Caribbean shows that these groups were viewed in an overwhelmingly similar fashion. South Asian and Caribbean people of color alike were at first tolerated then restricted, as their growing numbers were deemed to compromise British security and prestige. They were viewed as improvident and unintelligent but attitudes towards them were largely derived from those held towards their non-white parent(s). For Caribbean people of color, this meant that they were associated with the blackness that had acquired synonymy with "slave." Eurasians, on the other hand, were tainted with the "effeminacy" and timidity associated with natives of India. In both cases criticisms were made of sexual habits, parenting skills, and the abnegation of bourgeois femininity, masculinity, and morality represented by non-white peoples. White people—male and female—in these colonial contexts were also heavily criticized for undermining the gender ideals that were being so assiduously rooted in the male and female body within Britain.

These similarities make it clear that taxonomies of racial mixture depended overwhelmingly on the past or present existence of chattel slavery in a given context. This tight linkage of race, taxonomy, and slavery is important because it highlights three major features of race. First, it exposes the breadth and complexity of race (racial theory and racist practice) across a range of global settings, from slave societies to societies with slaves to others in which slavery played little or no part. Second, and more specific to race in the Americas, the argument originating with Eric Williams—that slavery made race—gains powerful reinforcement from considering the role that sexual reproduction played in both; in particular, interracial sex and its consequences shaped racial theories and the legal classifications to which they gave rise, both of which served to sustain systems of chattel slavery in the New World.[185] Finally, ethnogenesis and taxonomies of racial mixture highlight the ongoing importance of utility in both the origins and continuing existence of race. The fundamental usefulness of race is what has enabled it to overcome what would otherwise have been devastating exposés of its contradictions and inconsistencies. The sheer range of racial ideas, the arbitrariness of categories and their terminology, and the unreliability of their grounding conceit (that one can identify racial identity from empirical markers on the body), might, on their own or altogether, have sunk race as a worthwhile concept had it not been so very useful. In the last analysis, ethnogenesis matters because it sheds important light on the (sexual) origins of racial thought and allows us to glimpse the incredibly broad range of its cultural and institutional strategies of survival. These have sadly allowed the idea to persist despite the twentieth-century demolition of its scientific credentials.

CHAPTER SEVEN

Conclusion

The primary aim of *Generating Difference* has been to show that it is not possible fully to comprehend the development of racial thought in the long eighteenth century without appreciating the role played by sex in general and reproduction in particular. Two final examples encapsulate the principal conclusions to be drawn from the foregoing pages. The first recalls the portrayal of the abolitionist, James Stephen, in Cruikshank's *New Union Club* (1819) and concerns a Bristol woman, Sarah Elliott, who in 1737 was sentenced to stand in the pillory and spend three months in jail.[1] Her crime? She had attempted to extort money from Richard Cornwall, "a Christian Negro," by falsely claiming that she had a child by him. That they had been intimate is suggested by the fact that he did not reject her claim out of hand, but "the Black insisted on seeing the Child before he would condescend to her Demand, and told her, that if 'twas his Child, he shou'd know by the Colour of the Skin. The Woman artfully to deceive the Fellow, procur'd a borrow'd Child, with its Skin smutted over; but he calling for a wet Napkin, and rubbing the Child's Face, found it of a fair Complexion, quite different to his Specy."[2] Her brazen attempt to deceive Cornwall is all the more remarkable because Bristol, the leading provincial slaving port in these years, had a comparatively large black population that Elliott cannot have failed to encounter. Furthermore, as we have seen, she would have had recourse to other means beyond discoloring the child's face to argue for Cornwall's paternity, such as the power of her imagination. But in a testament to the increasing power of the body to signify identity, he demanded—and she accepted the need to provide—proof based on simple pigmentation rather than facial or other forms of resemblance, which led her to manipulate the child's appearance in spectacularly unconvincing fashion.

The second example occurred on the other side of the Atlantic exactly fifty years later. Bryan Edwards (1743–1800), whose *History, Civil and Commercial, of the British Colonies in the West Indies* (1793) did much to publicize the taxonomy of racial mixture in the Caribbean, returned to Jamaica in the first half of 1787 after a sojourn in Britain and was immediately embroiled in a lawsuit that exposed the limits of the system he would soon describe. In this case, James Any, "a Sambo man" and son of a woman named "Mulatto Bess," sued for his freedom on the basis of his mother's manumission in December 1751. The case turned on when Any was born, for the principle of *partus sequitur ventrem* ("that which is born follows the womb") applied, meaning that his condition followed that of his mother at the moment of his birth. Had he been born before December 1751, his mother's manumission would not extend to him, and he would legally be a slave; if after, he would have been born to a free woman and would also be free. There was no documentary evidence of his date of birth, but witnesses testified that he "had passed as a free man" for thirty years, been enrolled in the militia, paid taxes, and even received wages as a carpenter from the former attorney of the estate to which it was claimed he belonged.[3] Edwards was both a trustee of this estate and a reluctant defendant, but his brother had named him as such prior to his return to Jamaica. Despite both the court's publicly expressed reluctance "to [try] freedom in this way" and the fact that the decision lay with a jury of slaveholding men, the court found for Any.[4]

This case was a rare example from the British Caribbean of cases that were to be all too common in the antebellum South and which have been analyzed by historians such as Ariela Gross.[5] In this suit, despite mention of the racial categories "sambo" and "mulatto," the only role played by skin pigmentation was landing Any in court in the first place: had he been white, there was no way he could be enslaved and such a case would never have been brought. The pertinent division was therefore between white (always free) and non-white (potentially enslaved). The limits to the practical value of the Jamaican system of racial gradations were further exemplified by the reliance on testimonies that showed Any had performed his freedom sufficiently to convince a jury that he was indeed a free man.

These examples illustrate the role of reproduction in highlighting both the importance and the contradictions of race: it was as much performed as read or inscribed, and extracorporeal knowledge was required to decide questions of identity. As *Generating Difference* has shown, the association of reproduction and race was of long standing and had close and strong ties with the priorities of European states in the long eighteenth century, where a large population was a desideratum for all but particularly for an underpopulated Great Britain. Reproduction—which inevitably meant marriage—was the only viable means to this end, and a culture of pronatalism was a major aspect of British society and government policy. This held also for the slaveholding Caribbean, but too little was done to ensure the natural increase of the enslaved population, even after the flow of slaves from Africa was cut off in 1807. This had much to do with maintaining established patterns of slave management and the fear of what change, however slight, might bring. But it was also related to the solid association of marriage and reproduction: for their white masters, the sexual lives of slaves, casual and polygamous, could never result in numerous and healthy offspring, although planters failed to recognize that their own practices were the greatest hindrance. This was doubly ironic given the weight attached to the principle of population (happy, healthy, and prosperous people will increase; the opposite proves that something is seriously awry) by both sides in the abolition and emancipation debates. The ground shared by the transimperial culture of pronatalism marks one way in which Britain was closely and powerfully integrated into circuits of racial thought and policy that, until recently, were thought to be remote from the home islands during this period.

Indeed, as part two of the book has shown, the powerful relationship between reproductive sex and racial thought began at home. It was overwhelmingly here that naturalists, embryologists, and racial theorists put pen to paper and developed their ideas, using materials from across the world. The key concept that underpinned the Enlightened life sciences, species, was defined by the notion of a reproductive community and hence fundamentally based on sex. Indeed, species came to be so overly reliant on the interbreeding criterion that it was vulnerable in the face of phenomena that were difficult to account for, above all hybrid offspring. The taxonomic category defined by a reproductive community thus shifted across the century and in the thought of Linnaeus and Buffon—from species to genus—and with this came a weakness in natural history that unscrupulous polygenists such as Long did not fail to exploit. Polygenism may have remained a minority viewpoint (although perhaps not *as* minor in private as the reliance on published materials have led some historians of race to suppose), but the crisis that overwhelmed climatic monogenesis just as organized abolitionism began to bite produced newer forms of sociological and degenerationist monogenesis that nonetheless continued to rely on sex and reproduction. Cultural and biological aspects of sex also featured strongly in accounts of the world's peoples and act as a powerful reminder that racial thought remained a mix of somatic and cultural phenomena until well into the nineteenth century.

Yet the undeniable and growing biologization of racial thought across the eighteenth century was undoubtedly helped by developments in embryology and the broader understanding of reproduction and heredity. Both principal schools of thought concerning human diversity depended on an embryology that allowed for transformative change, and while preformationism was in the ascendant between the mid-seventeenth and mid-eighteenth centuries, racial thought in Europe made little (but by no means no) headway. The emergence of newer vitalist and teleological drives that shaped embryonic development and maintained the consistency of physical form and species identity, particularly in the work of Blumenbach, provided a spur to racial thought that enabled it to become a nineteenth-century juggernaut. By 1847 it was possible for Benjamin Disraeli to write, without apparent exaggeration, "All is race; there is no other truth."[6]

The most important shared feature of discourses of race and embryology was unquestionably the mixed-race datum. The fact that mixed-race people proved both the equality of parental contributions and the unity of the human species bound racial and reproductive thought together in a uniquely powerful way. It formed the basis of Richard Cornwall's demand to inspect his putative child and ensured that, for example, questions of parental resemblance and the inheritance of acquired characteristics would remain high on the agenda for both groups of theorists and that the phenomena of mixed race would be the principal battleground on which debates concerning human difference would be fought. Long's deliberate misuse of circumstantial demographic evidence to claim that "mulattoes" are inter se infertile is just one example of the ways in which the mixed-race datum exercised a profound influence on the shape of racial thought. Mixed offspring played a role in the discussion on a number of other phenomena, such as albinism, hermaphroditism, and the power of the maternal imagination, all of which contributed to the development of forms of identity based on the human body, above all race and sex.

The final part of *Generating Difference* shifted the focus beyond the Atlantic world to Asia and the Pacific, to explore the ways in which race and sexual reproduction interacted in Britain's wider empire in this period. It also demonstrated how examining racial thought and practice through a reproductive lens exposes the fundamental limits of the endeavor to inscribe identity unambiguously on the human form. Hermaphrodites, hijras, and Hawaiians (as well as other Pacific sexualities that were difficult for Europeans to comprehend) all show how these limits might apply to sex as well as to race. Indeed, the Tahitian ability to smell sexual difference had a parallel in European efforts to locate racial difference in odor, as part of the wider project to identify less ambiguous markers of human diversity than skin color. The focus of this project remained visual, concentrated on mensuration of the skeleton or skull, but notwithstanding the spectacular nature of Pacific encounters—visible on the London stage, in exhibitions of the work of the voyages' artists, or in remarkable pen portraits on the printed page—the region also highlighted the boundaries of what the human body could tell. The incipient racial division of the Pacific that was later encapsulated by the labels "Polynesian" and "Melanesian" laid down roots only after this division was confirmed by means of linguistic, not physical, evidence. Visible cues pointed the way, but extracorporeal data was necessary before a conclusive racial division could be made.

The use of language as a means of establishing genealogical relationships between peoples was the centerpiece of the method used by James Cowles Prichard, but it was the absence of a vocabulary of mixed race in India that highlights how genealogical knowledge was vital for deciding precisely the sorts of questions that race alone was intended to answer. Attitudes and policy toward mixed-race people in both the Caribbean and India were overwhelmingly similar. Perceptions of mixed-race people were substantially shaped by sentiments toward their non-white parent(s); they were subject to a similar set of prejudices (concerning imprudence, sexual incontinence, and tyrannical behavior) in both locations and experienced similar restrictions in their personal and work lives once the size of the mixed population had reached a critical mass. The white population in both regions was a tiny minority of the whole, which resulted in considerable anxiety regarding the security of colonial rule, and hence perhaps the reluctance of the court that confirmed Any's freedom to decide other such cases. Only one of these settings, the Caribbean, was a slave society, dominated by the institution of chattel slavery that permeated all social, economic, and legal relationships. India was, by contrast, a society with slaves, yet one in which white involvement with the institution (which itself took a wide variety of forms and degrees of permanence) was comparatively marginal. The absence of an elaborate taxonomy of mixed race in India points to the importance of chattel slavery for producing this variant of racial thought and racist praxis. At the same time, the need for genealogical information to decide questions of slave status in the Caribbean or eligibility for work or benefits in India highlights the fundamental flaw in all forms of identity that seek to use the physical body as an unambiguous signifier.

Reproduction therefore played a paradoxical role in the development of racial, and more broadly somatic, identity. An embryology that allowed for transformative change was a necessary but not sufficient precondition for the emergence of recognizably modern forms of race. But reproduction also exposed with potentially devastating clarity the boundaries of what the body might tell. That race not merely survived but went on to flourish in the nineteenth century was due, in part, to its utility as a technology of rule, both in Europe's colonies and at home. Although expressed increasingly in a scientific idiom, race never entirely shed its etymological origins as genealogy or lineage: whether as a "one drop" rule that greatly simplified matters by demanding only a single non-white ancestor or in the National Socialist fantasy of hereditary Jewishness that depended on ancestral religious practice, race was never a matter of what the physical body alone might signify.

The story told in these pages is thus one of continuity within change. Modern concepts of race undeniably developed over this period, and this book has gone to great lengths to show how sex, reproduction, and embryology played a large part in this development. Through their influence, race and biological sex became new forms of identity whose authority is predicated on the ability to reliably interpret the human form. But in the case of race, this identity is fundamentally limited: in all but the most obvious cases, extracorporeal knowledge (above all, of parentage and genealogy, but also of language and culture) is required to decide tricky questions of belonging. This enabled passing and a whole range of activities that demonstrated resistance to the new racial order that was imposed on colonial populations and, increasingly, within Europe. Emphasizing this continuity will, it is hoped, allow us to recognize that the war against racial thought was not won with the near-complete destruction of its scientific credentials after 1945; unless we tackle its surviving vestiges, this war may yet be lost.

Abbreviations

AA	Immanuel Kant, *Kant's gesammelte Schriften*, ed. Royal Prussian (later German) Academy of Sciences, 29 vols. Berlin, 1900–.
Bendyshe	Thomas Bendyshe, ed. *The Anthropological Treatises of Johann Friedrich Blumenbach*. London, 1865.
BL	British Library, London
Bodl.	Bodleian Library, Oxford
Buffon, *HN*	Georges-Louis Leclerc de Buffon. *Histoire naturelle, générale et particulière*, part of *Œuvres complètes*, ed. Stéphane Schmitt, 17 vols. Paris, 2007–. References are to volumes of the *Histoire naturelle* within this series. Unless otherwise indicated, English translations are taken from Buffon-Smellie.
Buffon-Smellie	Georges-Louis Leclerc, Comte de Buffon. *Natural History, General and Particular*, tr. William Smellie. 9 vols. 2nd ed. London, 1785.
Buffon, *Supp.*	Georges-Louis Leclerc de Buffon. *Histoire naturelle, générale et particulière . . . Supplément.* 7 vols. Paris, 1774–1789. Unless otherwise indicated, English translations are taken from Buffon-Smellie.
Blumenbach, *DG1*	Johann Friedrich Blumenbach. "On the Natural Variety of Mankind, ed. 1775," in Bendyshe, 65–143.
Blumenbach, *DG3*	Johann Friedrich Blumenbach. "On the Natural Variety of Mankind, ed. 1795," in Bendyshe, 145–276.
ECL	*Eighteenth-Century Life*
ECS	*Eighteenth-Century Studies*
EHR	*English Historical Review*

Forster, *Observations*	Johann Reinhold Forster. *Observations Made During a Voyage Round the World*, ed. Nicholas Thomas, Harriet Guest, and Michael Dettelbach. University of Hawaii Press, 1996.
GM	*Gentleman's Magazine*
HCSP	Sheila Lambert, ed. *House of Commons Sessional Papers of the Eighteenth Century*. 147 vols. Scholarly Resources, 1975–1976.
HRFP	Home-Robertson Family Papers, NRS
Jefferson, *Notes*	Thomas Jefferson. *Notes on the State of Virginia*, ed. William Peden. University of North Carolina Press, 1982.
JBS	*Journal of British Studies*
JHI	*Journal of the History of Ideas*
Kames, *Sketches*	Henry Home, Lord Kames. *Sketches of the History of Man*, ed. James A. Harris. 3 vols. Liberty Fund, 2007.
Lawrence, *Lectures*	William Lawrence. *Lectures on Physiology, Zoology and the Natural History of Man*. London, 1819.
Long, *HJ*	[Edward Long]. *The History of Jamaica*. 3 vols. London, 1774.
Lords SC	H. L. Papers, Report from the Select Committee on the State of the West India Colonies, 1831–1832, 127.
Monboddo, *OPL*	[James Burnett, Lord Monboddo]. *The Origin and Progress of Language*. 6 vols. Edinburgh, 1773–1792.
NLS	National Library of Scotland, Edinburgh
NRS	National Records of Scotland, Edinburgh
Nugent, *Journal*	Philip Wright, ed. *Lady Nugent's Journal*. University of the West Indies Press, 2002 [1966].
P&P	*Past and Present*
OED	*Oxford English Dictionary*
Pauw, *RP*	[Cornelius De Pauw]. *Recherches philosophiques sur les Américains*. 2 vols. Berlin, 1768–1769.
Prichard, *PHM*	James Cowles Prichard. *Researches into the Physical History of Man*, ed. George W. Stocking Jr. University of Chicago Press, 1973 [1813].
PTRS	*Philosophical Transactions of the Royal Society*
Raynal, *Two Indies*	Guillaume-Thomas Raynal. *A Philosophical and Political History of the . . . East and West Indies*, tr. J. O. Justamond. 8 vols. London, 1783.
RMS Diss.	Royal Medical Society Dissertations, UESC

RCPE	Royal College of Physicians of Edinburgh
SHPBeBMS	*Studies in the History and Philosophy of the Biological and Biomedical Sciences*
SINH Diss.	Society for Investigating Natural History, Dissertations, 1782–1806, MS Da.67, UESC
Smith, *Species*	Samuel Stanhope Smith. *An Essay on the Causes of the Variety of Complexion and Figure in the Human Species*, ed. Winthrop D. Jordan. Harvard University Press, 1965.
NA	The National Archives of the United Kingdom
UESC	Special Collections, University of Edinburgh
Virey, *NH*	[Julien-Joseph Virey]. *Natural History of the Negro Race*, tr. J. H. Guenebault. Charleston, SC, 1837.
W&MQ	*William and Mary Quarterly*
White, *Gradation*	Charles White. *An Account of the Regular Gradation in Man.* London, 1799.

Introduction • (Re)producing Bodies and Identities

1. Swift, *Gulliver's Travels*, 249.
2. Boucé, "Rape of Gulliver," 108–9; Boucé, "Sexual Beliefs," 36.
3. Swift, *Gulliver's Travels*, 253.
4. Passmann, "Mud and Slime," 1–17.
5. For an overview, see Outram, *Enlightenment*, ch. 8.
6. Marryat, *More Thoughts*, 96.
7. Marryat, *More Thoughts*, 99, 106. *The New Union Club* (BM 13249) has been extensively discussed. See Stephens and George, *Catalogue*, 9:910–12; Wood, *Blind Memory*, 165–72; Gattrell, *City of Laughter*, 480–82; Odumosu, *Africans in English Caricature*, 165–92.
8. Walvin, *Black and White*; Shyllon, *Black People*; Barker, *African Link*; Fryer, *Staying Power*; Dabydeen, *Hogarth's Blacks*; Gerzina, *Black England*; Boulukos, *Grateful Slave*; Chater, *Untold Histories*; Hanley, *Beyond Slavery*.
9. On Sancho, see in particular Sancho, *Letters*; Vincent Carretta, "Sancho, (Charles) Ignatius (1729?–1780)," *Oxford Dictionary of National Biography*, updated August 8, 2024, https://doi.org/10.1093/ref:odnb/24609; Hanley, *Beyond Slavery*, ch. 1. On the empire "at home," see Wilson, *Sense of the People*; Wilson, *New Imperial History*; Walvin, *Fruits of Empire*; Hall and Rose, *At Home with the Empire*; Molineux, *Faces of Perfect Ebony*; Livesay, *Children of Uncertain Fortune*.
10. Brubaker and Cooper, "Beyond 'Identity,'" 1–47.
11. Yuval-Davis, "Theorizing Identity," 272; Hall, "Who Needs 'Identity'?," 15–30.
12. Sidbury and Cañizares-Esguerra, "Genesis of Destruction," 245; cf. Hodson, "Weird Science," 227–32.
13. Taylor, *Sources of the Self*; Seigel, *Idea of the Self*; Wahrman, *Modern Self*.
14. Hall, "Spectacle of the 'Other,'" 234–38.

15. Literature on these topics is vast. Among the most important titles are Butler, *Gender Trouble*; Butler, *Bodies That Matter;* Said, *Orientalism*; Thomas, *Colonialism's Culture*; Douglas, *Science*. See also ch. 5 below.

16. See Porter, "Bodies of Thought," 82–108, updated as "History of the Body," 233–60, and the excellent and wide-ranging surveys in Kalof and Bynum, *Cultural History*, and Toulalan and Fisher, *Sex and the Body*.

17. Earle, *Body of the Conquistador*; Floyd-Wilson, *English Ethnicity*, 1–47; Siraisi, *Medicine*, 97–106.

18. Laqueur, *Making Sex*, 149.

19. Laqueur, *Making Sex*, 8.

20. Park and Nye, "Destiny Is Anatomy," 54; Cadden, *Sex Difference*; Siraisi, *Medicine*, 91–96, 110–13; King, *One-Sex Body*; Stolberg, "Anatomy of Sexual Difference," 274–99. Cf. Laqueur, "Sex in the Flesh," 300–306, and Schiebinger, "Skelettestreit," 307–13.

21. Martensen, "Transformation of Eve," 107–33; Roper, *Oedipus and the Devil*, 16–17. For an example of a history that highlights the experience of sexual difference, see Duden, *Woman Beneath the Skin*.

22. Harvey, "Sexual Difference," 202–23; Shoemaker, *Gender in English Society*, 1–14; Hitchcock, *English Sexualities*, 45–47; Fissell, *Vernacular Bodies*, 12–13, 248.

23. Park, "'One-Sex Body,'" 96–100.

24. Park and Nye, "Destiny Is Anatomy," 57 (emphasis mine).

25. Haraway, *Primate Visions*; Fausto-Sterling, *Sexing the Body*; Sedgwick, *Epistemology*; and the works of Butler cited in n15.

26. See Nicholson, "Interpreting Gender," 79–105.

27. Butler, *Bodies That Matter*; Scott, "Gender," 1053–75; Scott, *Gender*.

28. Nicholson, "Interpreting Gender," 83; Laqueur, "Rise of Sex," 802–12.

29. Sedgwick, *Epistemology*; Fausto-Sterling, *Sexing the Body*; Kates, *Monsieur d'Eon*.

30. See Clark, *Working Life*; Bloch, *Sexual Life*. The foundational work on sex in the eighteenth century is Stone, *Family, Sex and Marriage*.

31. For an illuminating survey, see Harvey, "Century of Sex?," 899–916. More recent works include Gowing, *Common Bodies*; Harvey, *Reading Sex*; Knott and Taylor, *Women, Gender, and Enlightenment*; Toulalan, *Imagining Sex*; Dabhoiwala, *Origins of Sex*.

32. See the work of Roy Porter, esp. his "Mixed Feelings," 1–27; Porter and Hall, *Facts of Life*. On (the impropriety of) Victorian austerity, see Foucault, *History of Sexuality*; Mason, *Victorian Sexuality*.

33. See, for example, Porter and Rousseau, *Sexual Underworlds*; Maccubbin, *'Tis Nature's Fault*.

34. Wrigley and Schofield, *Population History*, 417–20; Wrigley, "British Population," 57–95.

35. Davidoff and Hall, *Family Fortunes*. Cf. Vickery, "Golden Age," 383–414.

36. See Perry, "Colonizing the Breast," 204–34.

37. Exceptions relate largely to prostitution. See Henderson, *Disorderly Women*; Trumbach, *Sex*; Bullough, "Prostitution and Reform," 61–74.

38. See Winthrop Jordan's "Note on the Concept of Race" in his *White Over Black*, 583–85, and Kathleen Brown's acute discussion of Jordan's use of "race" in her essay "Native Americans," 79–100.

39. See Voegelin, *Race Idea*; Poliakov, *Aryan Myth*; Banton, *Idea of Race*; Banton, *Racial Theories*; Mosse, *Toward the Final Solution*; Stepan, *Idea of Race*; Hannaford, *Race*.

40. Dikötter, *Discourse of Race*; Harrison, *Climates and Constitutions*; Bernal, *Black Athena*; Eliav-Feldon, Isaac, and Ziegler, *Origins of Racism*; *Journal of Medieval and Early Modern Studies* 31, no. 1 (2001); Bethencourt, *Racisms*. There has been some skepticism toward the effort to locate race and racism in the premodern world: see esp. Seth, *Europe's Indians*, and Seth, "Origins of Racism," 343–68.

41. Stoler, "Racial Histories," 183–206.

42. Wheeler, *Complexion of Race*.

43. Fredrickson, *Racism*; Bethencourt, *Racisms*, chs. 2, 9; Goetz, *Baptism of Early Virginia*. "Proto-racism" is Benjamin Isaac's term of choice. See his *Invention of Racism*.

44. Fredrickson, *Racism*, 141; Appiah, "Racisms," 4–5.

45. See, for example, Loomba, *Shakespeare, Race, and Colonialism*; Wilson, *Island Race*, 12; Goetz, *Baptism of Early Virginia*; Ramey, *Black Legacies*.

46. Marryat, *More Thoughts*, 104.

47. Heng, *Invention of Race*, 3, 19.

48. See the particularly trenchant critique of Heng's work in Pearce, "Inquisitor and the Moseret," 145–90.

49. Isaac, *Invention of Racism*, 23.

50. Bindman, *"Race Is Everything,"* 11.

51. Floyd-Wilson, *English Ethnicity*; Dawson, *Bodies Complexioned*.

52. See Kidd, *Forging of Races*; Eigen and Larrimore, *German Invention*; Sebastiani, *Scottish Enlightenment*; Lettow, *Reproduction*.

53. Barker, *African Link*, 155–200; Stepan, *Idea of Race*, ch. 1; Bindman, *Ape to Apollo*; MacLeod, *American Revolution*, ch. 4; Wilson, *Island Race*, chs. 2, 4; Hudson, "'Nation' to 'Race,'" 247–64.

54. See Popkin, "Philosophical Basis," 245–62; Eze, *Race and the Enlightenment*; Eze, "Hume, Race," 691–98; Bernasconi, *Race*. The inspiration for the revisionist picture of the Enlightenment is unquestionably Adorno and Horkheimer, *Dialectic of Enlightenment*.

55. See Kidd, *British Identities*; Kidd, *Forging of Races*; Hodgen, *Early Anthropology*, 207–51.

56. On the "noble savage," see Fairchild, *Noble Savage*; Ellingson, *Myth*; White, *Tropics of Discourse*, 183–96.

57. Meek, *Social Science*; Sebastiani, "'Race,'" 75–96; Tomaselli, "Enlightenment Debate on Women," 101–24.

58. Wheeler, *Complexion of Race*, 181–92; Sebastiani, *Scottish Enlightenment*, chs. 2–3; Kitson, *Romantic Literature*, 5.

59. See esp. Popkin, "Medicine," 405–42; Fredrickson, *Racism*, 31–35; Martínez, *Genealogical Fictions*.

60. Kidd, *British Identities*, ch. 2.

61. Braude, "Sons of Noah," 103–42; Goldenberg, *Curse of Ham*.

62. Kidd, *British Identities*; Horsman, "Racial Anglo-Saxonism," 387–410, reprinted in Horowitz, *Race*, 77–100; MacDougall, *Racial Myth*.

63. Venturino, "Race et histoire," 19–38; Dorlin, *Matrice de la race*; Miramon, "Noble Dogs," 200–216; Doron, *L'homme altéré*, ch. 2; Schaub and Sebastiani, *Race et histoire*; Mc Inerney, *Nobility*; Cooley, *Perfection of Nature*; Stewart, "William Frédéric Edwards," 273–300.

64. Bush, *Slave Women*; Altink, *Representations*; Rennie, *Far-Fetched Facts*; Seth, *Difference and Disease*; Hogarth, *Medicalizing Blackness*; Bindman and Gates, *Image of the Black*; Nussbaum, *Torrid Zones*; Levine, *Gender and Empire*; Hyam, *Empire and Sexuality*; Mosse, *Nationalism and Sexuality*.

65. Foucault, *History of Sexuality*; Stoler, *Education of Desire*.

66. Laqueur, *Making Sex*, 155. See also Harvey, *Reading Sex*, 78–79, 139–45.

67. See, for example, Cody, *Birthing the Nation*, ch. 8; Harvey, "Sexuality and the Body," 78–99.

68. Stepan, "Race and Gender," 261–77; Schiebinger, *Mind Has No Sex?*, 211–13. For an example of the influence of this analogic relationship, see Wiegman, *American Anatomies*, ch. 2.

69. Schiebinger, *Nature's Body*, 116.

70. Stepan, "Citizenship," 29.

71. Nelson, *Enlightenment Biopolitics*; Nelson, "Making Men," 1364–94; Stoler, *Education of Desire*; Tuttle, *Conceiving the Old Regime*; Stein, "Birth of Biopower," 331–37.

72. Goveia, *Slave Society*; Fox-Genovese, *Plantation Household*; Beckles, *Natural Rebels*; Bush, *Slave Women*.

73. Altink, *Representations*; Morgan, "Slave Women," 231–53; Roberts, *Slavery and the Enlightenment*, 153–57; Burnard, *Mastery*; Morgan, *Laboring Women*.

74. Paugh, "Politics of Childbearing," 119–60; Paugh, *Politics of Reproduction*; Turner, *Contested Bodies*.

75. Morgan, "'Some Could Suckle,'" 167–92; Morgan, *Laboring Women*.

76. Brown, *Good Wives*; Fischer, *Suspect Relations*; Spear, *Social Order*; Newman, *Dark Inheritance*.

77. Molineux, *Faces of Perfect Ebony*; Wells, "Race Fixing," 134–38; Harvey, "Sexual Difference," 202–23; Wahrman, "Change and the Corporeal," 584–602.

78. White, *Gradation*, 135.

Chapter 1 • *"The King's Honor": Population and Pronatalism in Greater Britain*

1. Aldridge, "Population and Polygamy," 129–48; *Baxter v. Baxter* (1947) 2 All E.R. 886 (H.L.); Gibson and Begiato, *Sex and the Church*, ch. 4.

2. Tuttle, *Conceiving the Old Regime*; Cody, *Birthing the Nation*; Fissell, *Vernacular Bodies*.

3. Wrightson, *English Society*, ch. 3; Weil, *Political Passions*, pt. 1; Harvey, *Little Republic*, ch. 1; Amussen, *Ordered Society*, ch. 2.

4. See, for example, Poole, *Annotations*, vol. 1; Dodd, *Practice of Inoculation*, 3–4; Brown, *Self-Interpreting Bible*.

5. Gen. 13:9.

6. Cic., *Off.* 1.54.

7. Biller, *Measure of Multitude*, 49–52, pt. 3; Arist., *Pol.* 1326a26-7.

8. Arist., *Pol.* 1326b10; Pl., *Leg.* 5.737b–737e; Hutchinson, *Population Debate*, 9–14.

9. Plut., *Vit. Lyc.*, 14–16. For contemporary examples, see [Mandeville], *Modest Defence*, 49. Cf. [Temple], *Vindication*, 23–29.

10. Treggiari, "Social Status," 887–93; Pl., *Resp.* 5.460a–461e. For contemporary examples, see [Salmon], *Critical Essay*, 41–47; Hanway, *Rising Generation*, 1:139; *London Evening Post*, November 23, 1736.

11. Very broadly, contractarians argued that political authority derived from an original contract made by the governed, where patriarchalists claimed that it stemmed from God and had its analogue in the absolute authority of the father in the household.

12. Somerville, *Sex and Subjection*, ch. 8; Weil, *Political Passions*, chs. 1–2. The limits to the seeming universalism of contract theory have been subjected to powerful critique. See Pateman, *Sexual Contract*; Mills, *Racial Contract*.

13. Bodin, *Six Bookes*, 571; Filmer, *Patriarcha*, 68–69.

14. Hobbes, *De Cive*, 140 (X. 18). On Hobbes, see also Chapman, *"Leviathan* Writ Small," 76–90; Sagar, "Mushrooms and Method," 98–117.

15. Locke, *Two Treatises*, 183 (1:59); Locke, *Political Essays*, 255.

16. Hobbes, *Leviathan*, 181.

17. Slack, *Invention of Improvement*, 194–95; Levine, *Battle of the Books*. For an example of this debate in religious writing, see [Brett], *Conjugal Love*, 12.

18. Riley, *Population Thought*, 52–56; Tomaselli, "Moral Philosophy," S7–S29; Whelan, "Population and Ideology," 49–57; [Montesquieu], *Lettres persanes*, 2:123–74 (letters 102–12); Montesquieu, *Spirit of the Laws*, bk. 23; Vossii, *Variarum Observationum*, 1–68 cf. De Souligné, *Comparison*, 32–148; [Wallace], *Dissertation*; Hume, "Populousness," 377–464.

19. Whelan, "Population and Ideology."

20. Wrigley and Schofield, *Population History*, 531–35 (table A3.3); H.C. Papers, Abstract of Answers and Returns [. . .] (Enumeration Abstract), 1801–2 (9), 497.

21. Bonar, *Theories of Population*, ch. 1; Hutchinson, *Population Debate*, 24–26, 33–44.

22. Coke, *Treatise*, 28.

23. Hale, *Discourse*, sig. A2v; Hutchinson, *Population Debate*, 55–58.

24. Monboddo, *Antient Metaphysics*, 5:242–323; "Of the numbers of the People in Britain" [1788], Monboddo Papers, NLS, MS 24544, 5–42; Wrigley and Schofield, *Population History*, 528–29 (table A3.1); Cookson, "Political Arithmetic," 38.

25. See esp. Brewer, *Sinews of Power*; Braddick, *State Formation*, pt. 3.

26. Innes, *Inferior Politics*, 127–41; Hoppit, "Political Arithmetic," 516–40.

27. The writings and philanthropy of Jonas Hanway are a case in point. See, for example, his *Thoughts*; *Robert Dingley*; and *Letters*.

28. On mercantilism, see (among many others) Heckscher, *Mercantilism*; Magnusson, *Mercantilism*. For a selection of older perspectives on mercantilism, see Minchinton, *Mercantilism*. For more recent discussions of problems with the concept, see the introduction to Stern and Vennerlind, *Mercantilism Reimagined*, 3–22; and the forum in *W&MQ*, 3rd ser., 69 (2012): 3–70.

29. Pincus, *1688*; Pincus, "Rethinking Mercantilism," 3–34; Slack, *Invention of Improvement*, 196–97; [Pollexfen], *England and East-India*, 5–8; Child, *Discourse*, 122–27. Cf. Smith, *Wealth of Nations*, 1:365, 2:642–62.

30. Davenant, *Discourses*, 2. See also Petty, *Political Arithmetick*, sig. a3v–a[5]r.

31. Rusnock, *Vital Accounts*, 1–14; Innes, *Inferior Politics*, ch. 7; McCormick, "Quantification," 239–51; Slack, *Invention of Improvement*, 46–47.

32. Graunt, *Observations*, 51–52; [Petty,] *Treatise*, 5; Lansdowne, *Petty Papers*, 2:46–58; Davenant, *Discourses*, 17.

33. Graunt, *Observations*, 44; Petty, *Five Essays*, 21–31; Innes, *Inferior Politics*, 135–37, 150–54; Rusnock, *Vital Accounts*, chs. 1, 7.

34. Monboddo, *Antient Metaphysics*, 5:280.

35. Described in 2 Sam. 24 and 1 Chron. 21.

36. *Letter on Registering*, 8. On the 1753 census bill, see Glass, *Numbering the People*, 16–20; Buck, "People Who Counted," 32–35; Rusnock, *Vital Accounts*, 183–85; 14 Parl. Hist. Eng. (1753) 1317–65; *GM* 23 (1753): 499–502, 549–52, 597–99.

37. McCormick, *William Petty*, chs. 6, 8. Cf. Riley, *Population Thought*, which argues that population thought became pragmatic only after 1740.

38. Bell, *Dissertation*.

39. *Considerations on the Causes*; Kames, *Sketches*, esp. bk. 1; Brown, *Estimate*, 182–202. On the luxury debate, see Berry, *Idea of Luxury*; Berg and Eger, "Luxury Debates," 1–27.

40. Bell, *Dissertation*, 12. Cf. [Temple], *Vindication*.

41. *Populousness with Oeconomy*; Brown, *Estimate*, 189; Rogers, *Mayhem*, ch. 6; *GM* 21 (1751): 101.

42. Innes, *Inferior Politics*, 150–54; Whelan, "Population and Ideology," 53–54; Franklin, *Autobiography*, esp. 246–50 ("The Speech of Miss Polly Baker"), 251–60 ("Observations Concerning the Increase of Mankind"), 324 ("Information to Those Who Would Remove to America").

43. Franklin, *Autobiography*, 258.

44. Gail Bederman has argued that the 1798 edition should also be viewed as an intervention in the conservative attack on Godwin rather than simply the first edition of a great work. See her "Sex, Scandal, Satire," 768–95.

45. Malthus, *Essay*, ed. James, 2:206; Hollander, "Malthus's Population Principle," 187–235.

46. *Remarks on a Late Publication*, 46.

47. Malthus, *Essay*, 2nd ed., 531. Samuel Taylor Coleridge marked this passage in his copy of the *Essay*, and its removal from the third edition (1806) failed to dampen Godwin's rage even after more than a decade. See Potter, "Unpublished Marginalia," 1061–68; Godwin, *Of Population*, 19; James, *Population Malthus*, 60, 456.

48. "Summary of Politics: Jamaica Complaints," *Cobbett's Weekly Political Register*, February 16, 1805, 231; Gilbert, Introduction, 1:6.

49. Malthus, *Essay*, ed. James, 2:139; marginal comment in Malthus, *Essay*, 2nd ed., 538 (British Library shelfmark C.44.g.2). See Potter, "Unpublished Marginalia," 1066. For Coleridge's views on Malthus, see Winch, *Riches and Poverty*, 297–306.

50. James, *Population Malthus*, 138; Bill to amend Laws relating to Relief of Poor in England, 1821, Bill [489]; 5 Parl. Deb., H.C. (2nd ser.) (1821) 995; Cody, *Birthing the Nation*, 285–92.

51. Corfield, *Impact of English Towns*, 175–78; Borsay, *English Urban Renaissance*, 69–70; van Lieshout, "Waterscapes"; Riley, *Eighteenth-Century Campaign*, chs. 4, 5, 7; Hamlin, "Transformation of 'Nuisance,'" 189–204; Porter, "Great Wen," 61–75.

52. Hale, *Primitive Origination*, 227–28.

53. Andrew, *Philanthropy and Police*, ch. 2.

54. Owen, *English Philanthropy*, 38.

55. *Account of the Bristol Infirmary*, 2.

56. Cherry, "Voluntary General Hospitals," 59–75, 251–65; Wrigley and Schofield, *Population History*, 234–36; Andrew, *Philanthropy and Police*, 53–54; Langford, *Polite and Commercial People*, 134–41; Owen, *English Philanthropy*, 36–52.

57. For example, 16 of 194 in-patients (8.2%) died in Bristol's Infirmary (1737–1738), as did 52 (3%) of 1,708 in York (1740–1743), and 9 of 275 (3.3%) in Northampton (1765–1766). These figures, it should be noted, derive from promotional material designed to attract donors: *Account of the Bristol Infirmary*, 2; *Account of the Public Hospital*, 30; *Report of [. . .] the County-Hospital*, 2.

58. Levene, Reinarz, and Williams, "Child Patients," 15–33.

59. Cherry, "Voluntary General Hospitals," 253.

60. Owen, *English Philanthropy*, 46–47.

61. Many of these arguments are also made by Berridge, "Health and Medicine," 3:205–6.

62. Inoculation was a prominent theme in eighteenth-century literature. See Shuttleton, *Smallpox*, ch. 7; Nussbaum, *Limits*, ch. 4.

63. The modern concept of "immunity" was, of course, barely understood in the period; there was a general recognition that one could not suffer from smallpox twice, but even this was debated in the early decades of inoculation. Miller, *Adoption of Inoculation*, 26–44. See the collection of letters on this subject in *PTRS* 32 (1722–3): 262–69.

64. Variolation is a method of inoculation that involves infecting the patient with matter containing the disease-causing microorganism against which immunity is sought—for example, pus from the pocks of smallpox (*variola*). Vaccination entails the use of a vaccine, which may be formed from a weakened or inactive form of the microorganism or another related entity, such as cowpox (*variolae vaccinae*) to immunize against smallpox.

65. Miller, "Putting Lady Mary," 2–16; Voltaire, *Letters*, 73–82; Rusnock, *Vital Accounts*, 43–70.

66. Miller, *Adoption of Inoculation*, ch. 4; Margot Minardi, "Boston Inoculation Controversy," 47–76.

67. Bennett, "Inoculation," 199–223.

68. Job 2:7; Massey, *Sermon*.

69. *Hippocratic Writings*, 67; Miller, *Adoption of Inoculation*, 101–10; Minardi, "Boston Inoculation Controversy"; Nussbaum, *Limits*, 115.

70. Particularly after inoculation was endorsed by the Royal College of Physicians in 1755.

71. Brunton, "Pox Britannica," 148–50, 169, 172.

72. Davenport, Schwartz, and Boulton, "Adult Smallpox," 1289–1314; Owen, *English Philanthropy*, 51.

73. Bennett, "Inoculation," 223; Brunton, "Pox Britannica," 249. Even the chief spokesman for the claim that inoculation substantially influenced population growth, Peter Razzell, has reconsidered that view. See his *Essays*, ch. 1.

74. Andrew, *Philanthropy and Police*, 54–57.

75. *Guardian*, 366 (July 11, 1713); [Defoe], *Augusta Triumphans*, 10.

76. [Bray], *Memorial*, 16; Hanway, *Candid Historical Account*, 9–14.

77. Andrew, *Philanthropy and Police*, 54–57; McClure, *Coram's Children*, 33–36.

78. *Account of the Hospital for [. . .] Young Children*, iv.

79. Innes, *Inferior Politics*, 140–41; Andrew, *Philanthropy and Police*, 109–15; Nichols and Wray, *Foundling Hospital*, 17; George, *London Life*, 146–47.

80. See, for example, *Gazetteer and London Daily Advertiser*, March 17, 1756.

81. Stephens and George, *Catalogue*, 3:297–99 (no. 2423); Wilson, *Sense of the People*, 155–56.

82. Andrew, *Philanthropy and Police*, 127–34; Owen, *English Philanthropy*, 49.

83. *Account of the Lying-In Charity*, 16; *General State*, 2–3.

84. Mackenzie, *Historical Account*, 1:521.

85. Hume, "Populousness," 400; Owen, *English Philanthropy*, 55–56; McClure, *Coram's Children*, 76–114.

86. *Tendencies of the Foundling Hospital*, 11–19; Kames, *Sketches*, 2:532–33. This was not helped by the prominence of foundling hospitals in quasi-erotic literature promoting fornication. See MacLachlan, *Essay*.

87. McClure, *Coram's Children*, 21–23, 174; Nichols and Wray, *Foundling Hospital*, 21. On women in charity, see Prochaska, *Women and Philanthropy*, intro.; Gray, *History of English Philanthropy*, 160; Pinches, "Women as Agents," 65–85.

88. McClure, *Coram's Children*, 13.

89. Cressy, *Coming Over*, ch. 5; Lover of his Countrey, *Grand Concern*, 13–16; [Sheridan], *Discourse*, 232; Petyt, *Britannia Languens*, sig. G5v; Ridley, *Sermon* (1746), 4–6.

90. Cf. both Malthus, *Essay*, ed. James, 1:346; and Paley, *Principles*, 615–16.

91. *London Evening Post*, April 27, 1773; *General Evening Post*, September 25, 1773; Bailyn, *Voyagers to the West*, 36–49. Samuel Johnson was in Scotland while "emigration was . . . a common topick of discourse" and felt that it was "hurtful to human happiness." Boswell, *Journal*, 18.

92. Larkin, *Stuart Royal Proclamations*, 463, 556.

93. Cressy, *Coming Over*, ch. 5; Bailyn, *Voyagers to the West*, 49–57; 5 Geo. I, c. 27 (1718); 23 Geo. II, c. 13 (1750); H.C. Papers, First Report from Select Committee on Artizans and Machinery, 1824 (51), 49–51, 56–57, 315–16, 590, 601.

94. Francis Bacon, "Of the True Greatness [. . .] of States," in *Major Works*, 400; Locke, *Political Essays*, 322; Robbins, "General Naturalization," 168–77; Statt, *Foreigners and Englishmen*, 68, 70, 73, 93, 95, 100, 120, 194, 209–11.

95. Fortrey, *Englands Interest*, 5–13; *Conditions for New-Planters*; Ritchie, *Duke's Province*, 79. Older arguments in favor of immigration had not argued in populationist terms but rather on the basis of a skills shortage or for merchants likely to enrich the country. See Statt, *Foreigners and Englishmen*, 51–53.

96. Sowerby, *Making Toleration*, 61–62; Dabhoiwala, *Origins of Sex*, ch. 2.

97. [Richard Steele], *Spectator* 200 (October 19, 1711) in Bond, *Spectator*, 5:282–87.

98. Statt, *Foreigners and Englishmen*, 201–2; Riley, *Population Thought*, 74–76; Mandeville, *Fable of the Bees*, 1:194, 248, 2:350–51; McClure, *Coram's Children*, 24; Owen, *English Philanthropy*, 65–66.

99. Statt, *Foreigners and Englishmen*, 166; Wilson, *Sense of the People*, 110–12, 169–74; Wilson, *Island Race*, intro. and ch. 1. Among the exceptions may be counted local concord in British borderlands, such as with French fishermen in the Channel. See Morieux, "Diplomacy from Below," 83–125.

100. Colley, *Britons*, is the preeminent example. See also Newman, *English Nationalism*.

101. Defoe, *Tour*, 362; *Speech of Sir John Knight*, 12. This speech was burned by order of the Commons for its indiscreet attacks on William III's Dutch entourage, which guaranteed its popularity and resulted in it being endlessly recycled in eighteenth-century debates. See, for example, *Letter to Sir John Phillips*, 6–22. For a contemporary view on this parochialism, see Country Gentleman, *Reflections*.

102. Piety and pragmatism were often combined. See, for example, Ridley, *Sermon* (1746), 17–19.

103. Dickinson, "Poor Palatines," 464–85; Otterness, *Becoming German*, chs. 1–3.

104. [Swift], *Examiner* 45 (June 7, 1711).

105. Robbins, "General Naturalization," 171; Otterness, *Becoming German*, 41–42, 57–77, 106–12.

106. 13 Geo. II, c. 7 (1740).

107. 26 Geo. II, c. 26 (1753).

108. Perry, *Public Opinion*; Engelman, *Jews of Georgian England*, chs. 1–3; Harris, "London Evening Post," 1143–44; Crome, "'Jew Bill' Controversy," 1449–78.

109. Stone, *Road to Divorce*, chs. 4–5; Outhwaite, *Clandestine Marriage*; Probert, *Marriage Law*, 7–10, 21–67; Gibson and Begiato, *Sex and the Church*, ch. 4.

110. Outhwaite, *Clandestine Marriage*, chs. 3–4; Probert, *Marriage Law*, 218–19.

111. Leman, *Matrimony Analysed*, 70.

112. *Letter to the Public*, 21.

113. 15 Parl. Hist. Eng. (1753) 39–40.

114. On the "crisis" of 1748–1753, see Rogers, *Mayhem*. On the importance of promoting the reproduction of sailors, see Leman, *Matrimony Analysed*, 49; *Letter from a By-Stander*, 10–11; *Letter to the Public*, 30; K[eith], *Observations*, 21, 24; 15 Parl. Hist. Eng. (1753) 81. This point was labored when moves were made to repeal Hardwicke's Act in 1765. See *Reflections on the Repeal*, 33.

115. Probert, "Judicial Interpretation," 129–51. For a recent analysis that acknowledges that some people *were* improvidently married, see Griffin, "Conundrum Resolved?," 125–64.

116. See esp. K[eith], *Observations*, 4, 19–20; [Shebbeare], *Marriage Act*, 2:326–27. Cf. *Letter to the Public*, 14–15; Outhwaite, *Clandestine Marriage*, 119–20.

117. Paley, *Principles*, 592–93.

118. Probert, *Marriage Law*, 44; Blackstone, *Commentaries*, 1:442–44.

119. Thomas, "Double Standard," 195–216; Dabhoiwala, *Origins of Sex*, 99–110. Even Jeremy Bentham, who argued in defense of masturbation, sodomy, and even bestiality, remained ambivalent about adultery, arguing that it was of variable harm. See Bentham, "Sextus," 67 and n.

120. Capp, "Double Standard Revisited," 70–100; Boswell, *Life of Johnson*, 5:209.

121. Jordanova, "Popularization of Medicine," 76.

122. 15 Parl. Hist. Eng. (1753) 47.

123. Bannet, *Domestic Revolution*, 97–106.

124. Hole, *Practical Discourses*, 6:82 (pt. 3).

125. Such verses include Gen. 1:28, Ps. 127:3–5, 1 Tim. 2:15, and 1 Tim. 5:14. For interpretations that stressed upbringing, see Doddridge, *Family Expositor*, 5:453n; Gell, *Gell's Remains*, 2:392; Ridley, *Sermon* (1764), 15–16.

126. Humphreys, *Marriage*, 15; Paley, *Principles*, 287.

127. On the shared burden, see Humphreys, *Marriage*, 15–16. On wifely responsibility for education, see Poole, *Annotations*, 2:sig. Cccccc2r; Creffield, *Good Wife*, 24.

128. Polwhele, *Discourses*, 2:120.

129. See, respectively, [Brett], *Conjugal Love*, 12–13; and Cockburn, *Dignity*, 11–12; Preston, *Sermon*, 9–10.

130. Ridley, *Sermon* (1764), 12.

131. King, "Vitis Palatina," 282–302; Humphreys, *Marriage*, 14–17; Polwhele, *Discourses*, 2:110–35.

132. Fisher, *Honour*, 13; Hole, *Practical Discourses*, 6:14 (pt. 3); Cornwallis, "Bridal Bush," 27–28; Atterbury, *Sermons*, 2:62–63; Fordyce, *Sermons*, 1:34.

133. Fielding, *Tom Jones*, 72.

134. Edgeworth, *Belinda*, 42–43.

135. Wollstonecraft, *Posthumous Works*, 2:40–41.

136. Smollett, *Roderick Random*, 435.

137. On wife sales, see Richardson, *Pamela* (2001), 179–80; Thompson, *Customs*, 404–66; Stone, *Road to Divorce*, 143–48.

138. Richardson, *Pamela* (2001), 265, 269.

139. Goldsmith, *Vicar of Wakefield*, 9.

140. Sterne, *Tristram Shandy*, 474.

141. Fielding, *Tom Jones*, 201.

142. Franklin, *Autobiography*, 247.

143. Franklin, *Autobiography*, 246, 249.

144. Franklin, *Autobiography*, 247; Hall, *Benjamin Franklin*.

145. Dabhoiwala, *Origins of Sex*, ch. 1; Laqueur, *Solitary Sex*; Andrew, "'Adultery à-la-mode,'" 5–23; Stone, *Road to Divorce*, ch. 9.

146. [Berkeley], *Essay*, 7.

147. [Cooke], *New Theory*, 319.

148. *Marriage Promoted; Essay, or, Modest Proposal;* 6 & 7 W. & M., c. 6 (1695); Brooks, "Act of 1695," 31–53; Slack, "Government and Information," 33–68; Slack, *Invention of Improvement*, 181; Tuttle, *Conceiving the Old Regime*; King, *Rates and Duties*.

149. *Considerations upon Street-Walkers*, 7; Tucker, *Elements of Commerce*, 16–30, 39–40; *HCSP*, 9:237–40, 103:164; *H.C. Jour.*, February 14; February 25; March 19, 1750/1; March 26, 1751; "Whether by the encouragement of proper Laws the number of Births in Great Britain might not be greatly increased?," David Skene Papers, Aberdeen University Library, MS 37, fols. 11–12. The MPs were William Hay and Robert Nugent.

150. [Venette], *Tableau de l'amour*, 220–26; Venette, *Conjugal Love Reveal'd*, 125–30; Porter, "Spreading Carnal Knowledge," 233–55; Porter and Hall, *Facts of Life*, 33–64; Boucé, "Sexual Beliefs," 37–41.

151. Graham, *Lecture*, 5; Porter, "Sexual Politics," 199.

152. Porter, "Literature of Sexual Advice," 146.

153. Porter, "Touch of Danger," 206–32; Cody, *Birthing the Nation*, 200–210.

154. Grantham, *Marriage Sermon*, 10–11. Not everybody got the joke: Grantham's text was taken seriously enough to be listed as a genuine sermon on Gen. 29:25 in standard reference works of the eighteenth century. See Letsome, *Preacher's Assistant*, 5 (pt. 1); Cooke, *Preacher's Assistant*, 1:7.

155. Scott, *Fasti Ecclesiæ Scoticanæ*, 4:106.

156. McLaren, *Reproductive Rituals*, 75–81; Stone, *Family, Sex and Marriage*, 276; Hitchcock, *English Sexualities*, 53; Peakman, *Lascivious Bodies*, 51.

157. Riddle, *Contraception and Abortion*; Schiebinger, *Plants and Empire*.

158. A disturbing exception occurred in 1730, when a sow gelder attempted to spay his wife against her will. She forgave him but the court imposed a £400 bond for good behavior during his lifetime. *London Journal*, August 22, 1730.

159. [Carlile], *Every Woman's Book*; Place, *Illustrations and Proofs*; McCalman, *Radical Underworld*, 204–31.

160. James, *Population Malthus*, 385–87.

161. [Defoe], *Conjugal Lewdness*, 129, 152–62.

162. In the eighteenth century "abortion" meant both voluntary and involuntary miscarriage but is here used in its modern sense.

163. At least after quickening (the point at which the mother perceives the first movements of her fetus). See McLaren, *Reproductive Rituals*, 109.

164. See Cody, *Birthing the Nation*, 276–83; McLaren, *Reproductive Rituals*; Keller, "Embryonic Individuals," 321–48; Keller, *Generating Bodies*; Fissell, *Vernacular Bodies*; Gowing, *Common Bodies*.

165. 21 Jac. I, c. 27.

166. Cody, *Birthing the Nation*, 281; McLaren, *Reproductive Rituals*, 131; Rabin, "'Lewd Women'," 45–69; Kilday, *Infanticide*.

167. 21 Jac. I, c. 27.

168. See *Anno vicesimo primo*. Convictions under the statute were much higher in the period before *c.* 1730 than thereafter, not least because evidentiary problems made it difficult to prove infanticide, notwithstanding the presumption of the mother's guilt. See Cody, *Birthing the Nation*, 275; Hunter, "Signs of Murder," 266–90. See also Francus, "Monstrous Mothers," 133–56.

169. *Guardian*, 365–67 (July 11, 1713).

170. The number of acquittals does not mean that many women were not executed under the provisions of this statute. See Cody, *Birthing the Nation*, 281; McLaren, *Reproductive Rituals*, 130–35; Kilday, *Infanticide*, ch. 2.

171. Dabhoiwala, *Origins of Sex*, ch. 5; Henderson, *Disorderly Women*, 166–91; Bullough, "Prostitution and Reform," 61–74.

172. Cody, *Birthing the Nation*, 20. See also Andrew, *Philanthropy and Police*.

173. Dabhoiwala, *Origins of Sex*, ch. 4; Miller, *John Milton*; Nussbaum, *Torrid Zones*, 73–94; Hill, *World Turned Upside Down*, 183; Pearsall, *Polygamy*, esp. ch. 4; Andrew, *Debating Societies*, nos. 451, 655, 674, 675, 755; William Webb, "On the Effects of Domestication on Animals," SINH Diss., vol. 12 (1793–4), 397. The Old Bailey Sessions Papers record 445 instances of bigamy between 1680 and 1820. See *Old Bailey Proceedings Online* (version 9.0), accessed October 6, 2024, searched for all offenses listed as "bigamy" between 1680 and 1820, https://www.oldbaileyonline.org/search/crime?offence=bigamy&year_gte=1680&year_lte=1820#results.

174. [Delany], *Reflections*; Aldridge, "Population and Polygamy," 135–38.

175. O'Connell, "'Matrimonial Ceremonies'," 98–116.

176. See ch. 3.

177. Although some saw certain groups—such as Africans—as hyperfertile due to polygamy, its resultant hypofertility was the most common viewpoint. See Nussbaum, *Torrid Zones*, 78–79.

178. Millar, *Observations*, 78.

179. On Madan, see Trumbach, *Sex*, 188–90; Pearsall, *Polygamy*, 190–98.

180. Madan, *Thelyphthora*, 1:100 n.

181. Madan, *Thelyphthora*, 1:102–7, 2: 272–98.

Chapter 2 • *The Limits of Pronatalism: Slavery and Population in the British Caribbean*

1. John Fairbairn to George Home of Wedderburn, November 16, 1813, HRFP, GD267/5/24.

2. Fairbairn to Home, January 4, 1814, HRFP, GD267/5/24.

3. See, in particular, Sheridan, *Doctors and Slaves*, ch. 8; Morgan, *Laboring Women*; Altink, *Representations*; Paugh, *Politics of Reproduction*; Turner, *Contested Bodies*.

4. Smith, *Wealth of Nations*, 1:88.

5. Smith, *Wealth of Nations*, 1:90–91; Postlethwayt, *Universal Dictionary*, s.v. "manure"; Gray, *Happiness of States*; Rosen, *Medical Police*, 249.

6. Jer. 29:6; Augustine, *On the Catechising of the Uninstructed [De cathecizandis rudibus]*, ch. 21; Charbit, *Classical Foundations*, 43.

7. Montesquieu, *Spirit of the Laws*, 21–30, 427–56; [Wallace], *Dissertation*, 89–91; Hume, "Populousness," 383–98. See also Higman, "Demographic Theory," 164–94; Glacken, *Rhodian Shore*, 625–32; Drescher, *Mighty Experiment*, 36–38.

8. Lascelles et al., *Instructions*, 2; 9 Parl. Deb., H.C. (1st ser.) (1807) 121; Higman, "West India 'Interest,'" 17.

9. 3 Parl. Deb., H.C. (3rd ser.) (1831) 1410 (emphasis mine).

10. Lords SC, 828; Stephen, *Slavery*, 2:76; Macaulay, *Negro Slavery*, 39–40, 113; Whyte, *Zachary Macaulay*, 174.

11. 3 Parl. Deb., H.C. (3rd ser.) (1831) 1434–35, and 13 Parl. Deb., H.C. (3rd ser.) (1832) 88–90; Drescher, *Mighty Experiment*, ch. 3.

12. Stewart, *View*, 308–9.

13. "Paper—or address to Country—to interfere for better Treatment of negroes in W. Indies," Clarkson Papers, Huntington Library, San Marino, CA, CN56.

14. Wilberforce, *Appeal*, 34.

15. [Brougham], *Concise Statement*, 44–45.

16. Wilberforce, *Appeal*; Higman, *British Caribbean*, ch. 2. On the post-1807 intercolonial slave trade, see Eltis, "Traffic in Slaves," 55–64; Williams, "Slave Trade," 175–91. By 1823, all Caribbean colonies excluding Anguilla, British Honduras, the Bahamas, Bermuda, and the Cayman Islands had completed at least two registration returns.

17. There were precursors to Malthus's qualification of the desirability of population growth, particularly in an urban context, stretching back to Plato. See Pl., *Leg.* 737d; Stangeland, *Pre-Malthusian Doctrines*, 23–25.

18. Higman, "Slavery," 164–94; 8 Parl. Deb., H.C. (1st ser.) (1807) 993–94; Malthus, *Essay*, 3rd ed., 1:143–45, 2:505–59.

19. 9 Parl. Deb., H.C. (1st ser.) (1807) 118. See also Bashford and Chaplin, *New Worlds*, ch. 6; Paugh, *Politics of Reproduction*, ch. 5.

20. Mathison, *Notices*, 18.

21. Thomas Duncan to George Home, February 22, 1814, HRFP, GD267/5/12.

22. West Indian, *Notes*.

23. West Indian, "Notes in Defence of the Colonies," *Jamaica Journal*, June 5, 1824, 130 (in MSS Brit. Emp. s. 444, Bodl., vol. 39, fol. 10).

24. Burnley, *Opinions*, xi.

25. Higman, *British Caribbean*, 77.

26. Tadman, "Demographic Cost," 1539; MacLeod, *American Revolution*, 31–33; *Virginia Gazette*, June 30; July 28 (supp.); August 11, 1774.

27. On the causes of slave depopulation, see Tadman, "Demographic Cost," 1534–75; Ward, *West Indian Slavery*, ch. 5; Wood and Clayton, "Jamaica's Struggle," 287–308; Higman, *Population and Economy*, ch. 6; Higman, *Plantation Jamaica*, 291; Higman, *British Caribbean*, ch. 9; Morgan, "Slave Women," 231–53.

28. Cf. testimony offered by slaveowners to the House of Commons in 1790: *HCSP*, 71:216, 236, 238. See also Sheridan, *Doctors and Slaves*, 222; Blome, *Description*, 37; Schiebinger, *Plants and Empire*, 142–43; Higman, "Demography and Family Structures," 489, 497.

29. Testimony of Gilbert Francklyn, *HCSP*, 71:129–30; Williamson, *Observations*, 1:372; Burnard, *Mastery*, 215; Burnard, *Jamaica*, 84–86; Turner, *Contested Bodies*, 44–46.

30. *HCSP*, 71:117, 72:134.

31. *HCSP*, 71:97, 113–14, 116, 139, 163, 188, 211, 216, 220, 230, 238, 242–43, 246, 267, 274, 277, 294, 296–97; 72:9, 11, 30, 32, 39, 52, 71, 102–4, 130, 139–40, 149, 177, 179.

32. *HCSP*, 73:202, 213, 258, 308; 82:46, 58, 84, 89–90, 101, 110, 183, 187.

33. Belgrove, *Treatise*, 51–86; Hall, *Miserable Slavery*, 85; Diary, 1758, Thomas Thistlewood Papers, Beinecke Library, Yale University, OSB MSS 176, 171–84.

34. Thompson, "Drax's Instructions," 585.

35. Rachel Tudway to Thomas Grigg, February 28, 1717/18; Tudway to Edward Jones, April 18, 1719; Tudway to Thomas Fenton, November 29, 1723; November 15, 1728; Letterbook from England, 1717–29, Tudway of Wells MSS, Somerset Archives and Local Studies, DD/TD/15/8.

36. *HCSP*, 69:458–59.

37. Richard Holloway to Charles Tudway, October 10, 1759; Letterbook from Antigua, 1759–84, Tudway of Wells MSS, Somerset Archives and Local Studies, DD/TD/15/6.

38. What anxiety to the contrary as existed did not prevent the measure. See *HCSP*, 82:218.

39. See, in particular, the contributions by Oliver Ellsworth (CT) and George Mason (VA) on August 22, 1787. Madison, *Debates*, 443–44.

40. Lowe and Campbell, "Slave-Breeding Hypothesis," 401–12; Tadman, *Speculators and Slaves*, 122–25; Tadman, "Demographic Cost," 1557. A nuanced case for the existence of "slave breeding" and its importance in antebellum abolitionism and the later memory of slavery is offered in Smithers, *Slave Breeding*.

41. [Morgann], *Plan*; Brown, *Moral Capital*, 213–20.

42. Brown, *Moral Capital*, 228–30, 235–36.

43. Burke, "Sketch," 3:577. See also Marshall, *Edmund Burke*, ch. 7, esp. 189, 191–92, 193–94.

44. Burke, "Sketch," 3:577.

45. Burke, "Sketch," 3:578. This provision was not present in all versions of the code: see Marshall, *Edmund Burke*, 194. On coerced sexual relations between slaves, see Turner, *Contested Bodies*, 62–67.

46. Burke, "Sketch," 3:579, 580–81.

47. *Acts of Assembly*, 256–80.

48. *HCSP*, 67:197.

49. As shown in *Two Reports*, forwarded by the agent of Jamaica, Stephen Fuller, to the home secretary, Lord Grenville, in April 1790. Fuller to Grenville, April 2, 1790, Jamaica, Original Correspondence, NA, CO137/88, fols. 277r–298r.

50. *HCSP*, 67:213–14.

51. Only in the course of 1792 did events in France generate a pronounced conservative reaction in Britain. See Hilton, *Dangerous People*, 58.

52. 29 Parl. Hist. Eng. (1791) 359. The 1792 statute (reprinted in Edwards, *History*, 2:151–98) was passed in March, before the Commons vote in favor of gradual abolition on April 2. See 29 Parl. Hist. Eng. (1792) 1158; "An Act [. . .] for Consolidating [. . .] the several Laws relating to Slaves," [1792], Jamaica, Acts, 1790–1792, NA, CO139/47, fol. 106r.

53. *HCSP*, 82:215.

54. Ellis was later officially appointed by the Society of Merchants and Planters to represent them in the Commons. See Higman, "West India 'Interest,'" 17.

55. 33 Parl. Hist. Eng. (1797) 259–60.

56. Responsibility for colonial matters was vested in the Home Office between 1782 and 1801.

57. H.C. Papers, Correspondence with British W. India Colonies on Slave Trade, 1797–1800, 1803–4 (119), §§A and G; *Journals of the Assembly of Jamaica* 10 (1797–1802): 320–21; Sheridan, *Doctors and Slaves*, 229; Balcarres to Portland, January 5; March 22, 1800, Jamaica, Original Correspondence, NA, CO137/104, fols. 3r, 226r.

58. *HCSP*, 82:215; H.C. Papers, Select Committee on Extinction of Slavery in British Dominions, Report, Minutes of Evidence, Appendix, Index, 1831–32 (721), 288.

59. *Slave Law*, 5.

60. *Consolidated Slave Law*, iv.

61. The term "abortion" was used in this period to mean both the voluntary and involuntary termination of a pregnancy. See Schiebinger, *Plants and Empire*, 113–15.

62. Altink, *Representations*, 28.

63. Such as a purchasing strategy that concentrated on women so as to facilitate "breeding," as practiced by Simon Taylor, attorney of Chaloner Arcedeckne's Golden Grove estate in Jamaica. See Wood and Clayton, "Jamaica's Struggle," 287–308, and Higman, *Plantation Jamaica*, 219–20.

64. Thomas Clarke to Henry Cullen, April 20, 1788, Cullen Papers, RCPE, CUL1/3/192; Burnard, *Mastery*, 214–15; Morgan, *Laboring Women*, 87–92, 138; Turner, *Contested Bodies*, 75.

65. See, for example, Williamson, *Observations*, 1:371–73, 2:197–211; [Collins], *Practical Rules*, 151–74; Thomson, *Treatise*, 110–18.

66. The 1834 figure does not include the 91 slaves transferred to Waltham from L'Esterre Estate in 1833. A grand total of 247 slaves present at Waltham on July 31, 1834, thus represents the extent of the claim for compensation made by Home's heir, William Foreman Home. He received £6,226 16s. on November 2, 1835. See Slave Registers, Grenada, St. Mark's Parish, 1817, 1833, 1834, NA, T71/265/330–33, T71/320/130–38, T71/326/168–72, T71/329/100–108; H.C. Papers, Return of Sums awarded by Coms. Of Slave Compensation, 1837–38 (215), 98. Figures for Grenada as a whole are taken from H.C. Papers, Return of Slave Population in H.M. Colonies in W. Indies, 1815–22, 1824 (424), and Population Returns from Slave Colonies since 1833, 1835 (420), which list slave populations of 28,029 (1817) and 23,536 (1833).

67. Fairbairn to Home, November 1, 1814, HRFP, GD267/5/24.

68. Stephen, *Slavery*, 1:161n; H.C. Papers, Colonial Laws Relating to Importation and Protection of Slaves in W. India Colonies, 1788–1815, 1816 (226), 67–75.

69. "A List of Negroes Belonging to Waltham Estate, Island of Grenada October 1st 1804," HRFP, GD267/5/11/3(2).

70. See Wood and Clayton, "Jamaica's Struggle," 287–308; Altink, *Representations*, 29. Orlando Patterson has shown that 4 out of 260 women were listed as "indulged," having 6 children, on Green Park Estate, Jamaica, in 1 week in 1823. Of an adult enslaved population more than 260, this corresponds to 1.5 percent of all adults, or (assuming an equal sex ratio) 3 percent of all adult women. See Patterson, *Sociology of Slavery*, 60–61.

71. Turner, *Contested Bodies*, 104–5.

72. Lewis, *Journal*, 66, 140.

73. Sheridan, *Doctors and Slaves*, ch. 8; Altink, *Representations*, ch. 2; Turner, *Contested Bodies*; Paugh, *Politics of Reproduction*, ch. 3.

74. Cf. Beckles, *Natural Rebels*, 97. See also Adair, *Unanswerable Arguments*, 122; Sheridan, *Doctors and Slaves*, 245–46; Turner, *Contested Bodies*, ch. 6.

75. Schiebinger, *Plants and Empire*, 142–49.

76. Thistlewood Diary, 1767, 101, 174; Hall, *Miserable Slavery*, 145; Burnard, *Mastery*, 218–21; Henderson, *Account*, 117.

77. "List of Negroes," HRFP, GD267/5/11/3(2).

78. See also the testimony of Robert Thomas to the House of Commons in 1790: *HCSP*, 71:258.

79. Lords SC, 279.

80. [Schaw], *Journal*, 112–13. See also [Bancroft], *Essay*, 371–73 (quoted in Kames, *Sketches*, 2:563–4n).

81. Ward, *West Indian Slavery*, 38–60. On the sugar boycotts, see Everill, *Not Made by Slaves*, 49–54, 85–88; Midgley, *Women Against Slavery*, 35–40, and Turley, *English Antislavery*, 36, 79.

82. There continue to emerge counterarguments, notably that in Ryden, *West Indian Slavery*. See Drescher, *Econocide*, xxii.

83. Scholarship challenging Williams, *Capitalism and Slavery*, which asserted that the slave trade and slavery were only abolished once they had ceased to be profitable, is sizable. See especially Drescher, *Econocide*; Anstey, *British Abolition*; Eltis, *Economic Growth*. A fuller summary of the literature is available in Brown, *Moral Capital*, 12–23 nn.

84. Morgan, "'Some Could Suckle,'" 184–85, 189; Paugh, *Politics of Reproduction*, 94; [Brougham], *Concise Statement*, 60.

85. Lewis, *Journal*, 41.

86. For example, Cooper, *Facts*, 35–42; Mathison, *Notices*, 91–92; Wimpffen, *Voyage*, 290; Dancer, *Medical Assistant*, 262. On the possible origins of this stereotype in western African cultural taboos against moaning in labor, see Turner, *Contested Bodies*, 119–20.

87. *HCSP*, 69:211, 213, 277, 279; Thomson, *Treatise*, 111; Sheridan, *Doctors and Slaves*, 234–39. Cf. Altink, *Representations*, ch. 3.

88. Grainger, *West-India Diseases*, 14.

89. Fairbairn to Home, November 1, 1814, HRFP, GD267/5/24.

90. Lords SC, 274–77, 475, 482–83; Henderson, *Brief View*, 18.

91. Williamson, *Observations*, 2:131.

92. See the discussions of polygamy in chs. 1 and 3.

93. Long, *HJ*, 2:385. See also the testimony of John Marshall, *HCSP*, 73:390. The productive potential of polygamy was one argument against baptizing slaves that Morgan Godwyn sought to counter. See his *Negro's & Indians Advocate*, 137.

94. La Beche, *Present Condition*, 17–18. De la Beche's text enjoyed some influence, being cited in public anti-slavery meetings as a source of information on colonial slavery. See cutting from unknown newspaper [after December 13, 1825], Thomas Fowell Buxton Papers, Bodl., MSS Brit. Emp. s. 444, vol. 39 (Newspaper Cuttings, 1822–27), fol. 39.

95. *Address on the State*, 11.

96. See, for example, McNeill, *Observations*, 41.

97. Adair, *Unanswerable Arguments*, 131; Burnard, *Mastery*, 221–27.

98. Sheridan, *Doctors and Slaves*, 238–39; Lewis, *Journal*, 141; *HCSP*, 71:257, 72:29; Dancer, *Medical Assistant*, 270.

99. Lewis, *Journal*, 65.

100. The proslavery divine George Bridges even attributed the success of nonconformist missionaries in gaining slave converts to their deployment of music. See Bridges, *Annals of Jamaica*, 2:442n3. On music and its connections with racial theory, see Wells, "Masculinity and Its Other," 85–113.

101. Altink, *Representations*, 14–20; Wilson, *Man-Midwifery*, ch. 3; Cody, *Birthing the Nation*, ch. 2.

102. Altink, *Representations*, 14–20; Bush, *Slave Women*, 132–37; Beckles, *Natural Rebels*, 103; Morgan, "Slave Women," 250–53; Dancer, *Medical Assistant*, 267; Roughley, *Jamaica Planter's Guide*, 95–96.

103. Cooper, *Facts*, 42. For an example of midwifery in erotica, see *Spy on Mother Midnight*.

104. Nugent, *Journal*, 124. See also Williamson, *Observations*, 1:63–64.

105. Lords SC, 275.

106. Mathison, *Notices*, 29.

107. Journal of Jonathan Troup, 1788–90, Aberdeen University Library, MS 2070, fols. 27v–28v, 30r–v, 31v, 34r–35r.

108. [Collins], *Practical Rules*, pt. 1; Roughley, *Jamaica Planter's Guide*, ch. 2; Wilberforce, *Appeal*, 16–29; Turner, *Contested Bodies*, ch. 4. Cf. Schiebinger, *Plants and Empire*, 148–49.

109. [Stewart], *Account*, 200, 246. See also Bridges, *Annals of Jamaica*, 2:401.

110. [Alexander], *Address*, 13.

111. Wilberforce, *Appeal*, 18.

112. *HCSP*, 73:216.

113. Fairbairn to Home, February 4, 1815, HRFP, GD267/5/24.

114. See, for example, Edwards, *History*, 2:143.

115. La Beche, *Present Condition*, 17–18; *HCSP*, 69:279; Paugh, *Politics of Reproduction*, 82; Altink, *Representations*, 95.

116. Cooper, *Facts*, 9.

117. James Simpson, for example, testified to encouraging marriage on his estates. See H.C. Papers, Report, 1831–32 (721), 394.

118. *Plan for the Abolition*, 20.

119. Lords SC, 157; Henderson, *Brief View*, 20.

120. Sheridan, *Doctors and Slaves*, 243–44; Altink, *Representations*, 99, 106–15.

121. Thistlewood Diary, 1753, [101]; Hall, *Miserable Slavery*, 60; Burnard, *Mastery*, 156–62.

122. See, for example, *Marly*, 133–34, 148.

123. See the §22 of the Dominican, and §§9–11 of the Grenadan laws of 1788. *HCSP*, 67:110–16, 70:130–35. See also Altink, *Representations*, 78–87.

124. Rev. W. S. Austin, for example. See H.C. Papers, Report, 1831–32 (721), 187–88.

125. George Bridges and John Lindsay were two of the most prominent proslavery clerics. See Bridges, *Annals of Jamaica*; Higman, *Proslavery Priest*; John Lindsay, "A Few Conjectural Considerations upon the Creation of the Human Race, Occasioned by the Present British Quixottical Rage of Setting the Slaves from Africa at Liberty" [ca. 1788], Edward Long Papers, BL, Add. MS 12439.

126. Manyoni, "Extra-Marital Mating Patterns," 104–11; Altink, *Representations*, 93–95.

127. Cooper, *Facts*, 42–47. See also Paugh, *Politics of Reproduction*, chs. 1–2, and her "Politics of Childbearing," 119–60.

128. Altink, *Representations*, 93–95, 211n13.

129. Berlin, *Many Thousands Gone*, 130.

130. Burnard, *Mastery*, 202–4; Altink, *Representations*, 95–106.

131. As argued by Allan Maconochie in *Information*, 41. Paugh, *Politics of Reproduction*, ch. 2.

132. Eph. 6:5.

133. Porteus, *Letter to the Clergy*.

134. Elliott, *Empires*, 107; Higman, "Demography and Family Structures," 502; Peabody, "Slavery, Freedom," 610–12.

135. For a prominent example of an abolitionist text that called for the conversion of slaves but said nothing about teaching literacy, see Ramsay, *Treatment and Conversion*, 150–96.

136. *HCSP*, 69:469.

137. Davies, *Witchcraft*, 51, 302n147.

138. Adair, *Unanswerable Arguments*, 160–62n; [Gibbes], *Instructions*, 101.

139. Beilby Porteus, *Letter to the Governors*, 19–20.

140. Craton, *Testing the Chains*, ch. 19.

141. Home to Fairbairn, January 27, 1815, HRFP, GD267/5/37.

142. Home to Fairbairn, April 28, 1815, HRFP, GD267/5/37.

Chapter 3 • Gentes and Genitals: Sex in Enlightenment Racial Theory

1. White, *Gradation*, 134–35.

2. A partial but limited exception is Nelson, *Enlightenment Biopolitics*, esp. 140–49.

3. Douglas, *Science*, 107.

4. Wagner, *Eros Revived*, 191–200; Harvey, *Reading Sex*, chs. 2–3; Schiebinger, "Mammals, Primatology and Sexology," 184–209.

5. Arist., *Gen. an.* 731b4–732a11; Sloan, "Logical Universals," 103–6; Wilkins, *Species*, 15–21.

6. Lucr. 1.155–56, 1.159–60, 1.188–90.

7. Bacon, "New Atlantis," in *Major Works*, 482; Wilkins, *Species*, 39–45, 47–50, 58–60; Sloan, "Logical Universals," 106–8.

8. Arist., *Cat.* 2a13–2a18.

9. On Lawrence, see Luke, "Sir William Lawrence," 141–52; Wells, "Sir William Lawrence," 319–61.

10. Lawrence, *Lectures*, 98.

11. Lawrence, *Lectures*, 297.

12. Lawrence, *Lectures*, 549.

13. Ray, *Wisdom of God*, 5.

14. Buffon, *HN*, 4:416 (Buffon-Smellie, 3:405).

15. Rousseau, *Discourses*, 208.

16. See Wells, "Blurred Lines," 123–48.

17. Pallas, "Variation des animaux," 82ff. (pt. 2); Meiners, *Untersuchungen*, 1:46–47.

18. Frisch, *Beschreibung*, 6:14; Blumenbach, *De Generis*, 3rd ed., 68n; Prichard, *PHM*, 11n; Buffon, *HN*, 14:431–83; Lawrence, *Lectures*, 264.

19. Wilkins, *Species*, 70–74; Mayr, *Biological Thought*, 259; Banton, *Racial Theories*, 21.

20. Dekkers, *Dearest Pet*, 65, 73; Wilkins, *Species*, 44, 78, 86; Blumenbach, *DG1*, 73–81.

21. Buffon, *HN*, 14:431–32; Long, *HJ*, 2:356; Estwick, *Considerations*, 2nd ed., 73–74. Cf. Bernier, "Nouvelle division," 133–40, which used the language of "species" (*espèces*) in its logical sense.

22. Hunter, *Observations*, 2nd ed., 143.

23. Blumenbach, *Handbuch*, 23; Figal, *Heredity*, 51–55.

24. Kidd, *Forging of Races*, 2.

25. See the discussions of precisely this issue in Seth, "Origins of Racism," 343–68; Bindman, *Race Is Everything*, 11; and Bloch, *Historian's Craft*, 129–34.

26. For examples from Bendyshe's translation, see Blumenbach, *DG3*, 174, 193, 201, 204, 209, 238, 241, 244, 247, 254, 256, 265, 269, 275. On Blumenbach's dissatisfaction with Gruber's translation, see Dougherty, *Correspondence*, 4:283–4n2, and Blumenbach's review in *Göttingische Anzeigen*, 1889–90. Gruber even devoted a significant portion of one of his explanatory notes to outlining and endorsing Kantian terminology—including "races" (*Racen*)—in preference to Blumenbach's own. See Blumenbach, *Verschiedenheiten im Menschengeschlechte*, 259–61n. Cf. Lenoir, "Vital Materialism," 93n54, and Douglas, "Climate to Crania," 38–39.

27. Williams, *Keywords*, s.v. "culture," "nature." Douglas and Ballard, *Foreign Bodies*, is typical in its heavy concentration on tracing the development of the word "race" and its meanings.

28. See esp. Hudson, "'Nation' to 'Race,'" 247–64; Banton, *Racial Theories*; Doron, "Race and Genealogy," 75–109.

29. Hickey, *Memoirs*, 1:136. Hickey was writing from memory after his return to Britain in 1808; if he was interpolating a contemporary meaning of "race" into a conversation of forty years earlier, this only reinforces the point that the term remained highly versatile.

30. *Encyclopædia Britannica*, 19:5; Webster, *Dictionary*, s.v. "race"; Richardson, *New Dictionary*, s.v. "race."

31. "Es wäre ein Auftrag an einen geschäftslosen Menschen, zu entwickeln, in welchem Sinne jeder Schriftsteller dieses Wort gebraucht haben mag. Von den Reisebeschreibern, welche neuerlich die Bewohner der Südseeinseln geschildert haben, darf ich wohl sagen, daß sie ihre Zuflucht zu dem Worte Rasse nur da zu nehmen scheinen, wo es ihnen unbequ[e]m ward Varietät zu sagen." Forster, "Noch etwas," 159. See also Forster to Sömmerring, January 19, 1787, in Sömmerring, *Briefwechsel*, 507–10.

32. Edinburgh medical students were aware of its ambivalence by 1830. See Henry Cheek, "On the Varieties of the Human Race," RMS Diss., vol. 91 (1829–30), 302n.

33. Prichard, *Physical History*, 3rd ed., 1:109 (emphasis in original).

34. This term and its counterpart, polygenesis, are nineteenth-century coinages but are used here as they delineate these two bodies of thought better than any contemporaneous alternative.

35. Braude, "Sons of Noah," 103–42; Goldenberg, *Curse of Ham*; Malcolmson, *Skin Color*; Boyle, *Experiments*, 159–62; Browne, *Pseudodoxia Epidemica*, 1:520–21; Newton, *Dissertations*, 30; Grégoire, *Enquiry*, 25; *British Apollo*, 1:2; Tise, *Proslavery*, 116–20.

36. [Peyrère], *Prae-Adamitae*, translated as *Men before Adam*; *Co-Adamitae* (1732); *London Magazine*, July 1750, 318. On La Peyrère, see Popkin, *Isaac la Peyrère*.

37. The classic and most frequently cited example of hyperprecision in biblical chronology is James Ussher, bishop of Ulster, who calculated that Creation took place on the afternoon of October 23, 4004 BC. See Ussher, *Annals*, 1.

38. Stillingfleet, *Origines Sacrae*, 534.

39. Ramsay, *Essay*, 207, 234; Smith, *Species*, 8–9; Richard Millar, "How far can the Varieties of the human Species that are observable in the different Countries of the World, be accounted for from physical Causes?," SINH Diss., vol. 4 (1785–86), 164–65. Some polygenists argued that the possession of the "moral sense" was a racial characteristic. See Estwick, *Considerations*; Higman, *Proslavery Priest*, 198. On the moral sense, see esp. Shaftesbury, *Characteristics*, esp. 163–230; Hutcheson, *Inquiry*.

40. [Toland], *Two Essays*, 28; Brown, *Political Biography*, 29–30; Lodwick, "Concerning the Originall of Mankind [ca. 1675]," in *On Language*, 200; Poole, "Lodwick's Creation," 250.

41. Kames, *Sketches*, 1:25, 29–30, 2:559n3; White, *Gradation*, 13, 110, 117, 132; Virey, *NH*, 1, 22, 40–56. See also William Bourke, "On the Varieties of the Human Species," SINH Diss., vol. 12 (1793–94), 64; Atkins, *Voyage to Guinea*, 39; Atkins, *Navy Surgeon* (1734), app. 23–24.

42. Kames, *Sketches*, 1:53–67.

43. Bourke, "Varieties," 63.

44. Alexander Robertson, "Do the Varieties Which We Observe Among Mankind Arise from the Action of Moral and Physical Causes; or, Are There Different Races of

Men?," RMS Diss., vol. 40 (1798), 217; Edward Mease, "May We Attribute the Differences Which Are Observable in the Human Race, to the Effects of Moral & Physical Causes," RMS Diss., vol. 5 (1773), 9–27.

45. Isaac, *Invention of Racism*, 60–69.

46. Hippoc., *Aer.* 14, where the process of inheriting an artificially molded head shape is described. See also the discussion in chapter 4, pp. 141–43.

47. Among those to do so were Francis Bacon and Alexander Ross. See Loomba and Burton, *Race*, 214–15, 247–48.

48. See, for example, Arbuthnot, *Effects of Air*; Wilson, *Some Observations*, esp. pt. 3.

49. Buffon, *HN*, 2:98–162; Roger, *Life Sciences*, 463–68; Reill, *Vitalizing Nature*, 42–52.

50. John Gregory, "Lectures on the Institutes of Medicine" [1773], John Gregory Papers, UESC, Gen. 2106D, vol. 2, 93; Haller, "Vorrede," sig. bv; Goldsmith, *History*, 2:18.

51. Browne, *Pseudodoxia Epidemica*, 1:507–30; Boyle, *Experiments*, 151–63; Lewis, "William Petty's Anthropology," 282; Malcolmson, *Skin Color*, 31–37.

52. John Hunter (1728–93) was one exception. See his *Observations*, 2nd ed., 244; White, *Gradation*, 100. Another was Prichard, who, however, dropped the claim in the second (1826) edition of his *Researches*. See Prichard, *PHM*, 233–39.

53. Hdt. 3.101. This point was made by Toland in his *Two Essays*, 27.

54. Buffon, *HN*, 14:432–34; John Anderson, "Discourses of Natural and Artificial Systems in Natural History and [. . .] the Varieties in the Human Kind" [1774], OA/3/1, 30, John Anderson Papers, University of Strathclyde Archives and Special Collections, which also claims that some Philadelphians thought their black neighbors were becoming whiter.

55. Smith, *Species*, 106–7n. See also Jordan's introduction to Smith, *Species*, xlv–xlvi.

56. Blumenbach, *DG3*, 200–206; Nelson, *Biopolitics*, 58–59.

57. Two monogenists who sat on opposite sides of this debate were Prichard and Stanhope Smith. See Prichard, *PHM*, 181–87, and Smith, *Essay*, 20. Cf. Wahrman, *Modern Self*, 93–96, 108–9.

58. White, *Gradation*, 104.

59. See Prichard, *PHM*, lxxviii–lxxixnn114, 116.

60. Estwick, *Considerations*, 2nd ed., 74; Long, *HJ*, 2:356; Kames, *Sketches*, 1:17–18.

61. White, *Gradation*, 1–41; Virey, *NH*, 19–20, 30–31, 54–56; Meiners, *Untersuchungen*, 1:9, 11–12.

62. In private or semiprivate surroundings, such as a society whose proceedings were not published, it was another matter. Of 11 dissertations on human varieties read to the Edinburgh Society for Investigating Natural History between 1782 and 1800, 2 (or almost 20 percent) were unapologetically polygenist. Jenkins, "Race Before Darwin," 339–40, draws the opposite conclusion from these figures, emphasizing the near unanimity behind monogenesis.

63. Sebastiani, *Scottish Enlightenment*, ch. 2.

64. Robertson, *History of America*, 1:314 (emphasis mine).

65. Sebastiani, *Scottish Enlightenment*, 6–9, 92–95; Meek, *Social Science*; Wood, "Natural History," 89–123; O'Brien, *Narratives of Enlightenment*, 156–61; Wheeler, *Complexion of Race*, ch. 4.

66. Alexander, *History of Women*, 1:103.

67. See, for example, Goldsmith, *History*, 2:122.

68. See Klein, *Shaftesbury*; Carter, *Men*.

69. Millar, *Observations*, ch. 1.

70. "Le peu d'inclination, le peu de chaleur des Américains pour le sexe, démontroit indubitablement le défaut de leur virilité & la défaillance de leurs organes destinés à la régéneration: l'amour exerçoit à peine sur eux la moitié de sa puissance: ils ne connoissoient ni les tourments, ni les douceurs de cette passion, parceque [*sic*] la plus ardente & la plus précieuse étincelle du feu de la nature s'éteignoit dans leur ame tiede & phlegmatique." Pauw, *RP*, 1:42.

71. Pauw, *RP*, 1:42–53; [Webb], *Selections*, 2, 19–20; Dorlin, *Matrice de la race*, ch. 10.

72. See, for example, Raynal, *Two Indies*, 6:448–49; Kames, *Sketches*, 2:555–77.

73. Buffon, *HN*, 14:431–83; Kames, *Sketches*, 1:47–48.

74. Buffon, *HN*, 4:417–21; Jenkins, "Race Before Darwin," 342; Stepan, "Biological Degeneration," 97; Richards, *Romantic Conception*, 221.

75. Buffon, *Supp.*, 4:531; Jefferson, *Notes*, 53–65. On the negative connotations of "degeneration," see Johnson's definition of the term in Johnson, *Dictionary*; Roger, *Life Sciences*, 468, 674n266; Doron, *L'homme altéré*.

76. "variétés de l'espèce . . . se perpétuent de génération en génération, commes les difformités ou les maladies des pères & mères passent à leurs enfans; & . . . elle n'ont été confirmées & rendues constantes que par le temps & l'action continuée de ces mêmes causes." Buffon, *HN*, 3:573 (translation mine).

77. Blumenbach, *DG3*, 210–15 (quotation at 213).

78. Blumenbach, *DG3*, 221.

79. See, for example, his "Menschen-Racen," 1–14, translated in *Philosophical Magazine*, 3 (1799): 284–90.

80. See chapter 5, note 51. The basic exposition of this theory is contained in Kant's brief essay "Of the Different Races of Human Beings" (1775/1777), later elaborated and defended in his "Determination of a Concept of a Human Race" (1785) and "On the Use of Teleological Principles in History" (1788). These are respectively published in the standard German edition of Kant's works, AA, 2:427–43, 8:89–106 and 157–84, and in English translation in Kant, *Anthropology, History, and Education*, 82–97, 143–59, 192–218. The literature on Kant and race is sizable. See esp. Bernasconi, "Who Invented the Concept of Race?," in *Race*, 11–36; Hill and Boxill, "Kant and Race," 448–72; Mills, "Kant's *Untermenschen*," 169–93.

81. AA, 8:163–4; Kant, *Anthropology, History, and Education*, 199. Kant's deemphasis on race in the 1790s coincided with a wholesale reevaluation of the place of *Keime* in his biological thought. See Kleingeld, "Kant's Second Thoughts," 573–92; Sloan, "Preforming the Categories," 229–53.

82. Forster, "Noch etwas," 57–86, 150–66, unevenly translated in Mikkelsen, *Kant*, 143–67; AA, 8:172–6; Kant, *Anthropology, History, and Education*, 208–11 (quotation at 209).

83. "Lassen Sie mich lieber fragen, ob der Gedanke, daß Schwarze unsere Brüder sind, schon irgendwo ein einzigesmal die aufgehobene Peitsche des Sklaventreibers sinken hieß?" Forster, "Noch etwas," 154, 163 (quotation), 166. A good example of self-righteous monogenism was on display in Edinburgh only a few months before Forster wrote these words: see Millar, "Varieties," 164–65.

84. "ihre Vernunft, welche Lüsternheit und Begierde erkünstelt." Forster, "Noch etwas," 83.

85. Rousseau, *Discourses*, 138–39; Wokler, "Perfectible Apes," 112. Cf. J. C. Prichard, "Of the Varieties of the Human Race," RMS Diss., vol. 58 (1807–8), 96.

86. Jarrold, "On National Character," 346. See also William Webb, "On the Effects of Domestication on Animals," SINH Diss., vol. 12 (1793–94), 397.

87. "des animaux noirs qui paroissent moins dégénérés par la domesticité." Virey, "Du Lait," 470.

88. See also "Of the Degeneracy of Men in a State of Society," Monboddo Papers, NLS, MS 24515, fols. 113–52.

89. See, for example, Buffon, *HN*, 14:435–36, 442–43; James Watson Roberts, "Of the Degeneration of Animals," SINH Diss., vol. 4 (1785–86), 96–100; Lawrence, "Account," 211–13; Jacoby, "Slaves by Nature?," 89–99. Ellis et al., "Edward Long and Other Animals."

90. Blumenbach, *Beyträge*, 48 (translated in Bendyshe, 294).

91. Smith, *Essay*, 49n.

92. See Vico, *New Science*, 97–99 (paras. 332–37).

93. Hunt and Jacob, introduction, 8; Hunt, Jacob, and Mijnhardt, *Book That Changed Europe*, 221.

94. Bernard and Picart, *Ceremonies*, 5:136; Madam, *Thelyphthora*, 1:39–40, 51–52; 2:313–92. Prostitution- and adultery-as-polygamy was a popular trope in salacious literature of the period. See Wagner, *Eros Revived*, 133–61; *Bon Ton Magazine*, January 1796, 416–17.

95. Madan, *Thelyphthora*, 1:100n, 102–7; 2:272–98; Knights, *Devil in Disguise*, 125–38; Miller, *John Milton*, 48–59; Bernard and Picart, *Ceremonies*, 3:21–22, 47–48, 93; 4:122, 257, 475; 5:136, 308, 7:101.

96. Bernard and Picart, *Ceremonies*, 1:74, 236; 3:47–48, 184.

97. John Arbuthnot in 1710 argued that the sex ratio, which featured a constant proportion of slightly more males than females, proved not only that polygamy was unnatural but also the involvement of divine providence: the probability of such a constant ratio over any prolonged period of time was infinitesimal, and the surplus of males was absorbed by accidents and risky occupations. See his "Argument," 186–90.

98. Kames, *Sketches*, 1:282–83; Paley, *Principles*, 262–67; Malthus, *Essay*, ed. James, 1:74–120.

99. Bernard and Picart, *Ceremonies*, 4:448.

100. See chs. 1 and 2, above, and Barker, *African Link*, 124–25.

101. Forster, *Observations*, 298.

102. "Degeneracy of Men," fol. 130r.

103. Laqueur, *Solitary Sex*; [Smalbroke], *Reformation necessary*, 21; [Mandeville], *Modest Defence*, 8.

104. Buffon, *HN*, 2:513–15; Blumenbach, *DG3*, 170; Gwilliam, "Female Fraud," 518–48; Harol, *Enlightened Virginity*.

105. Gélis, "Refaire le corps," 7–28.

106. Buffon, *HN*, 3:538 (Buffon-Smellie, 3:175).

107. Blumenbach, *DG3*, 232; Schiebinger, "Anatomy of Difference," 387–405; Schiebinger, *Nature's Body*, 135.

108. Blumenbach, *DG3*, 233. Camper also used the fetus of a black African to prove that facial features were not produced by neonatal manipulation. See Meijer, *Race and Aesthetics*, 164–66.

109. Sömmerring, *Verschiedenheit des Negers*, 3–5; Lawrence, *Lectures*, 392.

110. Morgan, *Laboring Women*, 16–17; Morgan, "'Some Could Suckle,'" 167–92. For examples, see Goldsmith, *History*, 2:228; Bulwer, *Anthropometamorphosis*, 311–12.

111. *Voyages and Travels*, 76; Lithgow, *Totall Discourse*, 433, reprinted in the magazines *Phœnix Britannicus* (see Lithgow, "W. Lithgow's Description," 214) and *GM* 46 (1776): 73.

112. Lawrence, *Lectures*, 418; Hudson, "'Hottentots,'" 308–32; Schiebinger, *Nature's Body*, 161.

113. Forster, *Observations*, 181; William Webb, "Are the Diversities among Mankind the effect of Physical & Moral causes?," SINH Diss., vol. 13 (1794–95), 84–85.

114. Lawrence, *Lectures*, 416.

115. White, *Gradation*, 63.

116. Anderson, "Discourses," 39.

117. Blumenbach, *DG3*, 247–48; Millar, "Varieties," 180–81.

118. Darker nipples functioned both as a sign of pregnancy and as a marker of racial difference. See "MS Notes on Midwifery," Thomas Young Papers, RCPE, YOT/7, 117, and Kames, *Sketches*, 1:23. For the explicit association of pregnancy, pigmentation, and racial difference, see Sömmerring, *Verschiedenheit des Negers*, 47, and White, *Gradation*, 114.

119. Arist., *Hist. an.* 584b7–14. For other early modern examples, see Loomba and Burton, *Race*, 143, 189n70, 210, 256–57.

120. On Amerindian resistance to pain, see Raynal, *Two Indies*, 6:470–71; Kames, *Sketches*, 1:33–34; Prichard, *PHM*, 167–73; Sayre, *Sauvages Américains*, ch. 6; Chaplin, *Subject Matter*, 243–79.

121. Long, *HJ*, 2:380; Morgan, "'Some Could Suckle,'" 189–90; Dodd, *Sermon*, 2.

122. Case of Ann Galloway in "Cases from the Edinburgh Lying-In Hospital, 1793–4," RCPE, EGL/1, 13–16.

123. Goldsmith, *History*, 2:224.

124. Bland, *Comparative Parturition*, 31.

125. Ramsay, *Essay*, 225–28.

126. Schiebinger, *Mind Has No Sex?*, ch. 7.

127. Sömmerring, *Verschiedenheit des Negers*, 37; White, *Gradation*, 72–73; Virey, *NH*, 95; Blumenbach, *Beschreibung der Knochen*, 329n; Lawrence, *Lectures*, 462; Prichard, *PHM*, 67.

128. See Crawford, "Attitudes to Menstruation," 47–43; Crawford, *Blood*; Shail and Howie, *Menstruation*; Read, *Menstruation*.

129. Smith, "Body Embarrassed?," 26–46.

130. P. Turpin, "Question: The Theory of Menstruation," RMS Diss., vol. 3 (1770), 87.

131. Bulwer, *Anthropometamorphosis*, 390–91.

132. Blumenbach, *DG3*, 221; Beusterien, "Jewish Male Menstruation," 447–56; Pomata, "Menstruating Men," 109–52; Katz, "Shylock's Gender," 440–62. On Jewish male lactation, see Bulwer, *Anthropometamorphosis*, 318.

133. James Skinner, notes on midwifery lectures given by James Hamilton [1805–6], James Hamilton Papers, RCPE, SKJ/3, p. 9; "Lectures on Midwifery" [ca. 1760–1780], Thomas Young Papers, UESC, Dc.8.166, pp. 210–11; Thomas McKenzie, "What is the Nature and Cause of Catamenia?," RMS Diss., vol. 70 (1813–14), 15–27; Denman, *Introduction*, 1:159–98; Buffon, *HN*, 2:510.

134. Skinner, notes, p. 11; Turpin, "Question," 97; Blumenbach, *DG3*, 248.

135. Schapera, *Early Cape Hottentots*, 45; Merians, *Envisioning the Worst*, 132–34.

136. John MacFadzean, "An Essay to prove that there is but one Species of Man," SINH Diss., vol. 7 (1787–88), 183; Lawrence, *Lectures*, 423.

137. Blumenbach, *DG3*, 384; cf. Long, *HJ*, 2:383.

138. [Lawrence], "Generation."

139. "toutes les femmes naturelles du Cap sont sujettes à cette monstrueuse difformité, qu'elles découvrent à ceux qui ont assez de curiosité ou d'intrépidité pour demander à la voir ou à la toucher." Buffon, *NH*, 3:515 (translation mine). In later years he became doubtful and claimed always to have been so: "It always appeared to me contrary to every natural order" (qui m'a toujours paru contre tout ordre de nature). Buffon, *Supp.*, 4:501.

140. Schiebinger, *Nature's Body*, 170–71; Fausto-Sterling, "Gender, Race, and Nation," 19–48; Crais and Scully, *Sara Baartman*, 139–41.

141. Lawrence, *Lectures*, 425; Hudson, "'Hottentots,'" 308–32.

142. *OED*, s.v. "testis, n.2." This entry erroneously asserts that the classical philologist Alois Walde suggested an alternative etymology based on *testa* (pot, shell, etc.). This derives from a misreading of his *Lateinisches etymologisches Wörterbuch*, 777. On the darker semen of Africans, see Curran, *Anatomy of Blackness*, 125–30; Sömmerring, *Verschiedenheit des Mohren*, 27; and Le Cat, *Traité*, 58, which strongly influenced De Pauw (see Pauw, *RP*, 2:21, 27).

143. Admittedly, this meaning of "effeminate" was rapidly losing ground to its current sense by the middle of the eighteenth century. See Goldberg, *Sodometries*, 111; Smith, *Homosexual Desire*, 196–97.

144. Blumenbach, *DG3*, 249.

145. Pauw, *RP*, 1:42, 145. For an English translation of these sections, see also [Webb], *Selections*, 47.

146. Bernard and Picart, *Ceremonies*, 4:478–79; Buffon, *HN*, 3:515; Raynal, *Two Indies*, 1:305–12; Linnaeus, *Animal Kingdom*, 46; Curran, *Anatomy of Blackness*, 108–10.

147. White, *Gradation*, 62. The absence of the frenum from the penis of the "orang outang" was one of its key differences from humans noted by De Pauw. Pauw, *RP*, 2:56.

148. Female genitals were also part of discussions of skin color, even among prostitutes, as when either Sarah Jones or Mary Smith showed her vulva to the black sailor John Guy in 1736, "and told me it was as black as my Face, &c. &c." Trial of Sarah Jones and Mary Smith, September 8, 1736, *Old Bailey Proceedings Online* (version 9.0), accessed October 6, 2024, https://www.oldbaileyonline.org/record/t17360908-39.

149. Blumenbach, *DG1*, 109.

150. Virey, *NH*, 6.

151. White, *Gradation*, 115. This notion of neonatal ruddiness/lightness before subsequent darkening was very common. See, for example, Boyle, *Experiments*, 164–67; Kolben, *Present State*, 1:56; Kames, *Sketches*, 1:25–26; Jarrold, *Anthropologia*, 214, 243; Case of Mary Blair in "Cases," RCPE, EGL/1, pp. 17–20.

152. On Hunter, see Wilkinson, "'Other' John Hunter," 227–41.

153. Hunter, *Disputatio Inauguralis*, reprinted and translated in Bendyshe, 357–94 (quotation at 373).

154. Robertson, "Varieties," 207.

155. Kames, *Sketches*, 1:170–71; Blumenbach, *DG1*, 127; Blumenbach, *DG3*, 271–72; Malachi Blake, "Upon the Varieties of the Human Species," SINH Diss., vol. 10 (1790–91), 239–41; Virey, *NH*, 14; Lawrence, *Lectures*, 314–22; Schiebinger, "Anatomy of Difference," 391–92.

156. Among a vast number of references, see Goldsmith, *History*, 2:227; Lawrence, *Lectures*, 311, 546; Prichard, *PHM*, 80–85; Anderson, "Discourses," p. 29; White, *Gradation*, 97; Wells, *Two Essays*, 438–39; Estwick, *Considerations*, 2nd ed., 70–71.

157. MS Notes, "Negroe Cause [. . .] Janry 15th 1778," Dreghorn Collection of Sessions Papers, vol. 49, p. 12, Advocates Library, NLS; John Lindsay, "A Few Conjectural Considerations upon the Creation of the Human Race, Occasioned by the Present British Quixottical Rage of setting the Slaves from Africa at Liberty" [ca. 1788], Edward Long Papers, BL, Add. MS 12439, p. 30.

158. Moore, *Travels*, 7.

159. Blumenbach, *DG1*, 124; Blumenbach, *DG3*, 224. Recall Swift's linkage of red hair with lasciviousness, a common association in the period: Swift, *Gulliver's Travels*, 249; Boucé, "Rape of Gulliver," 108–9.

160. Prichard, *PHM*, 163; Stocking, "From Chronology to Ethnology," xvii.

161. "Les Américains du Nord, exposés à l'inclémence . . . à tous les changemens des saisons, ont aussi le visage fort hâlé; mais ils seroient beaucoup moins noirs, s'ils ne se frottoient avec des drogues & des graisses." Pauw, *RP*, 1:202.

162. Merians, *Envisioning the Worst*, ch. 4.

Chapter 4 • *Ex Ovo Omnia: Embryology, Sex, and Race*

1. Swift, *Gulliver's Travels*, 249.

2. Swift, *Gulliver's Travels*, 253–54.

3. Swift, *Gulliver's Travels*, 254–55.

4. Swift, *Gulliver's Travels*, 230–31.

5. *Some reasons humbly offer'd.* See also Swift, *Gulliver's Travels*, 255, 355n; Rawson, *God, Gulliver, and Genocide*, 230–32.

6. Examples: Bulwer, *Anthropometamorphosis*, 1–46; Hale, *Primitive Origination*, 245–88; Parsons, *Philosophical Observations*, 26, 29; [Wallace], *Various Prospects*, 62; Smith, *Species*, 11–23. Some defenses of spontaneous generation were interpreted as tending to support the phenomenon among humans. See Elford, *Short Essay*, 29–30, writing on Jackson, *Thirty Letters*, 2:56–70.

7. *Country Gentleman*, April 11, 1726, 159–63.

8. Sbiroli, *Libertine o madri illibate*; Harvey, "Sexual Difference," 202–23; Rousseau, *Notorious Sir John Hill*.

9. Digby, *Two Treatises*, 220; Hunter, *Observations*, 148.

10. [Maupertuis], *Earthly Venus*, 23.

11. Russell, *Like Engend'ring Like*, ch. 3. For breeders' own ideas in an earlier period and outside the British world, see Cooley, *Perfection of Nature*. On the expense of working with mammals, see Adelmann, *Marcello Malpighi*, 2:822.

12. A note on terminology: "semen" and "seed" are here taken as synonymous and are used instead of the Hippocratic σπερμα (sperm) to avoid confusion with the later discussion of spermatozoa.

13. Arist., *Gen. an.* 728b32–729a21.

14. Particularly for Aristotle. *Gen. an.* 767a36–768a22.

15. Arist., *Gen. an.* 767a36–768a22; Hippoc. *Nat. puer.* 6; Galen, *On Semen*, 2:5.

16. The word is Darwin's. See his *Variation*, 2:359. The fullest treatment of pangenesis remains Zirkle, "Early History," 91–151.

17. Arist., *Gen. an.* 721a31–722a1 (1:17); Galen, *On Semen*, 2:5.

18. Arist., *Gen. an.* 722a10–13 (translation: Aristotle, *Generation*, 55); Arist., *Hist. an.* 586a2–4.

19. On the analogy with cheese-making, see Needham, *History of Embryology*, 85.

20. Needham, *History of Embryology*, 91.

21. Translated as *Anatomical Exercitations* (quotation at sig. A3v).

22. On Digby's embryology, see Needham, *History of Embryology*, 121–30. For contemporary accounts of Harvey and Digby, see Aubrey, *Brief Lives*, ed. Bennett, 1:195–204, 324–29. Scholarly work on the intersection of politics and embryology includes Fissell, *Vernacular Bodies*; Cody, *Birthing the Nation*; Keller, *Generating Bodies*.

23. Although, as Jacques Roger shows, Digby's thought was tending toward iatro-chemical and mechanist ideas. See Roger, *Life Sciences*, 105–7.

24. Harvey, *Excercitations*, 196.

25. Stephanson, *Yard of Wit*, esp. 102–3; Lennox, "Comparative Study," 21–46; Pagel, *New Light*, 155.

26. Arist., *Gen. an.* 739b19–20; Arist., *Ph.* 198b36–199a1. For an eighteenth-century example, see [Cooke], *New Theory*, 299.

27. Essner, *Nürnberger Gesetze*, 32–40.

28. Johnson, "Reproduction in the Female," 642.

29. Banton, *Racial Theories*, 42–43; Burkhardt, "Lord Morton's Mare," 1–21; Harvey, *Disputations*, 38–39, 82, 179, 223. For a later example, influenced by Harvey, see Haighton, "Experimental Inquiry," 159–96.

30. "Countrey Revell," John Aubrey Papers, Bodl., MS Aubrey 21, fol. 15v; Aubrey, *Brief Lives*, ed. Bennett, 2:1037.

31. Ritvo, "Barring the Cross," 48, 56n49.

32. Culpeper, *Directory*, 87–89, 122–24; Lemnius, *Discoruse [sic] Touching Generation*, 107–8; *Aristotle's Master-Piece*, 6–18; Nicolas Venette, *Conjugal Love Reveal'd*, 141–42.

33. Highmore, *History*, 107–8.

34. Browne, *Pseudodoxia Epidemica*, 1:324–29.

35. Digby, *Two Treatises*, 214.

36. Harvey, *Disputations*, 417–18.

37. Culpeper, *Directory*, 83–84. Important texts on complexion include Floyd-Wilson, *English Ethnicity*, and Dawson, *Bodies Complexioned*.

38. Harvey, *Disputations*, 228.

39. Roger, *Life Sciences*, 113–23; Descartes, "Description," esp. 252–86. Translated in Descartes, *World and Other Writings*, 186–205.

40. Nicholas Russell and others have argued that both theories were an extension of metamorphosis: if embryonic development was a simple case of growth from parts existing at the moment of conception, then why could they not have existed before conception? See Russell, *Like Engend'ring Like*, 43–44; Keller, *Generating Bodies*, 138–39.

41. On the terminology of preformation/preexistence, see Bowler, "Preformation and Pre-Existence," 221–44; Keller, "Embryonic Individuals," 321–48.

42. Reill, *Vitalizing Nature*, 60.

43. Cole, *Early Theories*, 42–43. Malebranche's influence probably extended to convincing the influential Dutch microscopist Jan Swammerdam "that all creatures were present 'in the loins of their first parents.'" See Adelmann, *Marcello Malpighi*, 2:907 and n3.

44. Roger, *Life Sciences*, ch. 6; Pinto-Correia, *Ovary of Eve*, chs. 1, 3; various chapters in Smith, *Problem of Generation*, esp. Bitbol-Hespériès, "Monsters, Nature, and Generation," 57–58, and Aucante, "Descartes's Experimental Method," 70–75.

45. Graaf, *mulierum organis generationi inservientibus*; Haighton, "Experimental Inquiry," 159–96; Needham, *History of Embryology*, 163; Cobb, *Generation*, 168–72; Pinto-Correia, *Ovary of Eve*, 43–46.

46. *Onania*, 174; Blumenbach, "Specimen physiologiae," 113; Stollberg and Böcker, *Arthur Schopenhauers Mitschriften*, 131.

47. Among many examples, see [Cooke], *New Theory*.

48. Astruc, *Treatise*, 333–34.

49. It was also supported by the frequent observation that the proximal ostium is so narrow as not to permit the entry of a hog bristle. See, for example, Handley, *Mechanical Essays*, 39–40.

50. Coined by K. E. von Baer as a learned alternative to "Samenthierchen" (little animals of the seed). Burdach, *Physiologie als Erfahrungswissenschaft*, 1:90.

51. Eller, "Recherches," 14–15.

52. Leeuwenhoek, "Part of a Letter," 270–72; Astruc, *Treatise*, 330.

53. Pinto-Correia, *Ovary of Eve*, 282–84.

54. Dawson, *Nature's Enigma*; Roger, *Life Sciences*, 498–509; Pinto-Correia, *Ovary of Eve*, 166–72.

55. Roger, *Life Sciences*, 318.

56. Bowler, "Bonnet and Buffon," 259–81; Roger, *Buffon*, ch. 9; Wood, "Buffon's Reception," 176.

57. The preceding two paragraphs are indebted to Roe, *Matter, Life, and Generation*, and Larson, *Interpreting Nature*, 140–66.

58. Roe, *Matter, Life, and Generation*, ch. 4; Larson, *Interpreting Nature*, 162–64.

59. Blumenbach, *Essay*, 20.

60. Blumenbach, *DG3*, 194–96; Blumenbach, *De Generis*, 2nd ed., 1–2. On the claims made by Blumenbach for the *Bildungstrieb*, see, in particular, Lenoir, "Vital Materialism," 77–108; cf. Zammito, "Lenoir Thesis," 120–32; Zammito, *German Biology*, 206–14; Richards, "*Bildungstrieb*," 11–32; Richards, *Romantic Conception*, 216–29.

61. In addition to the references in the previous note, see Zammito, *Birth of Anthropology*, 302–7; Zammito, "Policing Polygeneticism," 35–54; Bernasconi, "Kant and Blumenbach's Polyps," 73–90; Mensch, "Kant and the Skull Collectors." The influence of Blumenbach in Edinburgh is clear from George Moore, "What Is the Most Plausible Theory of Generation?," RMS Diss., vol. 52 (1804–5), 107–8; James Skinner, notes on midwifery lectures given by James Hamilton [1805–6], James Hamilton Papers, RCPE, SKJ/3, p. 35.

62. Malpighi, *externo tactus organo*, 21–22.

63. McCormick, *William Petty*, ch. 6; Lewis, "William Petty's Anthropology," 261–88; Lewis, *William Petty*; Stuurman, "François Bernier," 1–21; Stuurman, *Invention of Humanity*, 301–2; [Toland], *Two Essays*, 15–28.

64. Atkins, *Treatise*, 204.

65. Atkins, *Treatise*, 205.

66. Blumenbach encountered Atkins after 1781 but evidently thought his contribution minor: he dismissively referred to him as "a man without a great name" (non magni nominis vir) in marginalia to his own copy of the 1781 edition of *De Generis* and did not mention him in the subsequent edition. See annotated copy of Blumenbach, *De Generis*, 2nd ed., Nachlass Blumenbach, Niedersächsische Staat- und Universitätsbibliothek Göttingen, MS Cod. Blumenbach 20, p. 61.

67. It did both for Kames, in whose *Sketches* "soil and climate" were listed as factors that monogenists believed shaped human diversity (1:35), while he argued different races were created for each "climate" and, implicitly, "soil" (1:47).

68. Atkins, *Navy Surgeon* (1742), 369.

69. Seth, *Difference and Disease*, 205; Barker, *African Link*, 88; Atkins, *Navy Surgeon* (1742), 368–69.

70. Blumenbach, *DG3*, 195; Mazzolini, *"Las Castas,"* 349.

71. It was unequivocally the strongest for the author of *A Philosophical Essay on Fecundation*, 44–45.

72. This tale also featured in other ancient texts, such as Plut., *Mor.* 563a3–6; Antig. Car., *Historiarum mirabilium*, ch. 122; Ar. Byz., *Historiae animalium epitome*, 2:272; Pliny, *HN*, 7:51; and in early modern medical and obstetric writings—for instance, Cardano, *Contradicentia medicorum*, bk. 2, tract 6, §17, reprinted in Cardano, *Opera omnia*, 6:642; Fienus, *Viribus imaginationis*, 190; and Sharp, *Midwives Book*, 123. See Zirkle, *Beginnings of Plant Hybridization*, 45–47; Sollors, *Neither Black nor White*, ch. 2.

73. MacCurdy, *Notebooks*, 173; López-Beltrán, "Human Heredity"; López-Beltrán, "Forging Heredity," 211–35.

74. Fissell, *Vernacular Bodies*, ch. 7; Cody, *Birthing the Nation*, chs. 3–4; Weil, *Political Passions*, chs. 1, 3; Rusnock, *Vital Accounts*; Hoffmeier, "Critique of Preexistence," 133–34.

75. [Cooke], *New Theory*, 233–34; Barrère, *Cause physique*, translated in Gates and Curran, *Who's Black and Why?*, 184–90; John Gregory, "Lectures on the Institutes of Medicine" [1773], John Gregory Papers, UESC, Gen. 2106D, vol. 2, pp. 90–107; Skinner, notes, pp. 33–41.

76. The phrase is Jean-François Pagès's (1798), quoted in López-Beltrán, "Forging Heredity," 231.

77. Examples of works discussing race and disease include Sheridan, *Doctors and Slaves*; Arnold, *Colonizing the Body*, 280–88; Bewell, *Romanticism and Colonial Disease*, esp. chs. 3, 6; Harrison, "'Tender Frame,'" 68–93; Harrison, *Climates and Constitutions*; Harrison, *Medicine*; Chaplin, *Subject Matter*, ch. 5; Charters, "Making the Body Modern," 214–31; Hogarth, *Medicalizing Blackness*; Seth, *Difference and Disease*.

78. See, for example, White, *Gradation*, 77–78, and Wells, *Two Essays*, 434.

79. Long believed both: that black people were subject to particular diseases yet immune to yellow fever. See annotated copy of Long, *HJ*, vol. 2, Edward Long Papers, BL, Add. MS 12405, fol. 273r.

80. Hogarth, *Medicalizing Blackness*, ch. 1; Rush, *Account*; Rush, "Observations," 289–97; cf. John Anderson ("Discourses of Natural and Artificial Systems in Natural History and [. . .] the Varieties in the Human Kind" [1774], OA/3/1, p. 34, John Anderson Papers, University of Strathclyde Archives and Special Collections) was especially keen to emphasize that color was *not* disease. William Bourke, "On the Varieties of the Human Species," SINH Diss., vol. 12 (1793–4), 59–68; "On the Different Races of Mankind" [after 1787], Edward Long Papers, BL, Add. MS 12438, p. 5; Hunter, "Inaugural Disputation," 386–87; Buffon, *HN*, 3:573. In an unpublished essay from *c.* 1790, Andrew Duncan drew the same parallel between latent hereditary dispositions and neonatal pigmentation: see his "On the Varieties of the Human Species Chiefly as Produced by Climate," Andrew Duncan, Jr. Papers and Letters, UESC, Dc.1.90, fols. 141v–142v. Robert Boyle was ambivalent on the point: see his *Experiments*, 153–54, 164–65.

81. Especially important are proto-eugenic works such as Vandermonde, *Manière de perfectionner*, discussed in Nelson, "Making Men," 1364–94, and his *Biopolitics*, 63–68; Pinto-Correia and Monteiro, "Support of Racial Mixture," 19–26; Lorenz, *Menschenzucht*, 164–67. Other examples include MacCurdy, *Notebooks*, 173; Harvey, *Disputations*, 42–43; cf. Highmore, *History*, 44; Dionis, *Anatomy*, 232; Turner, *Force*, 103–5; Blumenbach, *DG3*, 202–3, 207–18; Skinner, notes, pp. 35–37; Jarrold, *Anthropologia*, 49–50.

82. Highmore, *History*, 91–96; Sharp, *Midwives Book*, 120–24; *Essay on Conjugal Infidelity*, 11–12; William Thomson, notes on surgical lectures given by John Hunter [ca. 1778], UESC, DC.2.44¹, lecture 12 (unfol.); Goldsmith, *History*, 2:243–66; Prichard, *PHM*, 25–32. [Gott], *Divine History*, 438, argued that physical characteristics and diseases but not intellectual or moral endowments could be transmitted. On the power of nurses to influence children through their milk, see Mendelson and Crawford, *Women*, 29; Douthwaite, *Dangerous Experiments*, 165, 283n22; *Dialogue Between a Gentleman*, 39.

83. Harvey, *Disputations*, 65, 274–81; cf. Highmore, *History*, 29–32; Sharp, *Midwives Book*, 56. Aristotle's was a critique of *strict* pangenesis, according to which semen is a derivative of all parts of the body (i.e., that it consists of actual parts of the body).

84. Alexander Robertson, "Do the Varieties which we observe among Mankind arise from the action of Moral and Physical causes; or, are there different Races of Men?," RMS Diss., vol. 40 (1798), 201–17; Prichard, *PHM*, 194–204; Pauw, *RP*, 1:199, discussed in *Monthly Review*, 42 (1770): 524–25. Richards, *Romantic Conception*, 219–20.

85. Thomson, notes, lecture 12 (unfol.). See also Lawrence, *Lectures*, 452–53.

86. Goldsmith, *History*, 2:248–49; Darwin, *Zoonomia*, 491, 514–24. See also Wilson, "Eighteenth-Century 'Monsters,'" 1–25.

87. Drake, *Anthropologia Nova*, 1:334–35; Barrère, *Dissertation*, 23; John Usher, "On Conception," SINH Diss., vol. 3 (1784–85), 155; Thomas Smith, "On Generation," SINH Diss., vol. 6 (1786–87), 81; Moore, "Plausible theory," 99–108.

88. The discussion here is limited to physical attributes, but over the eighteenth century it was increasingly believed that some cultural and specific characteristics that had previously been accepted as features of humanity *ab origine* were acquired and subsequently inherited. For example, Lord Monboddo and the Italian surgeon Moscati respectively argued for the posterior acquisition of language and bipedalism. See Monboddo, *OPL*, bk. 2; Moscati, *Structur der Thiere*; Zammito, *German Biology*, 194–98.

89. Kames, *Sketches*, 1:25; White, *Gradation*, 117, 132; Smith, *Species*, 28–30, 80–81n; Monboddo, *Origin*, vol. 1, 2nd ed., 307; "Of the Ourang Outang, & whether he be of the Human Species," Monboddo Papers, NLS, MS 24537, fol. 36r.

90. Contributions to SINH Diss. by Alexander Macpherson (vol. 4 [1785–86], 67–70), Richard Millar (vol. 4 [1785–86], 161–85), John MacFadzean (vol. 7 [1787–88], 179–85, and vol. 8 [1788–9], 31–34), William Bourke (vol. 12 [1793–94], 59–68), William Webb (vol. 12 [1793–94], 381–97, and vol. 13 [1794–95], 53–90); Robertson, "Varieties," 201–17.

91. Other monogenists who also denied inheritance of acquired characteristics included William Lawrence (see Lawrence, *Lectures*, 96–97, 503–10) and the abolitionist poet Edward Rushton: see his essay "An Attempt to Prove That Climate, Food, and Manners, Are not the Causes of the Dissimilarity of Colour in the Human Species [1824]," in *Collected Writings*, 204. On Prichard, see Margaret Crump's biography, *James Cowles Prichard* (which sadly appeared too late to be consulted for this book).

92. Prichard, *Researches*, 2nd ed., 2:544–45; Augstein, *Prichard's Anthropology*, 113.

93. Augstein, "Prichard's Views," 34; "Register of Students, 1802–1810," James Hamilton Papers, RCPE, HJA/3/1. On Prichard's "preformism," see Augstein, "Prichard's Views," 257; Augstein, *Prichard's Anthropology*, 111–12.

94. Stocking, "From Chronology to Ethnology," lxxx.

95. Atkins, *Treatise*, 205; Lodwick, "Concerning the Originall of Mankind [ca. 1675]," in *On Language*, 200–201; [Toland], *Two Essays*, 26.

96. Godwyn, *Negro's & Indians Advocate*, 23–24; Jordan, *White Over Black*, 30–31.

97. The preeminent example is Bory de Saint-Vincent, who in 1825 identified fifteen species within the human genus, many of which contained two or more races. See Saint-Vincent, *L'homme*.

98. White, *Gradation*, 133. See also Long, *HJ*, 2:356, itself borrowed extensively from Estwick, *Considerations*, 2nd ed., 73–74. This sentiment was shared earlier by Voltaire in his "Relation," 187–200.

99. Kames, *Sketches*, 16–19; White, *Gradation*, 19–21, 127–28; Long, *HJ*, 2:358. Hence their opponents insisted on the common origin of dogs. See, for example, Ramsay, *Treatment and Conversion*, 223.

100. Kames's opposition to slavery is proven by his judgment in favor of the slave Joseph Knight in the case of *Knight v. Wedderburn* (1778). See MS Notes, "Negroe Cause [. . .] Janry 15th 1778," Dreghorn Collection of Sessions Papers, vol. 49, pp. 4–5, Advocates Library, NLS. White's position was more complex and is considered in greater detail in Wells, "Misplaced Politeness," 73–97.

101. There is now a sizable literature on the "orang outang" in the eighteenth century. For a useful guide, see Sebastiani, "'Monster,'" 80–99.

102. Long, *HJ*, 2:371 (emphasis in original).

103. Long referred to Monboddo, although not by name. See Long, *HJ*, 2:370–71n.

104. Long, *HJ*, 2:356–57, 363, 371. Long was somewhat ambivalent but never unequivocally stated that the "oran-outang" was human. Cf. Seth, "Materialism, Slavery," 764–72; Seth, *Difference and Disease*, ch. 6; Sebastiani, *Scottish Enlightenment*, 106; Sebastiani, "Challenging Boundaries," 127; Hall, "Whose Memories?," 146. Both Blumenbach and Lawrence initially believed that the *History* was produced by Samuel Estwick, and Lawrence continued to confuse them: Blumenbach to Banks, December 28, 1794, Joseph Banks Correspondence, BL, Add. MS 8098, fol. 221r; Lawrence, *Lectures*, 295n.

105. This term is used without quotation marks in this and the following paragraph for reasons of precision: less offensive alternatives ("mixed-race" or simply "mixed") are ambiguous (they could refer to a black-white *couple*) or circumlocutional.

106. Long, *HJ*, 2:335.

107. Long, *HJ*, 2:336.

108. On Long's plagiarism, see Barker, *African Link*, 48 and, as an example, Long, *HJ*, 2:353–54; cf. *Modern Part*, 14:21.

109. Long, *HJ*, 2:336; Barker, *African Link*, 52; Paugh, *Politics of Reproduction*, 75–76.

110. Barrère, *Dissertation*, 28; Bacon, *Four Sermons*, xiii–xiv; *Morning Chronicle*, March 22, 1788; Lawrence, *Lectures*, 295n; Nott, "Mulatto a hybrid," 252–56; Lewis, *Journal*, 55, 88; Broca, *Phenomena of Hybridity*, esp. 30; Stepan, "Biological Degeneration," 109–11; Hall, "Whose Memories?," 146.

111. The total population of Jamaica (slave and free) at this time was *c.* 142,000. Engerman and Higman, "Demographic Structure," 48; Higman, *Population and Economy*, 145–56; Livesay, *Children of Uncertain Fortune*, 24.

112. For example, Treytorens, "Observations," 16; Long, *HJ*, 2:332; Moreton, *West India Customs*, 123–32; Goveia, *Slave Society*, 216; Mohammed, "'Browning,'" 22–48; Newton, *Children of Africa*; Livesay, *Children of Uncertain Fortune*.

113. Ramsay, *Essay*, 239–40.

114. Annotated copy of Long, *HJ*, vol. 2, fol. 257r.

115. See, for example, Vogt, *Vorlesungen*, 242–43, translated as *Lectures on Man*, 436–37.

116. Blumenbach, *Essay*, 56; Zammito, *German Biology*, 213–14; Reill, *Vitalizing Nature*, 167–68.

117. Hunter, "Inaugural Disputation," 363–64.

118. William Lawrence, for example, rejected the literal application of "hybrid" to describe humans, even while describing the transmutation of white into black and vice versa. See Lawrence, *Lectures*, 293–97; Davidson, *Breeding*, 91.

119. See, for example, the letters by Humanitas (July 1) and R. W. (August 25) to the *Morning Chronicle* as part of a controversy at the onset of the parliamentary abolition campaign in 1788.

120. [Lawrence], "Generation." Examples of such committed polygenists include Civis (*Morning Chronicle*, February 5; August 19, 1788); McNeill, *Observations*, 42; White, *Gradation*, 129–30.

121. Blatchford, *Observations on Equivocal Generation*, 7. On the case, see *Kennedy v. Gifford*, 19 Wend. 296 (1838). Arist., *Gen. an.* 746a29–33 (2:7). For an early modern restatement, see Porta, *Natural Magick*, 34.

122. Buffon, *Supp.*, 3:16–17; Buffon-Smellie, 8:18–20n.

123. Buffon, *Supp.*, 3:1–38; Edwards, "Account of a Bird," 833–37.

124. Tyson, *Orang-Outang*, 51. Speech had been the sole prerogative of humans at least since Homer and remained decisive throughout this period. See Gera, *Ancient Greek Ideas*, 36n66, 182; Friedman, *Monstrous Races*, 29; Bertoletti, "Anthropological Theory," 115; Ashcraft, "Leviathan Triumphant," 141–81; Aarsleff, *Study of Language*, 13–43; Hodgen, *Early Anthropology*, 411.

125. Camper, "Account," 139–59; Ogilby, *Africa*, 558; Buffon, *HN*, 14:145–74; "Of the Ourang Outang, & whether he be of the Human Species," Monboddo Papers, NLS, MS 24537, fol. 18r. Camper's dissection was, however, of an orangutan (*Pongo pygmaeus*)

and not a chimpanzee, so his results are strictly incomparable with Tyson's. Meijer, *Race and Aesthetics*, 134–36; Tinland, *L'Homme Sauvage*, 119–29; Wokler, "Tyson and Buffon," 2309.

126. "daß auch verdorbne Europäer in den dortigen oder indianischen Pflanzorten . . . in einen heißen Klima eine Bastarterzeugung vom Orang-utang und einer Sklavinn für europäische Beobachter besorgten." Pallas, "Kurze Nachrichten," 156.

127. "nachtheiligen und tödtlichen Klima." Pallas, "Kurze Nachrichten," 156.

128. Zimmermann, *Geographische Geschichte des Menschen*, 1:118nh; Monboddo, *Origin*, vol. 1, 2nd ed., 334. See chapter 3, p. 94, above.

129. *Vénus physique* was widely cited in works of law, natural history, reference, theology, and even humor, and copies appeared in numerous catalogs of auctioned libraries across the century (examples include ESTC nos. T1340, N14880, T12289, T72215, T30077, T60463). Parsons's writings were less widely referenced, but he was read or personally known by many, particularly in scientific circles: examples include Blumenbach, Camper, Prichard, Virey, and White. Biographical details on Parsons can be found in Nichols, *Literary Anecdotes*, 5:472–89, and on Maupertuis in Beeson, *Maupertuis*; Terrall, *Man Who Flattened*.

130. Voltaire, "Relation," 187–200.

131. Maupertuis, *Earthly Venus*, 79; Maupertuis, *Vénus physique*, 155–56.

132. Mazzolini, "Albinos, Leucoæthiopes," 161–204.

133. *OED*, s.v. "albino"; Curran, "Rethinking Race History," 153–56. The most influential descriptions for eighteenth-century writers of American and African peoples with albinism were, respectively, Wafer, *New Voyage*, 134–38, and Purchas, *Purchas his Pilgrimes*, 2:980 (although see also Bulwer, *Anthropometamorphosis*, 108). The term "leucæthiop" was reintroduced in 1658 by Isaac Vossius (in his edition of Pomponius Mela's *De situ orbis*); he applied it in 1666 to the albinos of Luango, originally described by Andrew Battell. For labels other than "albino" or "leucæthiop," see Mazzolini, "Albinos, Leucæthiopes"; Mazzolini, "Blumenbach on Albinism," 113.

134. Blumenbach, *DG1*, 130–40; Blumenbach, *DG3*, 218–19, 260–61; Blumenbach, *Handbuch*, 63; Blumenbach, *Beyträge*, 119–26. For Blumenbach's paper on albinotic eyes, see Marx, *Zum Andenken*, 8 (translated in Bendyshe, 8).

135. Goldsmith, *History*, 2:240.

136. White, *Gradation*, 100; Prichard, *PHM*, 240–42.

137. See, for example, Maupertuis, *Earthly Venus*, 75–80; Buffon, *HN*, 3:541–46; Goldsmith, *History*, 2:240–42.

138. "Different Races of Mankind," pp. 4–5.

139. Hunter, *Observations*, 2nd ed., 244; Pauw, *RP*, 2:5–46; Le Cat, *Traité*.

140. "Cependant le Maure blanc a tous les traits du Maure noir." Le Cat, *Traité*, 111.

141. Speculation concerning an albinotic nation appeared in the *Histoire de l'Académie Royale des Sciences*: "Observations de phisique générale," 15–17; "Diverses observations anatomique," 12–13. It also featured in Raynal, *Two Indies*, 4:7; Parsons, "Account," 45–53. It was still being discussed in the nineteenth century: see *Chambers' Edinburgh Journal*, November 18, 1843, 347–48. Spanish discussions are summarized in Katzew, *Casta Painting*, 46–48. Discussions of other groups are contained in J. C. Prichard, "Of the Varieties of the Human Race," RMS Diss., vol. 58 (1807–8), 112–14; Lawrence, *Lectures*, 303–4, 448; Monboddo, *Origin*, vol. 1, 2nd ed., 307–8.

142. Ogilby, *Africa*, 509; Buffon, *Supp.*, 4:556; Virey, *Histoire naturelle*, 2:154–55; Virey, *NH*, 91.

143. *Modern Part*, 16:293n; Buffon, *Supp.*, 4:556, 567; James Cunningham, "On the Colour of Hair," SINH Diss., vol. 1 (1782–83), 220–22.

144. Le Cat, *Traité*, 107–8; Parsons, "Account," 46. The literature on the maternal imagination is now sizable. Key works in English include Wilson, "'Out of Sight,'" 63–85; Huet, *Monstrous Imagination*; Daston and Park, *Wonders*, ch. 5; Finucci, "Maternal Imagination," 41–77; Mazzoni, *Maternal Impressions*; Fissell, *Vernacular Bodies*, ch. 7; Crawford, *Marvelous Protestantism*, ch. 5; Smith, "Imagination," 80–99; Turner, "Birth Anomaly," 217–37.

145. Chaplin, *Subject Matter*, 139.

146. Jacob, *Logic of Life*, 26.

147. Turner, *Morbis Cutaneis*, 113; Fissell, *Vernacular Bodies*, 207–8; Kahn, "Versehen der Schwangeren," 325.

148. Jerome, *Hebrew Questions*, 203–4; Jerome, "Liber Hebraicarum," 985nd; Erasmus, *Omnium operum*, 2:99 (Tome 4). James Blondel recognized this misuse of Hippocrates: Blondel, *Power*, 11, 35–37.

149. Gen. 30:31–42. Examples of the use of the pseudo-Hippocratic example include Turner, *Morbis Cutaneis*, 113; Mauclerc, *Dr. Blondel Confuted*, 14; Eller, "Recherches," 3–19; Paré, *On Monsters and Marvels*, 38–39.

150. See, for example, Tornielli, *Annales Sacri*, 1:225–26; Lafitau, *Customs*, 1:62–67; *Athenian Oracle*, 2nd ed., 1:29–30; *British Apollo*, 1:2; Smith, *Species*, 115–16n; Molineux, *Faces of Perfect Ebony*, 94–95.

151. Wells, "Sex and Racial Theory," ch. 3; Harvey, *Impostress Rabbit Breeder*; Cody, "Doctor's in Labour," 175–96; Wilson, "'Out of Sight,'" 63–85.

152. "Imaginationi matris id universum assigno, qvæ partem vehementis desideratam animo fixo compehendens ejudem colorem fœtui impressit." Quoted in Müller-Wille, "Cabbage, Tulips, Ethiopians," 13n31.

153. "Quæ omnia, rudimentum futuri fœtus neutiquam in uno tantum sexu delitescere, evincunt." Linnaeus, *Sponsalia plantarum*, 26.

154. Smollett, *Peregrine Pickle*; Sterne, *Tristram Shandy*; Goethe, *Elective Affinities*. See also Rousseau, "Pineapples," 79–109; Stephanson, "*Tristram Shandy*," 93–108.

155. Edward Mease, "May we attribute the differences which are observable in the Human Race, to the Effects of Moral & Physical Causes," RMS Diss., vol. 5 (1773), 9–27.

156. John Lindsay, "A Few Conjectural Considerations upon the Creation of the Human Race, Occasioned by the Present British Quixottical Rage of setting the Slaves from Africa at Liberty" [ca. 1788], Edward Long Papers, BL, Add. MS 12439; Higman, *Proslavery Priest*, ch. 8.

157. Smith, *Species*, 115n.

158. Quilletus, *Callipædia*, 33. Originally published in Latin in 1655, the poem was translated into French in 1665, and English editions were published in 1710, 1712, 1718, 1720, 1728, 1729, 1732, 1750, 1760, 1761, 1768, 1769, 1771, and 1776.

159. Gee, *Trade and Navigation*, 42–43; Turner, "Birth Anomaly," 217–37.

160. Unsigned review of *Miraculum Naturae*, by Johann Swammerdam, *PTRS*, 7 (1672): 5000. (Original text in Swammerdam, *Miraculum Naturae*, 29–30.)

161. Lemnius, *Secret Miracles of Nature*, 11–12.

162. For example, Sharp, *Midwives Book*, 185–86; Turner, *Force*, 73; [Bianchini], *Essay*, 22; Gregory, "Lectures," vol. 2, pp. 96–97; Fissell, *Vernacular Bodies*, 130.

163. Lindsay, "Conjectural Considerations," pp. 103–4.

164. John Goodsir, notes on midwifery lectures given by Alexander Hamilton [1797], UESC, Gen. 59D, p. 35.

165. "A Full, True, and Particular Account of a Lady Who Longed for Charcoal" [ca. 1820], Bristol Broadsides Collection, Bristol City Library, B13013, no. 19.

166. Skinner, notes, pp. 43–45; Wells, *Two Essays*; "Scraps of English Folklore," 79.

167. This word is used in preference to the modern "intersex" because it is the contemporaneous term and because the latter encompasses a wide range of (often chromosomal) conditions that are not visibly manifested in the genitals or secondary sexual characteristics (the only way a "hermaphrodite" was identified in the early modern era).

168. Wilson, *Signs and Portents*, 134; Gilbert, "Seeing and Knowing," 150–70; Guerrini, "Advertising Monstrosity," 109–27; Guerrini, "Hermaphrodite of Charing Cross," 28–51.

169. Parsons, *Enquiry*, liv.

170. Cf. Reis, *Bodies in Doubt*, 11, in which Caspar Bauhin's words are misread as Parsons's own. See Parsons, *Enquiry*, xxxiv.

171. Parsons, *Philosophical Observations*, 78–79, 198–205; Parsons, *Enquiry*; Parsons, *Prælecturi*, 13; Parsons, "Short Account," 650–52; Parsons, "Hermaphrodite," 142–45.

172. For a discussion of the Gnostic view, see Duval, *Traité des hermaphrodits*, 272; Gilbert, *Early Modern Hermaphrodites*, ch. 1; DeVun, "Heavenly Hermaphrodites," 132–46. This idea was rejected by (among others) Bulwer, *Anthropometamorphosis*, 396–97; Hale, *Primitive Origination*, 316. Works emphasizing the growing harshness toward hermaphrodites include Mosse, *Nationalism and Sexuality*, ch. 5; Porter, "Bodies of Thought," 82–108; Mendelson and Crawford, *Women*, 19–21; Gilbert, *Early Modern Hermaphrodites*.

173. Cadden, *Sex Difference*, ch. 4; Daston and Park, "Sexual Ambiguity," 117–36. On the cellular womb, see Cadden, *Sex Difference*, 93; Kudlien, "Seven Cells," 415–23, and the following examples: Culpeper, *Directory*, 34; Sharp, *Midwives Book*, 38; *Aristotle's Master-Piece*, 8–9; Venette, *Conjugal Love Reveal'd*, 191–92.

174. See Venette, *Generation*, 2:397–427.

175. "Ces sortes de personnes sont plutôt un espéce d'Eunuque que d'Hermaphrodite, leur verge ne leur servant de rien & les régles ne leur venant jamais." Venette, *Generation*, 2:402. This was semiplagiarized in [Jacob], *Tractatus*, reproduced in Pettit and Spedding, *British Erotica*, 2:19.

176. Parsons, *Enquiry*, 144–54; Blumenbach *DGI*, 141; Lawrence, "Account," 183; Guicciardi, "Hermaphrodite et le prolétaire," 58.

177. Parsons, *Enquiry*, 7.

178. Home, "Account," 157–78; Hunter, "Account," 527–35; Hunter, *Observations*, 45–55; Parsons, *Philosophical Observations*, 5, 79.

179. Pauw, *RP*, 2:84.

180. William Webb, "Are the Diversities among Mankind the effect of Physical & Moral causes?," SINH Diss., vol. 13 (1794–95), 83; Parsons, *Enquiry*, 147.

181. For example, the discussion of an "Ethiopian" (gypsy) hermaphrodite in Colombo, *De Re Anatomica*. See Moes and O'Malley, "Realdo Colombo," 527–28.

182. Parsons, *Enquiry*, 147.

183. Guicciardi, "Hermaphrodite et le prolétaire," 66.

184. Home, "Account," 163.

185. Ovington, *Voyage to Suratt*, 497; Merians, *Envisioning the Worst*, 132.

186. Pauw, *RP*, 2:104–5.

187. Pauw, *RP*, 2:83–117; Guicciardi, "Hermaphrodite et le prolétaire," 56–68; Lafitau, *Customs*, 1:57–58; Bernard and Picart, *Ceremonies*, 3:111, 4:500.

188. Sadeur, *New Discovery*; Fausett, *Images of the Antipodes*, ch. 1.

189. Brown, "Politics of Sexual Difference," 171–93; Brown, *Good Wives*, ch. 3; Gilbert, "Seeing and Knowing," 150–70; Reis, *Bodies in Doubt*, ch. 1.

190. Mendelson and Crawford, *Women*, 19; Tissot, *Onanism*, 46.

191. *Supplement to the Onania*, 165; Wagner, *Eros Revived*, 19–20.

192. Coke, *Institutes*, fo. 8r.

193. Trial of Katherine Jones, September 3, 1719, *Old Bailey Proceedings Online* (version 9.0), accessed October 6, 2024, https://www.oldbaileyonline.org/record /t17190903-50; Sudai, "Sex Ambiguity," 478–513.

194. Daston and Park, "Sexual Ambiguity," 117–36.

195. Faselius, *Elements*, 21. Original: Faselius, *Elementa Medicinae Forensis*, 16 (§40).

196. Coke, *Institutes*, fo. 7v; Blackstone, *Commentaries*, 2:201, 246–47.

197. Gross, *What Blood Won't Tell*.

198. [Formey], *Souvenirs d'un citoyen*, 1:218–19.

199. [Maupertuis], *Réflexions philosophiques*; Condillac, *Essay*; Rousseau, "Essai sur l'origine des langues [ca. 1754]," in *Œuvres complètes*, 5:371–429, translated in Rousseau, *Discourses*, 247–99; Parsons, *Remains of Japhet*; Leibniz, "Brevis designatio Meditationum de Originibus Gentium ductis potissimum ex indicio linguarum [1710]," in *Opera Omnia*, vol. 4, pt. 2, 186–98; Prichard, *PHM*.

Chapter 5 • *"This Race Benign": Race and Reproduction in the Pacific, 1760–1820*

1. Bougainville, *Pacific Journal*, 229.

2. "Leur Odorat est très fin, et tel, qu'ils distinguent à L'odeur une femme d'un homme. Un passager . . . en avait amené une avec luy déguisée en homme; il reconnurent son sexe, malgré son déguisement, et l'indiquèrent par des signes très énergiques." "Observations de M. de la Condamine Sur L'insulaire de la Polynesie, amené de lisle de Tayti en France par M. de Bougainville" [1769], Rés. G.1.443 (Suppl. 1–18), pièce no. 5, Bibliothèque Nationale de France. See also Martin-Allanic, *Bougainville navigateur*, 2:965–66. On Baret, see Dunmore, *Monsieur Baret*, and Ridley, *Discovery of Jeanne Baret*.

3. *Supplément au Voyage*, 239–50.

4. Modern investigators have established that histochemical compatibility between potential partners is communicated on an olfactory level and that the number of male and female test subjects who can distinguish sex from the odors of sweat or breath is statistically significant. See Blake and Sekuler, *Perception*, 499–500, and Sergeant, "Female Perception," 25–45.

5. *GM* 42 (1772): 109; Martin-Allanic, *Bougainville navigateur*, 2:971; Bougainville, *Voyage Round the World*, 264–65; Alexander, *Omai, Noble Savage*, 37; Rensch, "Early European Perceptions," 407–10.

6. See ch. 3.

7. The revised edition (1785) substituted *Negers* (negro) for *Mohren*.

8. Forster, *Observations*, 155.

9. See Mary Louise Pratt's discussion of the "anti-conquest" in her *Imperial Eyes*, ch. 3.

10. Rennie, *Far-Fetched Facts*; Fausett, *Images of the Antipodes*.

11. Sturma, *South Sea Maidens*, ch. 1; Cheek, *Sexual Antipodes*, ch. 4.

12. The undiscovered great southern continent, which Ptolemaic geography taught must exist; the confirmation or elimination of this possibility was one of the chief goals of Cook's second (1772–1775) voyage.

13. Bacon, "New Atlantis," 457–89; Varaisse, *Sevarites*; Cheek, *Sexual Antipodes*, 125–26.

14. Sadeur, *New Discovery*, 32, 68–72; Sadeur, *Southern Land, Known*, 51–55; Barre, "Equality," 82.

15. Patot, *James Massey*, 20–21, 109–10, 128, 271; Rosenberg, *Tyssot de Patot*, ch. 8.

16. *Monthly Review*, 4 (1750): 157; Arthur, *Virtual Voyages*, 56–77; Morris, *John Daniel*, 93, 143; Cheek, *Sexual Antipodes*, 132–35; Lipski, "Hybridity and Conflicted Discourse," 119–39; [Neville], *Isle of Pines*.

17. *Travels of Hildebrand Bowman*, 83, 89. See also *Description of New Athens*, reprinted in Claeys, *Utopias*, 45; Lamb, *Preserving the Self*.

18. See, for example, Forster, *Voyage*, 2:470.

19. Hawkesworth, *Account*, 2:207.

20. Hawkesworth, *Account*, 2:128.

21. In addition to describing the "Point Venus" ritual as a Tahitian religious cere-mony, akin to Christian "matins" or "vespers" (2:128), Hawkesworth had also doubted the workings of providence in the introduction (1:xix–xxi); Wesley, *Extract*, 7; Guest, "Great Distinction," 44–45.

22. Montagu, *Letters and Friendships*, 1:279.

23. Pearson, "Hawkesworth's Alterations," 52.

24. See Lamb, *Preserving the Self*, esp. 3–15.

25. Commerson, "Lettre de M. Commerson," 197–207 passim; Bougainville, *Voyage autour du monde*, 209; Bougainville, *Pacific Journal*, 48, 61–63, 74. See also Smith, *European Vision*, 41–43.

26. Sturma, *South Sea Maidens*, ch. 2; Wilson, "Thinking Back," 352.

27. See Boswell, *Ominous Years*, 308–10; Boswell, *Life of Johnson*, 3:7–8; Cook, *South Pole*, 1:187–88.

28. Beaglehole, *Journals*, 1:128; Sturma, *South Sea Maidens*, 25–28; Thomas, *Discoveries*, 203–4.

29. Knox, *Essays, Moral and Literary*, 2:33.

30. Oliver, *Ancient Tahitian Society*, 2:913–64; Rennie, *Far-Fetched Facts*, 93; Smith, *European Vision*, 43.

31. *New Discoveries*, 90; *Public Advertiser*, July 3, 1773; see esp. Cody, *Birthing the Nation*, chs. 1, 8; Wahrman, *Modern Self*, chs. 1, 2; Dabhoiwala, *Origins of Sex*, 215–31.

32. Hawkesworth, *Account*, 2:207; Cook, *South Pole*, 1:358.

33. Cook, *South Pole*, 1:358.

34. A confused account of the *arioi* is contained in Becket, *Journal*, 47. See also Hegarty, "Unruly Subjects," 189; Beaglehole, *Journals*, 1:clxxxvi–cxc.

35. West, "Limits of Enlightenment Anthropology," 147–60.

36. Forster, *Voyage*, 2:129–30.

37. Forster, *Voyage*, 2:131.

38. Oliver, *Ancient Tahitian Society*, 1:342; Forster, *Resolution Journal*, 3:523–24; Beaglehole, *Journals*, 2:420, 842–43.

39. Hoare, *Granville Sharp*, 147–52.

40. Roderick, "Sir Joseph Banks," 67–89; Bewell, "Botany and Sexual Controversy," 173–93. See also *Town and Country Magazine*, 5 (1773): 457–59.

41. [Preston], *Seventeen Hundred and Seventy-Seven*, 25.

42. Voltaire, *Lord Chesterfield's Ears*, 20; Voltaire, "Oreilles," 125–207.

43. Jimack, *Diderot: Supplément*, 11. See also Curran, "Logics of the Human," 158–72.

44. "nous nous sommes aperçus au premier coup d'œil que vous nous surpassiez en intelligence; et, sur-le-champ, nous avons destiné quelques-unes de nos femmes et de nos filles les plus belles à recueillir la semence d'une race meilleure que la nôtre." Diderot, "Supplément au voyage de Bougainville," in *Œuvres philosophiques*, 500; Mander, "Female Figure," 109–10.

45. "ce tribut levé sur ta personne, sur ta propre substance." Diderot, "Supplément," 500. The classical work of this school is Moorehead, *Fatal Impact* (2000 [1969]). An early example is [Fitzgerald], *Injured Islanders*. The shift away from "fatal impact" theses has been discussed in the introduction to Lamb, Smith, and Thomas, *Exploration and Exchange*; Lee Wallace, *Sexual Encounters*, 19; Edmond, *Representing*, ch. 1.

46. Orr, "'Southern Passions,'" 212–31.

47. "Les adieux du viellard." Diderot, "Supplément," 465–72.

48. See Russell, "'Entertainment of Oddities,'" 48–70. In the absence of any other measures, Pacific voyagers estimated the Tahitian population based on the island's naval strength. See Cook, *South Pole*, 1:348–50; Forster, *Voyage*, 2:64–65; *Monthly Review*, 57 (1777): 20; Guest, *Empire*, 14–15.

49. Sturma, *South Sea Maidens*, ch. 5.

50. See Hall, *Benjamin Franklin*; *GM* 17 (1747): 175–76, 211; Franklin, "Observations Concerning the Increase of Mankind," in *Autobiography*, 251–60.

51. On animals and breeding, see the work of Ritvo, esp. "Possessing Mother Nature," 413–26.

52. On the "stupidity" of native peoples, see, for example, Forster, *Voyage*, 1:491–93. On the Spanish deposit of only males, see *Monthly Review*, 71 (1784): 283; Cook and King, *Pacific Ocean*, 2:23.

53. Fara, *Sex, Botany, and Empire*, ch. 6; Edmond, *Representing*, ch. 1; Thomas, *Colonialism's Culture*, ch. 3; Guest, "Great Distinction," 36–58.

54. Sturma, *South Sea Maidens*, ch. 2.

55. Beaglehole, *Journals*, 1:520–21.

56. Levine, "Sexuality, Gender, and Empire," 134–55.

57. Gagnon, "Others Have Sex," 23–38; Forster, *Resolution Journal*, 2:308–9; Thomas, *Discoveries*, xxv–xxvi. Diderot's "old man" voiced much the same sentiment: see his "Supplément," 469–70.

58. Edmond, *Representing*, ch. 3; Sturma, *South Sea Maidens*, ch. 3; Smith, *European Vision*, ch. 5.

59. Dening, *Mr Bligh's Bad Language*, 55–87.

60. Burke, "Philosophical Enquiry," 219–20, 255.

61. Hudson, "'Hottentot Venus,'" 19–41; Fredrickson, *Racism*, 59–60.

62. Montaigne, "Of Physiognomy," *Complete Works*, 809. See also Porter, *Windows of the Soul*; Graham, "Lavater's Physiognomy in England," 561–72.

63. See Bindman, *Ape to Apollo*.

64. Smith, *European Vision*, ch. 1. The emergence of this standard did not mean that editorializing and invention were absent from even the most rigorous of texts. See Edwards, *Story of the Voyage*, 219–25. See also Porter, "Exotic as Erotic," 117–44; Thomas, *Colonialism's Culture*, ch. 3. Cf. Hegarty, "Unruly Subjects," 183–97; Lamb, "Death of John Hawkesworth," 97–113.

65. See the essays in Lincoln, *Science and Exploration*.

66. See Porter, "Exotic as Erotic," 117–44; Pratt, *Imperial Eyes*, 15–37; Douglas, "Slippery Word," 1–29; Wilson, "Thinking Back," 362.

67. Smith, *European Vision*, 1–7, 127–32; Bindman, *Ape to Apollo*, 123–50.

68. See Obeyesekere, *Cannibal Talk*, 1–23. Cf. Thomas, *Discoveries*, 104–7.

69. *St. James's Chronicle*, August 4, 1774. See also "J[oseph?]. B[anks?].," letter to the editor, *General Evening Post*, August 6, 1774.

70. *London Evening Post*, July 22, 1775.

71. Cook, *South Pole*, 1:243–44.

72. Guest, *Empire*, 12–13, 133.

73. Boswell, *Ominous Years*, 341; Schaffer, "Inscription Devices," 106.

74. Beaglehole, *Journals*, 2:294. On "indigenous countersigns," see Douglas, *Science*.

75. The native had ostentatiously licked his fingers, "as if afraid to lose the least part, either grease or gravy, of so delicious a morsel." (Beaglehole, *Journals*, 2:818.) See also Obeyesekere, *Cannibal Talk*, 30–32; Salmond, *Cannibal Dog*, 223–26.

76. See Wilson, *Island Race*, ch. 5; Douglas, "Ethno-Historical Text," 65–99.

77. Daston and Galison, *Objectivity*, 17.

78. Kames, *Sketches*, 1:24, 29, 32; Kames to Lind, March 7, 1772, Single Letters and Small Collections, 1700–97, NLS, MS 10782, fols. 113–14.

79. Marx, *Zum Andenken*, reprinted in Bendyshe, 21. Blumenbach also collected Pacific artifacts following the Cook voyages. See Gascoigne, "German Enlightenment," 164–65; Blumenbach, *DG1*, 99; Douglas, "'Novus Orbis Australis,'" 106–9.

80. Blumenbach, *DG1*, 99–100n4; Hoare, *Tactless Philosopher*, 144–46. Jennifer Mensch attributes this change rather to Georg's influence: see her "Kant and the Skull Collectors," 205.

81. D'Urville, "Sur les îles," 5; Sturma, *South Sea Maidens*, 10; Edmond, *Representing*, 8. On Forster's *Observations* as preracial, see Douglas, "Slippery Word"; Douglas, "'Novus Orbis Australis,'" 102–6. Margaret Jolly has argued that all eighteenth-century voyage accounts were preracial. See her "'Ill-Natured Comparisons,'" 331–64.

82. Gascoigne, *Joseph Banks*, ch. 4; Fulford, Lee, and Kitson, *Literature*, 33–45, 148.

83. On the *Endeavour*'s as "Banks's voyage," see, for example, *Public Advertiser*, July 3 and 17, 1773. See also Wilson, "Pacific Modernity," 68.

84. On the figure of the "macaroni," see McNeil, *Pretty Gentlemen*; Cohen, "Grand Tour," 241–57; Wright, *Caricature History*, 257–62. For specific comment on macaronis and the Pacific, see Scobie, *Celebrity Culture*, 40–43; Fara, *Sex, Botany, and Empire*, 6–9; Gascoigne, *Joseph Banks*, 61–62.

85. Letters to Banks from the following: Camper (January 23, 1786), Buffon (June 23, 1783), Bougainville (June 25, 1784), Franklin (August 21, 1784), Manchester Literary and Philosophical Society (June 24, 1801) in Joseph Banks Correspondence, BL, at (respectively) Add. MSS: 8096, fols. 257–58; 8095, fols. 153–54, 244, 282–83; 33980, fols. 297–98. Carter, *Sir Joseph Banks*, 42, 96–97, 114, 168–69, 292, 376; Gascoigne, *Joseph Banks*, ch. 4. On the importance of the Forsters for emergent racial theories, see Gascoigne, "German Enlightenment," 141–71, and Guest, *Empire*, 11–13.

86. On the importance of Göttingen as an entrepôt for English texts, see Fabian, "English Books," 1780–82; Biskup, "University of Göttingen," 128–60.

87. Gascoigne, "German Enlightenment," 166–67; Blumenbach to Banks, June 20, 1787, Joseph Banks Correspondence, BL, Add. MS 8096, fols. 383–84. Banks's notes for his reply are contained at fol. 384r. On Camper's rivalry with Hunter, see Camper to Sömmerring, September 11, 1786, in Sömmerring, *Briefwechsel*, 413–20.

88. Blumenbach to Banks, June 9, 1790, Joseph Banks Correspondence, BL, Add. MS 8097, fol. 261; Carter, *Sir Joseph Banks*, 372; Blumenbach to Banks, May 8, 1792, Joseph Banks Correspondence, BL, Add. MS 8098, fols. 8–9.

89. Blumenbach, *DG3*, 149–54.

90. Dougherty, *Correspondence*, 2:185n5. Camper to Banks, January 23, 1786; February 19, 1787, Joseph Banks Correspondence, BL, Add. MS 8096, fols. 257–58, 411–12.

91. Turnbull, "British Anatomists," 26–50; Fara, *Sex, Botany, and Empire*.

92. "Thoughts on the Manners of Otaheite written in Holland for the amusement of the Prince of Orange" [1773], Sir Joseph Banks Papers, National Library of Australia, MS 9/4.

93. Goldsmith, *History*, 2:72–73.

94. Gill, *Exposition*, 61.

95. See Burke, "Philosophical Enquiry," 276; White, *Gradation*, 135. See also Armstrong, "'Effects of Blackness,' 213–36.

96. Virey, *NH*, 30; cf. White, *Gradation*, 135.

97. Browne, *Pseudodoxia Epidemica*, 1:520.

98. See, for example, Jefferson, *Notes*, 138.

99. See Hudson, "'Hottentots,'" 308–32.

100. Johnson, *Dictionary*, s.v. "Complexion." For examples of this use of "complexion," see *Athenian Oracle*, 2nd ed., 1:436; Bigland, *Historical Display*, 94–95.

101. Johnson, *Dictionary*, s.v. "Complexion"; Glacken, *Rhodian Shore*, 12.

102. On the relation between skin color and complexion, see Wheeler, *Complexion of Race*, 22–28. On a visual "common sense," see Stoler, "Racial Histories," 186–87.

103. Malcolmson, *Skin Color*.

104. Boyle, *Experiments*, 120, 151–67. See also [Newton], *Opticks*, 69.

105. Mitchell, "Essay," 102–50. For a sensitive and insightful reading of Mitchell, see Delbourgo, "Newtonian Slave Body," 185–207.

106. Mitchell, "Essay," 103.

107. Mitchell, "Essay," 109 (emphasis mine).

108. He spoke in terms of "a Suffocation of the Rays of Light." Mitchell, "Essay," 123.

109. See chapter 3, pp. 118–19.

110. Hawkesworth, *Account*, 2:187–88; Cook, *South Pole*, 2:34–37; Forster, *Observations*, 153–71; Blumenbach, *DG3*, 275.

111. Beaglehole, *Journals*, 1:399.

112. Buffon, *HN*, 3:454 (Buffon-Smellie, 3:104); for an early translation of Pyrard, see Purchas, *Purchas his Pilgrimes*, 2:1646–71.

113. "Collections for a History of Jamaica," Edward Long Papers, BL, Add. MS 18270, 43; Long, *HJ*, 2:353.

114. Schotte, *Synochus Atrabiliosa*, 104–5; Sömmerring, *Verschiedenheit des Negers*, 41. See also Seth, *Difference and Disease*, 267–71. For biographical data on Schotte, see Meusel, *Lexikon*, 12:422–23.

115. Jefferson, *Notes*, 138–39; Virey, *NH*, 31.

116. Smith, *Essay*, 32–36.

117. Lawrence, *Lectures*, 461.

118. White, *Gradation*, 56–57.

119. Tench, *Complete Account*, 179; Wells, *Two Essays*, 430; Tullett, "Grease and Sweat," 307–22; Kettler, *Smell of Slavery*, ch. 2.

120. Goldsmith, *History*, 2:180; Prichard, *PHM*, 54.

121. Raynal, *Two Indies*, 6:437.

122. Grégoire, *Enquiry*, ch. 7.

123. D'Elia, "Benjamin Rush," 414n8.

124. "On the Different Races of Mankind" [after 1787], Edward Long Papers, BL, Add. MS 12438, 3; Sömmerring, *Verschiedenheit des Negers*, 67; Virey, *NH*, 67; White, *Gradation*, 57, 80; Lawrence, *Lectures*, 461, 467; Prichard, *PHM*, 53–55; Smith, *Species*, 172–74.

125. Schiebinger, "Anatomy of Difference," 396; Schiebinger, *Nature's Body*, 140–42.

126. Roger, *Buffon*, 293–97; Cuvier, *Règne Animal*, 1:81–88; see also Cuvier, "Cuvier's Animal Kingdom," 341–51.

127. Schiebinger, *Mind Has No Sex?*, 211–13; Forster, "Noch etwas," 76–77.

128. Guenebault (Virey's translator) copied his from White rather than extracting them from Sömmerring. See White, *Gradation*, cxxxix–clxvi; Virey, *NH*, 57–75; Wells, "Misplaced Politeness," 73–97.

129. Sömmerring, *Verschiedenheit des Negers*, 3–72 (§§4–66); White, *Gradation*, cxxxix–clxvi.

130. White, *Gradation*, 41–56.

131. Schiebinger, "Anatomy of Difference," 399; Schiebinger, *Nature's Body*, 156–58.

132. Blumenbach, *DG3*, 239; White, *Gradation*, cxlvi.

133. White, *Gradation*, 63–64.

134. Virey, *NH*, 25.

135. Silenus was the tutor and companion of the Greek wine god, Dionysius, and was associated with drunkenness and musical creativity. Hogarth, *Analysis of Beauty*, 128.

136. Hogarth, *Analysis of Beauty*, 131.

137. Hogarth, *Analysis of Beauty*, 126.

138. [Camper], *Works*, 14, 1–12; see also Meijer, *Race and Aesthetics*.

139. [Camper], *Works*, v–vi.

140. [Camper], *Works*, 42.

141. [Camper], *Works*, 45–46.

142. Meijer, *Race and Aesthetics*, 3, 88–94.

143. [Camper], *Works*, 45–51; Meijer, *Race and Aesthetics*, 1–6, 55–61, and 94–96, where she proves convincingly that the facial angle was not intended as a tool for racial taxonomy and that this association was based on a misreading of his work. For a more negative assessment of Camper, see Mosse, *Toward the Final Solution*, 21–24.

144. On such naivety, see Wells, "Misplaced Politeness," 73–97.

145. White, *Gradation*, 125.

146. [Camper], *Works*, 42.

147. See the controversy between Stephen Jay Gould and Thomas Junker in *Isis*, 89 (1998) over a reworking of Blumenbach's illustration of his typology that appeared in Gould, *Mismeasure*, rev. ed., 409. See Junker, "Blumenbach's Racial Geometry," 498–501, and Gould, "Mental and Visual Geometry," 502–4. I am grateful to George Newberry for this reference. For a contemporaneous example, see Jenkins, "Race Before Darwin," 343–44 and fig.

148. Blumenbach, *DG3*, 237; Blumenbach, *De Generis*, 3rd ed., 204. Note Bendyshe's mistranslation of "national" (*gentilitium*) as "racial."

149. Lavater, *Essays on Physiognomy*, 2:208.

150. Lavater, *Essays on Physiognomy*, 2:225–26.

151. Blumenbach, *DG3*, 229; Bindman, *Ape to Apollo*, 92–122, 118.

152. See chapter 3, note 79. Lavater, *Essays on Physiognomy*, 3:99–103. A full modern translation is in Kant, *Anthropology, History, and Education*, 82–97.

153. Bindman, *Ape to Apollo*, 92–122; Mosse, *Toward the Final Solution*, 24.

154. Lavater, *Essays on Physiognomy*, 1:24.

155. On language, see Aarsleff, *Locke to Saussure*; Aarsleff, *Study of Language*; Hudson, *Writing and European Thought*.

156. Locke, *Essay*, 158–61 (2.11.8–13).

157. Locke, *Essay*, 160 (2.11.12).

158. Buffon, *HN*, 3:532–33 (Buffon-Smellie, 3:172).

159. Vico wrote in 1724 that "a poverty of words naturally makes men sublime in expression, weighty in conception, and acute in understanding much in brief expression." Vico, *First New Science*, 149. Aarsleff, *Study of Language*, 14–15; Pagden, *Fall of Natural Man*, 15–26.

160. Lafitau, *Customs*, 2:266.

161. Condillac, *Essay*, 113–19. On Condillac, see Aarsleff, *Study of Language*, 13–43; Jooken, "Two Distinct Views," 307–27.

162. Kames, *Sketches*, 1:47–51.

163. Jooken, "Monboddo and Adam Smith," 263–76.

164. Meek, *Social Science*, 99–130.

165. Kames, *Sketches*, 1:214–52 ("miraculous" appears on 214).

166. On Monboddo, see Cloyd, *James Burnett*.

167. On Monboddo and his work, see Hammett, "Lord Monboddo," 63–80; Hammett, "*Origin and Progress*"; Wokler, "Apes and Races," 145–68; Wokler, *L'homme physique*, 121–38; Jooken, "Linguistic Conceptions"; Jooken, "'Gaggling of Geese,'" 54–77.

168. Monboddo, *OPL*, 1:12.

169. Rousseau, "Discourse on Inequality," in *Discourses*, 148.

170. Monboddo, *OPL*, 1:142–43.

171. Monboddo, *OPL*, 1:5 (emphasis in original).

172. Both Rousseau and Monboddo made use of "Peter the Wild Boy"—a feral child found in 1724 in the vicinity of Hameln in the Electorate of Hanover—in this regard. See Rousseau, "Discourse on Inequality," in *Discourses*, 190–91; Monboddo, *Antient Metaphysics*, 3:335–78, 4:25–34, 5:1–10; Sebastiani, "Monboddo's 'Ugly Tail,'" 45–65.

173. See Barnard, "*Orang Outang*," 95–112.

174. Diderot, "Le rêve de d'Alembert," in *Œuvres philosophiques*, 385; Monboddo, *OPL*, 1:175.

175. Language and the body could be intrinsically related, as in the work of Franciscus Mercurius Van Helmont (1614–1699), who thought that Hebrew characters were diagrams showing the position of the speech organs in making the sound designated by each letter. These ideas enjoyed a brief renaissance after 1750. See Hudson, *Writing and European Thought*, 84.

176. Kames, *Sketches*, 1:49n.

177. White, *Gradation*, 37.

178. This essay was first published in 1748, and the footnote was added in an expanded version of 1753–1754. See Hume, *Essays*, 197–215. See also Popkin, "Hume's Racism," 211–26; Eze, "Hume, Race," 691–98.

179. Edward Long also devoted ten pages of his *History of Jamaica* to Williams, whom he attempted to show was haughty, contemptuous of his slaves, cruel, prejudiced, and had delusions of grandeur by referring to himself as white. Long referred elsewhere to Hume's footnote. See Long, *HJ*, 2:376, 475–85; cf. Dabhoiwala, "Man of Parts," and Hall, *Lucky Valley*, 438–45.

180. Gibbon also connected the two. See Gibbon, *Decline and Fall*, 4:616.

181. Winckelmann, *Art of Antiquity*, 118; Lavater, *Essays on Physiognomy*, 3:104–5.

182. Commerson, "Lettre de M. Commerson," 199.

183. Hawkesworth, *Account*, 2:228; Rensch, "Early European Perceptions," 403–14.

184. Bougainville, *Voyage Round the World*, 264–65.

185. Of outstanding importance to this project was Leibniz, who wrote several essays in which he used language to classify and account for the origins of many of the world's peoples. See, for example, his "Brevis designatio Meditationum de Originibus Gentium ductis potissimum ex indicio linguarum [1710]," in *Opera Omnia*, vol. 4, pt. 2, 186–98, and translations of this and other essays in Leibniz, *Esprit de Leibnitz*, vol. 2. See also Horsman, "Racial Anglo-Saxonism," 387–410; Droixhe, *La Linguistique*; Kidd, *British Identities*; Buzon, "Leibniz," 383–400; Stewart, "Mother Tongue," 71–107.

186. This was done as early as the unauthorized account of the *Endeavour* voyage in Becket, *Journal*, esp. 105–6.

187. Gascoigne, "German Enlightenment," 150–51. Banks had stumbled on but ultimately rejected this technique. See Hawkesworth, *Account*, 3:473–77; Forster, *Observations*, 182–90; Rensch, *Language*, 66.

188. Prichard, *PHM*, 244.

189. Prichard, *PHM*, 246–47.

190. Augstein, *Prichard's Anthropology*, 4, 61; Rensch, "Appendix II," 383; Beaglehole, *Journals*, 1:514–19; Lifschitz, *Language and Enlightenment*, 104–9; Combe and Buchan, "'Savage and Brutal Nations,'" 29–41; Stewart, "James Cowles Prichard," 76–91.

191. Forster, *Observations*, 174.

192. Forster, *Observations*, 184 (emphasis mine).

193. Forster, *Observations*, 190.

194. See McCormick, *Omai*; Fullagar, *Savage Visit*, ch. 6; Guest, "Ornament and Use," 317–44.

195. Thomas, *Discoveries*, 260–63.

196. See Mole, *Byron's Romantic Celebrity*, ch. 1. Cf. Braudy, *Frenzy of Renown*, which argues instead that fame has been a constant of Western history. See also Mole, *Romanticism and Celebrity Culture*.

197. McCormick, *Omai*, 111–12, 114, 124–28, 159, 333.

198. Garrick to George Colman the Elder, August 29, 1775, in Garrick, *Letters*, 3:1031–32; Joppien, "Loutherbourg's Pantomime," 83; Huse, "Noble Savage," 303–16; Dening, *Mr Bligh's Bad Language*, 262–76; Worrall, *Harlequin Empire*, 139.

199. Wilson, *Island Race*, 63–70; Guest, *Empire*, 1–27.

200. Lawrence, *Lectures*, esp. 568–72.

201. Rousseau and Porter, introduction; Smith, *European Vision*, 80–84, 326.

202. *St. James's Chronicle*, August 4, 1774, June 29, 1775; "Ode of Anacreon," *Morning Chronicle*, August 29, 1776; *Middlesex Journal*, December 1, 1774; *Lloyd's Evening Post*, July 1, 1776. On British perceptions of Mai, see Hackforth-Jones, "Mai/Omai in London," 13–30.

203. See, for example, *Omiah's Farewell*, ii.

204. Cf. Thomas Blake Clark, who argued that Mai got whiter the more he was represented. See his *Omai*, 83.

205. *London Chronicle*, August 9, 1774. See also McCormick, *Omai*, 100, 105, 109, 121; Solander to Dr. James Lind, August 19, 1774, in Duyker and Tingbrand, *Daniel Solander*, 333–35. For an account of nose flattening in the Pacific, see Cook, *South Pole*, 1:366–67; Forster, *Observations*, 332. For a play on this supposed feature of Omai and syphilitic noses, see [Preston], *Seventeen Hundred and Seventy-Seven*, 12.

206. [O'Keeffe], *Short Account*, 16. See also McCormick, *Omai*, 335.

207. For mention of "woolly" hair in a Pacific context, see Jolly, "'Ill-Natured Comparisons,'" 331–64; "Of the Countries in the South Seas," Monboddo Papers, NLS, MS 24536, fols. 85v–86r.

208. *Middlesex Journal*, December 21, 1775; *Morning Chronicle*, September 20, 1775.

209. Hoare, *Memoirs*, 149. A measure of the favored status of Pacific islanders is perhaps offered by Hume's refinement of his "racist footnote" to focus solely on "Negroes" rather than all non-Europeans, as contained in the posthumous (1777) edition of his *Essays*. See Wheeler, *Complexion of Race*, 186.

210. Kiernan, "Noble and Ignoble Savages," 86–116; Wilson, "Thinking Back," 345–62; Guest, *Empire*, 38.

211. See, for example, *Oxford Magazine*, June 1770, 228–29.

212. Carter, *Men*; Margaret Jolly, "Eroticism and Exoticism," 99–122.

213. Wallace, "Too Darn Hot," 232–42, and Wallace, *Sexual Encounters*; Morris, "Aikāne," 21–54; Morris, "Same-Sex Friendships," 71–102; Porter, "Exotic as Erotic," 117–44.

214. Cook, *South Pole*, 2:66–67.

215. Ellis, *Polynesian Researches*, 1:202.

216. See especially Wahrman, *Modern Self*, 45–82.

217. Thomas, introduction; Caplan, introduction, xvi–xvii; Scutt and Gotch, *Skin Deep*, 21–37.

Chapter 6 • Colonial Ethnogenesis and the Sexual Making of Race

1. Stephen to Zachary Macaulay, May 28, 1828, Macaulay Papers, Huntington Library, San Marino, CA, MY 810.

2. *Marly*, 219. On this novel, see Williamson, introduction, and Salih, *Representing Mixed-Race*, ch. 2.

3. *Marly*, 218.

4. Stepan, "Citizenship," 29.

5. See Beckles, *Natural Rebels*, 134.

6. On Long and Edwards's intended readership, see Hall, *Civilising Subjects*, 107; Kitson, *Romantic Literature*, 220n6.

7. Burnard, *Mastery*, 121–26; *GM*, 83, pt. 1 (January–June 1813): 659. See also Hall, *Lucky Valley*.

8. Richard B. Sheridan, "Edwards, Bryan (1743–1800)," *Oxford Dictionary of National Biography*, updated January 3, 2008, https://doi.org/10.1093/ref:odnb/8531.

9. See Higman, *Population and Economy*, ch. 7, esp. 153. See also Long, *HJ*, 2:260–61; Edwards, *History*, 2:15–16.

10. Long, *HJ*, 2:335–36; White, *Gradation*, 129. Long's *History* was widely read, and Matthew Lewis in 1816 twice mentioned his conclusions concerning mixed-race people. See Lewis, *Journal*, 55, 88.

11. This was also the case with the mid-eighteenth-century Spanish sources used by De Pauw, who in turn influenced Buffon's 1777 revisions of his article on human varieties. See Mazzolini, *"Las Castas,"* 351.

12. Edwards, *History*, 2:16.

13. See Jordan, *White Over Black*, esp. ch. 4; Jordan, "American Chiaroscuro," 183–200; MacLeod, *American Revolution*, ch. 2. For a desperate attempt to stoke fears of miscegenation in the last years of slavery, see Davis, *West Indies*, 26. For an example dating from the late eighteenth century, see [Tobin], *Cursory Remarks*, 118–19n; cf. Olaudah Equiano, letter to the editor, *The Public Advertiser*, January 28, 1788, reprinted in Equiano, *Interesting Narrative*, 330–32; Wheeler, *Complexion of Race*, 285.

14. Slaves inherited their slave status from their mother, which provided a further reason why sexual relationships between invariably free white women and slaves were so furiously anathematized.

15. Higman, *British Caribbean*, 3, 77.

16. Higman, *British Caribbean*, 77.

17. Although this association was not ironclad in the Iberian Peninsula itself. See Guasco, *Slaves and Englishmen*, 94.

18. Most of these rights were phased out in the early decades of the eighteenth century. See Heuman, *Between Black and White*, 3–19; Curtin, *Two Jamaicas*, 42–47. Petitions were rare but usually successful, as they were submitted by elite people of color who could afford the astronomical cost. See Livesay, *Children of Uncertain Fortune*, 42–52; Newman, *Dark Inheritance*, 97–98.

19. See Burnard, *Jamaica*, 134; Heuman, *Between Black and White*, pt. 1; Steel, "Philosophy of Fear," 12–14; Newman, *Dark Inheritance*, ch. 3; Livesay, *Children of Uncertain Fortune*, ch. 7.

20. *Marly*, 83; Wimpffen, *Voyage*, 114; Davis, *West Indies*, 91; Lewis, *Journal*, 55.

21. Journal of Jonathan Troup, 1788–90, Aberdeen University Library, MS 2070, fols. 107v–108r; Kriz, "Marketing Mulatresses," 209; [Stewart], *Account*, 297; Long, *HJ*, 2:335.

22. Edwards, *History*, 2:21–23; Long, *HJ*, 2:331–32.

23. Hyperfecundity was by far the rarer view, but it was not unknown. See also Garrigus, "Colour Line," 36–37; Altink, *Representations*, 74; Nugent, *Journal*, 69; Higman, *Population and Economy*, 154; Troup Journal, fol. 107v; Lewis, *Journal*, 55, 87–88; Wilberforce, *Appeal*, 8–9; Williamson, *Observations*, 1:42–43.

24. McNeill, *Observations*, 43; Lewis, *Journal*, 39, 179–81; La Beche, *Present Condition*, 34–35; Goveia, *Slave Society*, 224–25; Patterson, *Sociology of Slavery*, 64, 161–62; Moreton, *West India Customs*, 123–32; Edwards, *History*, 2:21.

25. Grégoire, *Enquiry*, 123; Thomas Clarke to Henry Cullen, April 20, 1788, Cullen Papers, RCPE, CUL1/3/192; *Considerations on Negro Slavery*, 20; Cooper, *Facts*, 24–25; [Stewart], *Account*, 297, 304; Wells, *Two Essays*, 434.

26. Long retailed both positive and negative (but mostly negative) stereotypes about mulattoes. See Long, *HJ*, 2:335.

27. Burnard, *Jamaica*, ch. 5.

28. [Brougham], *Concise Statement*, 42. See also Fredrickson, *Racism*, 39.

29. Sheridan, *Doctors and Slaves*, ch. 1; Higman, *British Caribbean*, 322–24.

30. [Collins], *Practical Rules*, 35.

31. See, for example, Wilberforce, *Appeal*, 8–9.

32. [Morgann], *Plan*, 15 (emphasis mine).

33. Stephen, *Slavery*, 1:31n.

34. See [Heyrick], *Immediate, Not Gradual Abolition*; *Concise View*, 18–19; [Alexander], *Address*, 13; Williamson, *Observations*, 1:49; Barker, *African Link*, ch. 4.

35. Nugent, *Journal*, passim but particular concentrations at 23–54, 100–126, 198–219. See also Higman, "Lady Nugent's Social History," 78–79.

36. Morgan Godwyn recognized it already in 1680: see his *Negro's & Indian's Advocate*, 35–36.

37. Waltham Day Book, 1832–1834, Home-Robertson Papers, HRFP, GD267/25/113.

38. William Alexander to John Stevens, June 25, 1750, Alexander Papers, New-York Historical Society, MS8, Series 2, Box 6.

39. See *Speech of [. . .] George Canning*, 26; 10 Parl. Deb., H.C. (2nd ser.) (1824) 1103. On the reference to *Frankenstein*, see Malchow, *Gothic Images*, ch. 1.

40. See, for example *Speeches of Mr. Barrett*; *First Report*, 15–16 (in MSS. Brit. Emp. s. 6, Bodl.); *Address on the State*, 14–15; *Considerations on Negro Slavery*, 7; Burnley, *Opinions*, 38.

41. On the loquaciousness of slaves, see Lewis, *Journal*, 95–96; Nugent, *Journal*, 71; Edwards, *History*, 2:83; La Beche, *Present Condition*, 36. On their passions and spite, see, for example, [Senior], *Jamaica*, ch. 4.; Bridges, *Annals of Jamaica*, 2:398–436; Mathison, *Notices*, 2; Stewart, *Past and Present*, 251–52.

42. See Lewis, *Main Currents*, 116–23; Ward, *West Indian Slavery*, ch. 3; Davis, *Age of Revolution*, 113–19; Petley, "Creole World View," 100–101.

43. See, for example, *Liverpool Mercury*, April 28, 1826; *British and Colonial Weekly Register*, June 5, 1824; *Exposure of the Arguments*.

44. Morgan, "Slaves and Livestock," 47–76; Burnard, "Slave Naming Patterns," 325–46.

45. Lewis, *Journal*, 100; Higman, *Population and Economy*, 1–5; M. L. Lewis to John Plummer, October 11, 1816, MS 694, National Library of Jamaica, summarized in Ingram, *Manuscript Sources*, 95.

46. See Mathison, *Notices*, 100; [Estwick], *Considerations*, 41–42; Blakley, *Empire of Brutality*.

47. [Stewart], *Account*, 236; Long, *HJ*, 2:403–4.

48. [Stewart], *Account*, 235.

49. Cooper, *Facts*, 9; Bridges, *Annals of Jamaica*, 2:400.

50. See Long, *HJ*, 2:444.

51. This phrase was first coined in 1772. See *OED*, s.v. "negro."

52. Higman, *Population and Economy*, 144–45.

53. Patterson, *Sociology of Slavery*, 33–51; Burnard, *Mastery*, ch. 1; Altink, *Representations*, ch. 3; Bush, *Slave Women*, ch. 4.

54. Higman, *Population and Economy*, 147; Altink, *Representations*, 75; Bush, *Slave Women*, 110–18.

55. Hence high rates of absenteeism. See Patterson, *Sociology of Slavery*, 33–51; Curtin, *Two Jamaicas*, ch. 1; [Stewart], *Account*, ch. 16.

56. Higman, *British Caribbean*, 115–21.

57. On slave resentment, see Bush, *Slave Women*, 110–18; *Anti-Slavery Monthly Reporter*, October 31, 1825, 36; Cooper, *Facts*, 35–42; Roughley, *Jamaica Planter's Guide*, 77. See also Olaudah Equiano's complaint of a racial double standard in his *Interesting Narrative*, 104. On white sexual temptation, and the need to avoid it, see esp. Long, *HJ*, 2:330–34.

58. There were 15,776 white people in Jamaica in 1844, a significant fall from 1830, but still half the number estimated by contemporaries. See Martin, *History*, 1:92; Higman, *British Caribbean*, 77; Heuman, *Between Black and White*, 9.

59. H.C. Papers, Select Committee on Extinction of Slavery in British Dominions, Report, Minutes of Evidence, Appendix, Index, 1831–32 (721), 303, 489; Equiano, *Interesting Narrative*, 104.

60. See Higman, *British Caribbean*, 156; Goveia, *Slave Society*, 216; and ch. 4, above.

61. See the evidence of Reverend John Barry on this point in Lords SC, 497–98; also Moreton, *West India Customs*, 123–32; Beckles, "Property Rights," 697; Wilberforce, *Appeal*, 22.

62. See esp. Goveia, *Slave Society*, 218; Mohammed, "'Browning,'" 22–48; Long, *HJ*, 2:332; Moreton, *West India Customs*, 124–25; Luffman, *Brief Account*, 114–16; McNeill, *Observations*, 41–43; Stewart, *View*, 327–28. Privilege bills encouraged women of color to couple with white men. See Livesay, *Children of Uncertain Fortune*, 41.

63. In old sugar colonies, with larger slave populations, whites would have been responsible for fewer colored births than in the new sugar colonies. In Tobago (1819), whites were responsible for 92.3 percent of colored births. See Higman, *British Caribbean*, 156.

64. On class/rank in Jamaican society, see Jones, "Contesting the Boundaries," 195–232; Patterson, *Sociology of Slavery*, 33–51; Curtin, *Two Jamaicas*, 47–51; Burnard, *Mastery*, ch. 1. On the blame of lower whites, see [Stewart], *Account*, ch. 12; Stewart, *View*, ch. 11; Davis, *West Indies*, 68–70; La Beche, *Present Condition*, 36–40; Altink, *Representations*, 81–82. Lady Nugent, although not proslavery, remonstrated with young white men for their behavior on August 10, 1803. See Nugent, *Journal*, 172. On setting a poor example, see M'Mahon,

Jamaica Plantership, 134; Phillippo, *Jamaica*, 148; McNeill, *Observations*, 41; Stewart, *View*, 310; published letter, *New Times*, n.d. [April 1824?], Thomas Fowell Buxton Papers, Bodl., MSS Brit Emp. s. 444, vol. 38 (Newspaper Cuttings, 1821–24), fol. 16. Lower-status white people were also blamed for excessive cruelty. See annotated copy of Long, *HJ*, vol. 2, Edward Long Papers, BL, Add. MS 12405, fol. 217r.

65. For abolitionist attacks on planters, see *HCSP*, 82:139, 182, 184. See also Moreton, *West India Customs*, 77; Wimpffen, *Voyage*, 277–78; Luffman, *Brief Account*, 114–16. See also extract of a letter to Bristol merchant [June 28, s.a.], Lucy Townsend Scrap Book, Bodl., MSS Brit. Emp. s. 4, fol. 72; published letter, *New Times*, n.d. [April 1824?]; Wilberforce, *Appeal*, 16–21, 76; [Brougham], *Concise Statement*, 59–60.

66. See also Petley, "Creole World View," 98–99; Burnard, *Mastery*, ch. 3.

67. See, for example, Burnard, "Tropical Hospitality," 202–23; Petley, "Gluttony," 85–106; Altink, *Representations*, 54; Prince, *History of Mary Prince*, 82n15. On masculinity in this period, see Begiato, *Manliness in Britain*; Harvey, *Little Republic*; McCormack, *Public Men*; the forum in *Journal of British Studies*, 44 (2005): 274–362; Carter, *Men*; Hitchcock and Cohen, *English Masculinities*.

68. Altink, *Representations*, esp. ch. 3.

69. Phillippo, *Jamaica*, 149; Lords SC, 456–91.

70. See La Beche, *Present Condition*, 17–18; [Martin], *Jamaica*, 36, 42–43; Davis, *West Indies*, 80–88; Gladstone, *Facts*, 4–5.

71. Lords SC, 499–500.

72. [Collins], *Practical Rules*, 155.

73. *Considerations on Negro Slavery*, 5; La Beche, *Present Condition*; Beaumont, *Compensation*; Steel, "Philosophy of Fear," 1–20.

74. On the shifting balance of these stereotypes across time, see Altink, *Representations*, 5–6. Barbara Bush has argued that slaves often deployed these stereotypes to their own advantage. See her "'Sable Venus,'" 761–89.

75. Two notable exceptions are Henrice Altink (*Representations*) and Hilary Beckles (*Natural Rebels*). On stereotypes, see these and Bush, "'Sable Venus,'" 761–89; Beckles, "Property Rights," 692–701.

76. Comparatively few slaves of color worked in the fields, typically being domestics if female or tradesmen if male.

77. For an early analysis of "Quashee," see Patterson, *Sociology of Slavery*, 174–81.

78. Patterson, *Sociology of Slavery*, 174–81; Bush, *Slave Women*, 52–63.

79. [Schaw], *Journal*, 113–14; Cooper, *Facts*, 27–28; Stewart, *View*, 171–72; Beckles, "White Women," 66–82.

80. Schaw, *Journal*, 112–13. On Schaw, see Coleman, "Janet Schaw," 169–93, and Bohls, *Women Travel Writers*, 46–65.

81. Buchanan, *Colonial Ecclesiastical Establishment*, 67–68.

82. Hawes, *Poor Relations*, 79–80. These prejudices were also held toward people of color in the Caribbean. See, for example, Troup Journal, fols. 107v–108r.

83. *Anglo-India*, 1:362–414.

84. Similar motives led to the imposition of comparable restrictions in the Dutch East Indies. See Stoler, *Education of Desire*, 43ff.

85. Hawes, *Poor Relations*, ch. 1; Collingham, *Imperial Bodies*, ch. 2; Sramek, *Gender, Morality*, ch. 2; Williamson, *East India Vade-Mecum*, 1:468.

86. Draft memorandum [ca. 1820], Military Collections, 1804–1821, India Office Records, BL, IOR/L/MIL/5/376, fol. 120r.

87. Harrison, "'Tender Frame,'" 68–93; Harrison, *Climates and Constitutions*, esp. 102–10; Arnold, "Bodily Difference," 254–73; Bewell, *Romanticism and Colonial Disease*, 35–40.

88. Orme, *Historical Fragments*, 457–72; Dubois, *Description*, 201.

89. Grose, *Voyage*, 209–10. See also *India Gazette*, May 31, 1790; Raynal, *Two Indies*, 1:95–96.

90. Although it was not always invoked: See Tzoref-Ashkenazi, "German Voices," 199.

91. See, for example, H.C. Papers, Select Committee on State of Affairs of East India Company Report, Minutes of Evidence, 1831–32 (735-I), 298–99, 312–13; *Calcutta Chronicle*, January 17; August 7, 1788; Dubois, *Description*, 202–3.

92. Stanhope, *Genuine Memoirs*, 141; Austen-Leigh, "Philip Dormer Stanhope," 165–67 and replies (213, 376).

93. Munro, *Narrative*, 67; Dubois, *Description*, 190, 196; Orme, *Historical Fragments*, 419; Grose, *Voyage*, 237.

94. Orme, *Historical Fragments*, 431.

95. Munro, *Narrative*, 19–20.

96. Dubois, *Hindu Manners*, 312; Forbes, *Oriental Memoirs*, 1:359.

97. Hardgrave, *Balthazar Solvyns*, 264. On *hijras* in general, see Nanda, *Neither Man nor Woman*, and Reddy, *With Respect to Sex*.

98. For a classic statement associating these regions with despotic government, see Montesquieu, *Spirit of the Laws*, 229–333 (pt. 3).

99. Yule and Burnell, *Hobson-Jobson*, s.v. "zenana"; Stanhope, *Genuine Memoirs*, 25; Williamson, *East India Vade-Mecum*, 1:413.

100. Dubois, *Description*, 191.

101. Williamson, *East India Vade-Mecum*, 1:369–70; Dubois, *Description*, 132–46; Dubois, *Hindu Manners*, 207. Among texts that argued for the existence of Hindu polygamy are [Towers], *Dialogues Concerning the Ladies*, 82–83, and Mill, *History of British India*, 1:298–301. According to the seventeenth-century traveler Jean Baptiste Tavernier, early marriage was also a means of preventing sodomy. See his *Six Voyages*, pt. 2, 181.

102. Orme, *Historical Fragments*, 408; Dow, *History of Hindostan*, xviii.

103. Williamson, *East India Vade-Mecum*, 1:390; Grose, *Voyage*, 220.

104. Dubois, *Hindu Manners*, 312.

105. Other examples feature in Grose, *Voyage*, 307–9; Orme, *Historical Fragments*, 165; Stanhope, *Genuine Memoirs*, 101; Williamson, *East India Vade-Mecum*, i. 451; Dow, *History*, xix; Dubois, *Hindu Manners*, 313–14. An influential contrary voice was Charles Grant's: see H.C. Papers, Observations by C. Grant on State of Society among Asiatic Subjects of Great Britain, 1792, 1812–13 (282) [hereafter Grant, "Observations"], 30.

106. Ghosh, *Social Condition*, 75.

107. Grose, *Voyage*, 222, 231.

108. Williamson, *East India Vade-Mecum*, 1:386. See also Stanhope, *Genuine Memoirs*, 38–39.

109. Grose, *Voyage*, 377; Orme, *Historical Fragments*, 408, 465; Dubois, *Hindu Manners*, 94.

110. Montesquieu said much the same of the Chinese in *Spirit of the Laws*, 126–28 (bk. 8, ch. 21).

111. Dow, *History of Hindostan*, xviii.

112. Orme, *Historical Fragments*, 464.

113. Grant, "Observations," 29. Another positive comment on Indian parents is in Orme, *Historical Fragments*, 431.

114. Grant, "Observations," 30; Dubois, *Description*, 190–91; Grose, *Voyage*, 302–7. See also Forbes, *Oriental Memoirs*, 2:48.

115. Dubois, *Hindu Manners*, 309. See also Dubois, *Description*, 192.

116. Mill, *History of British India*, 1:330.

117. See esp. Dubois, *Hindu Manners*, 500; Williamson, *East India Vade-Mecum*, 1:357; Buchanan, *Colonial Ecclesiastical Establishment*, 126–30.

118. Williamson, *East India Vade-Mecum*, 1:362–68; Grant, "Observations," 28; Mill, *History of British India*, 1:294–302; Dow, *History of Hindostan*, xiv, xviii–xix; Orme, *Historical Fragments*, 459, 465; Grose, *Voyage*, 143; Dubois, *Description*, 132–46; Dubois, *Hindu Manners*, 231.

119. See the work of Andrea Major, esp. her *Pious Flames* and *Sovereignty*; Banerjee, *Burning Women*; Fisch, *Burning Women*.

120. Stanhope, *Genuine Memoirs*, 100.

121. Grose, *Voyage*, 144; Major, *Pious Flames*, 58–63; Teltscher, *India Inscribed*, 51–68; Fisch, *Burning Women*, 341–43.

122. Mill, *History of British India*, 1:275, 293–301. On the treatment of women as an index of civilization, see esp. Tomaselli, "Enlightenment Debate on Women," 101–24; Knott and Taylor, *Women, Gender, and Enlightenment*; Sebastiani, *Scottish Enlightenment*, ch. 5.

123. That is, 250 of 4,000 Europeans (6.25 percent). Williamson, *East India Vade-Mecum*, 1:453.

124. Munro, *Narrative*, 57–58.

125. Williamson, *East India Vade-Mecum*, 1:451–52.

126. Hawes, *Poor Relations*, 1–20; Bickell, *West Indies*, 108; Long, *HJ*, 2:327–28.

127. Sen, "Colonial Aversions," 49–82; Ghosh, *Sex and the Family*, 1–34.

128. Williamson, *East India Vade-Mecum*, 1:457.

129. Love, *Vestiges of Old Madras*, 1:533.

130. Ghosh, *Sex and the Family*, esp. introduction and ch. 1.

131. Chatterjee, "Colouring Subalternity," 92.

132. In Spanish America, too, the label "mestizo" soon became shorthand for "illegitimate." See Schwartz and Salomon, "New Peoples," 483; Katzew, *Casta Painting*, 40; Fuchs, "Mimetic Racism," 11–14.

133. Despite the proven integration of illegitimate mixed-race children from the Caribbean into "Atlantic family" networks, the attempt to argue that they were legitimate was not made. Livesay, *Children of Uncertain Fortune*, 9–14.

134. Mill, *History of British India*, 1:292; Williamson, *East India Vade-Mecum*, 1:356, 386, 452.

135. *Original Papers*, 47n.

136. See Dubois, *Hindu Manners*, 205–35.

137. Munro, *Narrative*, 49.

138. Tucker, *Tucker's Note Book*, 63; *Constructions of the Regulations*, 292 (no. 806, July 26, 1833).

139. Groves, *66th Berkshire Regiment*, ch. 2.

140. Torrens to Secretary, Board of Control, August 21, 1820, Military Collections, 1804–1821, India Office Records, BL, IOR/L/MIL/5/376, fol. 233v (emphasis in original).

141. Nechtman, *Nabobs*; Lawson and Phillips, "'Execrable Banditti,'" 225–41; Spear, *Nabobs*; Holzman, *Nabobs*.

142. Holzman, *Nabobs*, 97–98.

143. [Price], *Right Horse*, 61–62; Foote, *Nabob*.

144. Foote, *Nabob*, 38.

145. *Town and Country Magazine* 8 (1776): 289–91, 345–47; [Clarke], *Nabob*, 3.

146. Nechtman, *Nabobs*; Lawson and Phillips, "'Execrable Banditti,'" 225–41.

147. Price, *Saddle*, 67–68; Griffith, *Wife in the Right*, 68.

148. Major, *Slavery*, 86–87; cf. A. P. Coleman, "Ochterlony, Sir David, first baronet (1758–1825)," *Oxford Dictionary of National Biography*, updated October 4, 2008, https://doi.org/10.1093/ref:odnb/20492; Hickey, *Memoirs*, 4:140–41.

149. Hickey, *Memoirs*, 3:276.

150. Scholarship on this is more advanced for the nineteenth century. See Chilton, *Agents of Empire*; Courcy, *Fishing Fleet*; MacMillan, *Women of the Raj*. For the eighteenth century, see Marshall, "British Society," 107; Spear, *Nabobs*, 13; Ghosh, *Social Condition*, 72–91, esp. 62–65; Hickey, *Memoirs*, 2:101; Thackeray, *Vanity Fair*, 704.

151. Further examples of enslaved women represented thus in visual caricature are James Gillray, *Philanthropic Consolations* (1796, BM 8793), Isaac Cruikshank and George Woodward, *A Morning Surprise* (1807, BM 10885), and George Cruikshank, *The New Union Club* (1819, BM 13249). On these, see Stephens and George, *Catalogue*, 7:245, 8:120, 9:910–12. On the representation of black women in eighteenth-century English caricature, see Molineux, *Faces of Perfect Ebony*, ch. 6; Odumosu, *Africans in English Caricature*, 110–21; Newman, *Dark Inheritance*, ch. 5.

152. 25 Parl. Hist. Eng. (1786) 1398.

153. See *HCSP*, 59:428.

154. Stephens and George, *Catalogue*, 6:336–37; Wilson, *Island Race*, 98–101; Gattrell, *City of Laughter*, 333–34.

155. Foote, *Nabob*, 9.

156. *Parker's Morning Advertiser and General Intelligencer*, August 19, 1784 (emphasis in original).

157. Griffith, *Wife in the Right*, 90.

158. Saint Vincent, acquired in 1763, is one notable exception.

159. Ward, *West Indian Slavery*, ch. 6. See also Craton, *Testing the Chains*, esp. pt. 4.

160. Slave-trading was (unsuccessfully) forbidden in 1774; slavery itself was delegalized in 1843 but only made a criminal offence in 1862. See Major, *Slavery*, 7–11, 52–53.

161. Major, *Slavery*, 3–22; Chatterjee, "Colouring Subalternity," 49–97; Finn, "Domestic Slavery," 181–203; Everill, *Not Made by Slaves*, 49–54, 85–88. For a contemporary example, see H.C. Papers, Select Committee, 1831–32 (735-I), 303–4.

162. Balfour, *Cyclopaedia of India*, 3:674.

163. See Visaria and Visaria, "Population (1757–1947)," 464–66; Higman, *British Caribbean*, 77.

164. Bayly, *Caste*, 104.

165. Arnold, "Bodily Difference," 264; Grose, *Voyage*, 142, 236; Dow, *History of Hindostan*, cxx; Scrafton, *Reflections*, 13, 21; Dubois, *Description*, 198–99.

166. See Bayly, *Caste*, esp. ch. 3; Robb, *Concept of Race*, esp. 165–259. For examples, see Dubois, *Description*, 198; Raynal, *Two Indies*, 1:75; Forbes, *Oriental Memoirs*, 1:40, 52–53. For dissenting voices, see Heber, *Narrative*, 1:23; Ward, *View*, 3:185.

167. Elliott, *Empires*, 286–87; Katzew, *Casta Painting*; Martínez, *Genealogical Fictions*; Bethencourt, *Racisms*.

168. With the possible exception of Cuba after the 1760s.

169. Bethencourt, *Racisms*, 174; Dantas, "Humble Slaves," 115–40.

170. *Original Papers*, 30.

171. Proposed Regulations for the Bombay Military Fund, enclosed in committee for feasibility study of a military fund for the Bombay Presidency to Sir Evan Nepean, March 15, 1816, Board's Collections, 1817–18, India Office Records, BL, IOR/F/4/538/12948/3.

172. On racism and rationality, see esp. Goldberg, *Racist Culture*, ch. 6.

173. A point powerfully made by Stoler, "Racial Histories," 186–87.

174. Broca, "phénomènes d'hybridité," 3:397; Broca, *Phenomena of Hybridity*, 29.

175. Risley mentions the terms "mulatto," "quadroon," and "mestizo," but only in relation to America. Risley, *People of India*, 179.

176. Collingham, *Imperial Bodies*.

177. Higman, *British Caribbean*, 155. Examples of the use of "Yellow" in colonial literature include Waller, *Voyage*, 106; Wimpffen, *Voyage*, 49; Nugent, *Journal*, 12. "Yellow" was also used in metropolitan discourses. See, for example, cutting of published letter from James McQueen to T. F. Buxton, *Glasgow Courier*, October 20, 1825, Thomas Fowell Buxton Papers, Bodl., MSS Brit. Emp. s. 444, vol. 39 (Newspaper Cuttings, 1822–27), fol. 34.

178. Slave Registers, Grenada, 1817–34, NA, T71/265–66, 269, 272, 274, 276, 279, 284–85, 288, 290, 296, 299, 301, 303, 305, 311, 313, 315, 317, 320, 326, 329.

179. Katzew, *Casta Paintings*, 44. These terms mean "hold yourself suspended in mid-air," "wolf," and "coyote," respectively.

180. Dayan, *Haiti*, 118; Garrigus, *Before Haiti*, 44; Garraway, "Family Romance," 227–46; Saint-Méry, *Description*, 1:68–96, esp. 1:86; Proisy, *État des Finances*, table X.

181. Not merely in India: "Portuguese" had come to signify "Jew" in northern Europe by the seventeenth century. See Bethencourt, *Racisms*, 138.

182. For an example of this circularity, see H.C. Papers, Select Committee, 1831–32 (735-I), 297. See also Munro, *Narrative*, 49; Forbes, *Oriental Memoirs*, 85; Spear, *Nabobs*, 13, 46–47, 61–62; Hawes, *Poor Relations*, 59, 73–92.

183. On the need for genealogical knowledge to distinguish between slave and free, see Kriz, "Marketing Mulatresses," 206–7.

184. Higman, *British Caribbean*, 155.

185. Williams, *Capitalism and Slavery*, 7.

Chapter 7 • Conclusion

1. Court of Quarter Sessions, Docket Book, 1733–1737, Bristol Archives, JQS/D/8, March 21, 1736/7.

2. *Samuel Farley's Bristol Newspaper*, February 12, 1736/7.

3. *James Any v Edwards* (Grand Court of Jamaica, 1787). Grant, *Cases Adjudged*, 329.

4. *Barry et al. v Fenn* (Grand Court of Jamaica, 1785). Grant, *Cases Adjudged*, 226. On these cases, see also Catterall, *Judicial Cases*, 5:357, 359.

5. Gross, *What Blood Won't Tell*.

6. Disraeli, *Tancred*, 1:303.

BIBLIOGRAPHY

1. Manuscripts

UNITED KINGDOM

Bristol, Bristol Archives
  Court of Quarter Sessions Records

Bristol, City Library
  Bristol Broadsides Collection

London
  British Library
    Additional Manuscripts
      Joseph Banks Correspondence
      Edward Long Papers
    India Office Records
      Board of Control Collections. 1817–18.
      Military Department Compilations and Miscellaneous Records. 1817–1821.

  National Archives of the United Kingdom
    Jamaica, Acts. CO139.
    Jamaica, Original Correspondence, Secretary of State. CO137.
    Slave Registration Returns. T71.

Oxford, Bodleian Library
  John Aubrey Papers
  Thomas Fowell Buxton Papers
  Rawlinson Manuscripts
  Lucy Townsend Papers

Taunton, Somerset Archives and Local Studies
  Tudway of Wells Manuscripts

Edinburgh
  National Library of Scotland
    Dreghorn Collection of Court of Sessions Papers, Advocates Library
    Monboddo Papers
    Single Letters and Small Collections. 1700–1797.

National Records of Scotland
    Home-Roberston Family Papers
Royal College of Physicians of Edinburgh
    "Cases from the Edinburgh Lying-In Hospital, 1793–4."
    Cullen Papers
    Alexander and James Hamilton Papers
    James Hamilton Papers
    Thomas Young Papers
University of Edinburgh, Special Collections
    Andrew Duncan, Jr. Papers and Letters.
    John Goodsir. "Notes of Midwifery from Dr Hamilton." 1797.
    John Gregory. "Lectures on the Institutes of Medicine." 1773.
    Royal Medical Society Dissertations. 1751–1968.
    Society for Investigating Natural History, Dissertations. 1782–1806.
    William Thomson. "Lectures on Surgery by John Hunter." ca. 1778.
    Thomas Young. "Lectures on Midwifery, vol. 1." ca. 1760–1780.

Glasgow, University of Strathclyde Archives and Special Collections
    John Anderson. "Discourses of Natural and Artificial Systems in Natural History
        and [. . .] the Varieties in the Human Kind." 1774.

Aberdeen, University Library
    David Skene Papers
    Journal of Jonathan Troup. 1788–1790.

FRANCE
    Paris, Bibliothéque Nationale de France
        MS Supplement to Charles de Brosses, *Histoire des navigations aux terres australes.*
        2 vols. Paris, 1756. Shelfmark: Rés. G.1.443 (Suppl. 1–18).

GERMANY
    Göttingen, Niedersächsische Staats- und Universitätsbibliothek
    Johann Friedrich Blumenbach Papers

UNITED STATES
    New Haven, Yale University, Beinecke Library
    Thomas Thistlewood Papers

    New York, New-York Historical Society
    Alexander Papers

    San Marino, CA, Huntington Library
    Clarkson Papers
    Macaulay Papers

AUSTRALIA
    Canberra, National Library of Australia
    Sir Joseph Banks Papers

2. Printed Primary Sources

An Account of the Bristol Infirmary. Bristol, [1738].

An Account of the Hospital for the Maintenance and Education of Exposed and Deserted Young Children. London, 1749.

Account of the Lying-In Charity. London, 1791.

An Account of the Public Hospital for the Diseased Poor in the County of York. York, 1743.

An Account of the Rise and Establishment of the Infirmary [. . .] erected at Edinburgh. Edinburgh, [1730].

Acts of Assembly. Passed in the Island of Jamaica; From 1770, to 1783, inclusive. Kingston, 1786.

Adair, James M. *Unanswerable Arguments Against the Abolition of the Slave Trade.* London, 1790.

Address on the State of Slavery in the West India Islands. Leicester, 1830.

[Alexander, William]. *Address to the Public, On the Present State of the Question Relative to Negro Slavery, in the British Colonies.* York, 1828.

Alexander, William. *The History of Women.* 2 vols. London, 1779.

Anglo-India, Social, Moral, and Political. 3 vols. London, 1838.

Anno vicesimo primo Jacobi Regis, &c. An Act to prevent the Destroying and Murthering of Bastard Children. London, 1680.

Arbuthnot, John. "An Argument for Divine Providence." *PTRS* 27 (1710–12): 186–90.

Arbuthnot, John. *An Essay Concerning the Effects of Air on Human Bodies.* London, 1733.

Aristotle, *Generation of Animals.* Translated by A. L. Peck. Harvard University Press, 1943.

Aristotle's Master-Piece; or, the Secrets of Generation. London, 1690.

Astruc, John. *Treatise on all the Diseases Incident to Women.* London, 1743.

Athenian Oracle. 2nd ed. 3 vols. London, 1704.

Atkins, John. *The Navy Surgeon: or, A Practical System of Surgery.* London, 1734.

Atkins, John. *The Navy Surgeon: or, A Practical System of Surgery.* London, 1742.

Atkins, John. *A Treatise on [. . .] Chirurgical Subjects.* London, [1724].

Atkins, John. *A Voyage to Guinea, Brasil, and the West-Indies.* London, 1735.

Atterbury, Lewis. *Sermons on Select Subjects.* 2 vols. London, 1743.

Aubrey, John. *Brief Lives.* Edited by Kate Bennett. 2 vols. Oxford University Press, 2016.

Bacon, Francis. *The Major Works.* Edited by Brian Vickers. Oxford University Press, 1996.

Bacon, Thomas. *Four Sermons.* London, 1750.

Balfour, Edward. *Cyclopaedia of India.* 3rd ed. 3 vols. London, 1885.

[Bancroft, Edward]. *An Essay on the Natural History of Guiana.* London, 1769.

Barre, François Poulain de la. "Of the Equality of the Two Sexes [1673]." In *Three Cartesian Feminist Treatises*, edited by Marcelle Maistre Welch. University of Chicago Press, 2002.

Barrère, Pierre. *Dissertation sur la cause physique de la couleur des negres.* Paris, 1741.

Beaglehole, J. C. *The Journals of Captain James Cook on His Voyages of Discovery.* 4 vols. Cambridge University Press, 1955–74.

Beaumont, Augustus Hardin. *Compensation to Slave Owners Fairly Considered.* London, 1826.

Becket, Thomas [pseud.]. *A Journal of a Voyage Round the World.* London, 1771.

Belgrove, William. *A Treatise upon Husbandry or Planting.* Boston, 1755.

Bell, William. *A Dissertation on [. . .] What Causes Principally Contribute to Render a Nation Populous?* Cambridge, 1756.

Bellamy, Daniel. *The Family Preacher.* 2 vols. London, 1776.

Bendyshe, Thomas, ed. *The Anthropological Treatises of Johann Friedrich Blumenbach.* London, 1865.

Bentham, Jeremy. "Sextus." In *Of Sexual Irregularities, and Other Writings on Sexual Morality*, edited by Philip Schofield, Catherine Pease-Watkin, and Michael Quinn. Oxford University Press, 2014.

[Berkeley, George]. *An Essay Towards Preventing the Ruine of Great Britain.* London, 1721.

Bernard, Jean-Frédéric, and Bernard Picart. *The Ceremonies and Religious Customs of the Various Nations of the Known World.* 7 vols. London, 1733–39.

Bernier, François. "Nouvelle division de la terre." *Journal des Sçavans* (1684): 133–40.

[Bianchini, Giovanni Fortunato]. *Essay on the Force of Imagination in Pregnant Women.* London, 1772.

Bickell, Reverend R. *The West Indies as They Are.* London, 1825.

Bigland, John. *An Historical Display of the Effects of Physical and Moral Causes on the Character and Circumstances of Nations.* London, 1816.

Blackstone, William. *Commentaries on the Laws of England.* 4 vols. Oxford, 1765–69.

Bland, Robert. *Observations on Human and on Comparative Parturition.* London, 1794.

Blatchford, Thomas W. *Observations on Equivocal Generation.* Albany, 1844.

Blome, Richard. *A Description of the Island of Jamaica.* London, 1678.

Blondel, James. *The Power of the Mother's Imagination over the Foetus Examin'd.* London, 1729.

Blumenbach, Johann Friedrich. *Beyträge zur Naturgeschichte.* Göttingen, 1790.

Blumenbach, Johann Friedrich. "Comparison between the Human Race and that of Swine." *Philosophical Magazine* 3 (1799): 284–90.

Blumenbach, Johann Friedrich. *An Essay on Generation.* London, 1792.

Blumenbach, Johann Friedrich. *De Generis Humani Varietate Nativa.* 2nd ed. Göttingen, 1781.

Blumenbach, Johann Friedrich. *De Generis Humani Varietate Nativa.* 3rd ed. Göttingen, 1795.

Blumenbach, Johann Friedrich. *Geschichte und Beschreibung der Knochen des menschlichen Körpers.* Göttingen, 1786.

Blumenbach, Johann Friedrich. *Handbuch der Naturgeschichte.* 5th ed. Göttingen, 1797.

Blumenbach, Johann Friedrich. "On the Natural Variety of Mankind, ed. 1775." In Bendyshe, *Anthropological Treatises,* 65–143.

Blumenbach, Johann Friedrich. "On the Natural Variety of Mankind, ed. 1795." In Bendyshe, *Anthropological Treatises,* 145–276.

Blumenbach, Johann Friedrich. Review of *Über die natürlichen Verschiedenheiten im Menschengeschlechte,* by Johann Friedrich Blumenbach, trans. Johann Gottfried Gruber. *Göttingische Anzeigen von gelehrten Sachen* 2 (1798): 1889–90.

Blumenbach, Johann Friedrich. "Specimen physiologiae comparatae inter animantia calidi sanguinis vivipara et ovipara." *Commentationes Societatis Regiae Scientiarum Gottingensis* 9 (1789): 108–28.

Blumenbach, Johann Friedrich. *Über die natürlichen Verschiedenheiten im Menschengeschlechte.* Translated by Johann Gottfried Gruber. Leipzig, 1798.

Blumenbach, Johann Friedrich. "Über Menschen-Racen und Schweine-Racen." *Magazin für das Neueste aus der Physik und Naturgeschichte* 6, no. 1 (1789): 1–14.

Bodin, Jean. *The Six Bookes of a Common-Weale.* Translated by Richard Knolles. London, 1606.

Bond, Donald F., ed. *The Spectator*. 5 vols. Oxford University Press, 1987.

The Book of Common-Prayer And Administration Of the Sacraments. London, 1662.

Boswell, James. *Boswell: The Ominous Years, 1774–1776*. Edited by Charles Ryskamp and Frederick A. Pottle. McGraw-Hill, 1963.

Boswell, James. *Journal of a Tour to the Hebrides*. London, 1785.

Boswell, James. *Life of Johnson*. Edited by George Birkbeck Hill and L. F. Powell. 2nd ed. 6 vols. Oxford University Press, 1964.

Bougainville, Louis-Antoine de. *The Pacific Journal of Louis-Antoine de Bougainville, 1767–1768*. Translated and edited by John Dunmore. Hakluyt Society, 2002.

Bougainville, Louis-Antoine de. *Voyage autour du monde*. Paris, 1771.

Bougainville, Louis-Antoine de. *Voyage Round the World*. Translated by J. R. Forster. London, 1772.

Boyle, Robert. *Experiments and Considerations Touching Colours*. London, 1664.

[Bray, Thomas]. *Memorial Concerning The Erecting [. . . of] an Orphanotrophy*. London, 1728.

[Brett, John]. *Conjugal Love and Duty*. Dublin, 1757.

Bridges, Reverend George Wilson. *The Annals of Jamaica*. 2 vols. London, 1828.

The British Apollo, Containing Two Thousand Answers to Curious Questions in Most Arts and Sciences. 3rd ed. 3 vols. London, 1726.

Broca, Paul. "Des phénomènes d'hybridité dans le genre humain." *Journal de la physiologie de l'homme et des animaux* 2 (1859): 601–25; 3 (1860): 392–439.

Broca, Paul. *On the Phenomena of Hybridity in the Genus Homo*. Edited by C. Carter Blake. London, 1864.

[Brougham, Henry]. *A Concise Statement of the Question Concerning the Abolition of the Slave Trade*. London, 1804.

Brown, John. *Estimate of the Manners and Principles of the Times*. 2nd ed. London, 1757.

Brown, John. *Self-Interpreting Bible*. London, 1792.

Browne, Thomas. *Pseudodoxia Epidemica*. Edited by Robin Robbins. 2 vols. Oxford University Press, 1981.

Buchanan, Claudius. *Colonial Ecclesiastical Establishment*. 2nd ed. London, 1813.

Buffon, Georges-Louis Leclerc, Comte de. *Histoire naturelle, générale et particulière*. Œuvres complètes, edited by Stéphane Schmitt. 18 vols. Honoré Champion, 2007–.

Buffon, Georges-Louis Leclerc, Comte de. *Natural History, General and Particular*. Translated by William Smellie. 2nd ed. 9 vols. London, 1785.

Bulwer, John. *Anthropometamorphosis: Man Transform'd: or, the Artificiall Changling*. London, 1653.

Burdach, Karl Friedrich. *Die Physiologie als Erfahrungswissenschaft*. 6 vols. Leipzig, 1826–40.

Burke, Edmund. "A Philosophical Enquiry into the Origin of Our Ideas of the Sublime and Beautiful." In *Writings and Speeches*, 1:185–320.

Burke, Edmund. "Sketch of a Negro Code." In *Writings and Speeches*, 3:562–81.

Burke, Edmund. *Writings and Speeches of Edmund Burke*. Edited by Paul Langford. 9 vols. Oxford University Press, 1981–2015.

Burnley, William. *Opinions on Slavery and Emancipation*. London, 1833.

Caesar, James. *The Great Happiness of a Faithful Princess in Child-Bearing*. London, 1716.

Camper, Peter. "Account of the Organs of Speech of the *Orang Outang*." *PTRS* 69 (1779): 139–59.

[Camper, Petrus]. *The Works of the Late Professor Camper.* Translated by T. Cogan. London, 1794.

Cardano, Girolamo. *Contradicentia medicorum.* 1548.

Cardano, Girolamo. *Opera omnia.* 10 vols. Lyon, 1663.

[Carlile, Richard]. *Every Woman's Book: or, What Is Love?* 4th ed. London, 1826.

Catterall, Helen Tunnicliff, ed. *Judicial Cases Concerning American Slavery and the Negro.* 5 vols. Carnegie Institution, 1926–37.

Child, Josiah. *A Discourse About Trade.* London, 1690.

Cicero, *On Duties* [*De Officiis*]. Edited by M. T. Griffin and E. M. Atkins. Cambridge University Press, 1991.

[Clarke, Richard]. *The Nabob: or, Asiatic Plunderers.* London, 1773.

Cleland, John. *Memoirs of a Woman of Pleasure.* Edited by Peter Sabor. Oxford University Press, 1999.

Co-Adamitae. London, 1732.

Cockburn, John. *The Dignity and Duty of a Married State.* London, 1708.

Coke, Edward. *The First Part of the Institutes of the Laws of England.* 2nd ed. London, 1629.

Coke, Roger. *A Treatise Wherein is demonstrated, That the Church and State of England are in Equal Danger with the Trade Of it.* London, 1671.

[Collins, David]. *Practical Rules for the Management and Medical Treatment of Negro Slaves, in the Sugar Colonies.* London, 1803.

Colombo, Realdo. *De Re Anatomica.* Venetiis, 1559.

Commerson, Philibert de. "Lettre de M. Commerson." *Mercure de France* (November 1769): 197–207.

A Concise View of Colonial Slavery. London, 1830.

Condillac, Etienne Bonnot de. *Essay on the Origin of Human Knowledge.* Edited by Hans Aarsleff. Cambridge University Press, 2001.

The Conditions for New-Planters In the Territories of His Royal Highness the Duke of York. Boston, 1665.

Considerations on Negro Slavery. Edinburgh, 1824.

Considerations on the Causes of the Present Stagnation of Matrimony. London, 1772.

Some Considerations upon Street-Walkers. London, 1726.

The Consolidated Slave Law. 2nd ed. [Kingston], 1827.

Constructions of the Regulations and Acts Issued by the Court of Sudder Dewanny Adawlut. Calcutta, 1855.

Cook, James. *A Voyage Towards the South Pole, and Round the World.* 2 vols. London, 1777.

Cook, James, and James King. *A Voyage to the Pacific Ocean.* 3 vols. London, 1784.

[Cooke, John]. *The New Theory of Generation.* London, 1762.

Cooke, John. *The Preacher's Assistant.* 2 vols. Oxford, [1783].

Cooper, Thomas. *Facts Illustrative of the Condition of Negro Slaves in Jamaica.* London, 1824.

Cornwallis, Henry. "The Bridal Bush [1709]." In *Conjugal Duty,* 22–32. London, 1732.

Country Gentleman. *Reflections upon Naturalization, Corporations, and Companies.* London, 1753.

Creffield, Edward. *A Good Wife a Great Blessing.* London, 1717.

Culpeper, Nicholas. *A Directory for Midwives.* London, 1651.

Cuvier, Georges-Léopold-Chrétien-Frédéric-Dagobert. "Cuvier's Animal Kingdom [. . .] (1840)." In *Theories of Race*, edited by Peter J. Kitson. Vol. 8 of *Slavery, Abolition and Emancipation: Writings in the British Romantic Period*. Pickering & Chatto, 1999.

Cuvier, Georges-Léopold-Chrétien-Frédéric-Dagobert. *Le Règne Animal*. 4 vols. Paris, 1817.

Dancer, Thomas. *The Medical Assistant*. Kingston, 1801.

Darwin, Charles. *Variation of Animals and Plants Under Domestication*. 2 vols. London, 1868.

Darwin, Erasmus. *Zoonomia: or, The Laws of Organic Life*. London, 1794.

Davenant, Charles. *Discourses on the Publick Revenues, and on the Trade of England*. London, 1698.

Davis, Anthony. *The West Indies*. London, 1832.

[Defoe, Daniel]. *Augusta Triumphans*. London, 1728.

[Defoe, Daniel]. *Conjugal Lewdness: or, Matrimonial Whoredom*. London, 1727.

Defoe, Daniel. *Moll Flanders*. Edited by G. A. Starr and Linda Bree. Oxford University Press, 2011.

Defoe, Daniel. *Tour Through the Whole Island of Great Britain*. Edited by Pat Rogers. Penguin, 1971.

[Delany, Patrick]. *Reflections upon Polygamy*. 2nd ed. Dublin, 1739.

Delany, Patrick. *Twenty Sermons on Social Duties*. 2nd ed. London, 1747.

Denman, Thomas. *Introduction to the Practice of Midwifery*. 2 vols. London, 1794.

Descartes, René. "Description du corps humain." In *Œuvres de Descartes*, edited by Charles Adam and Paul Tannery, 11:218–90. 12 vols. Léopold Cerf, 1897–1910.

Descartes, René. *The World and Other Writings*. Edited by Stephen Gaukroger. Cambridge University Press, 2004.

A Description of New Athens in Terra Australis Incognita. London, 1720.

A Dialogue Between a Gentleman and a Lady, Relating Chiefly to The Nursing and Bringing up of Children. London, 1698.

Diderot, Denis. *Œuvres philosophiques*. Edited by P. Vernière. Garnier Frères, 1964.

Digby, Kenelm. *Two Treatises*. Paris, 1644.

Dionis, [Pierre]. *The Anatomy of Human Bodies Improv'd*. London, 1716.

Disraeli, Benjamin. *Tancred; or, the New Crusade*. 3 vols. London, 1847.

"Diverses observations anatomiques." *Histoire de l'Académie Royale des Sciences* (1744): 11–15.

Dodd, William. *Practice of Inoculation*. London, 1767.

Dodd, William. *A Sermon Preach'd in the Parish Church of St. Andrew, Holborn, On Friday, April 26, 1754*. London, 1754.

Doddridge, Philip. *The Family Expositor*. 6 vols. London, 1739–56.

Dougherty, Frank William Peter, ed. *The Correspondence of Johann Friedrich Blumenbach*. Revised and edited by Norbert Klatt. 6 vols. Norbert Klatt Verlag, 2006–15.

Dow, Alexander. *The History of Hindostan*. London, 1772.

Drake, James. *Anthropologia Nova*. 2 vols. London, 1707.

Dubois, Abbé J. A. *Description of the Character, Manners, and Customs of the People of India*. London, 1817.

Dubois, Abbé J. A. *Hindu Manners, Customs and Ceremonies*. Translated and edited by Henry K. Beauchamp. 3rd ed. Oxford University Press, 1906.

Duval, Jacques. *Traité des hermaphrodits*. Paris, 1880 [Rouen, 1612].

Duyker, Edward, and Per Tingbrand, eds. *Daniel Solander: Collected Correspondence, 1735–1782*. Miegunyah, 1995.

Edgeworth, Maria. *Belinda*. Edited by Kathryn Kirkpatrick. Oxford University Press, 1994.

Edwards, Bryan. *The History, Civil and Commercial, of the British Colonies in the West Indies*. 2 vols. London, 1793.

Edwards, George. "Account of a Bird [. . .] Bred between a Turkey and Pheasant." *PTRS* 51 (1760): 833–37.

Edwards, John. *Theologia Reformata*. 2 vols. London, 1713.

Elford, William. *Short Essay on the Propagation and Dispersion of Animals and Vegetables*. London, 1786.

Eller, Johann Theodor. "Recherches sur la force de l'imagination des femmes enceintes sur le foetus, à l'occasion d'un chien monstrueux." *Histoire de l'Académie Royale des Sciences et des Belles Lettres de Berlin* (1756): 3–19.

Ellis, William. *Polynesian Researches*. 2nd ed. 4 vols. London, 1831.

Encyclopædia Britannica. 7th ed. 22 vols. Edinburgh, 1830–42.

Equiano, Olaudah. *The Interesting Narrative and Other Writings*. Edited by Vincent Carretta. Penguin, 2003.

Erasmus, *Omnium operum divi Eusebii Hieronymi Stridonensis*. 5 vols. Basel, 1516.

Essay on Conjugal Infidelity. London, 1727.

An Essay, or, Modest Proposal, Of a Way To encrease the Number of People. London, [1693?].

[Estwick, Samuel]. *Considerations on the Negroe Cause*. London, 1772.

Estwick, Samuel. *Considerations on the Negroe Cause*. 2nd ed. London, 1773.

Exposure of the Arguments, Why Murder and Blasphemy, as Essential to West Indian Prosperity, Ought Not to Be Forbidden. Aberdeen, 1825.

Faselius, Johann Friedrich. *Elementa Medicinae Forensis*. Jena, 1767.

Faselius, Johann Friedrich. *Elements of Medical Jurisprudence*. Translated by Samuel Farr. London, 1788.

Fielding, Henry. *Tom Jones*. Edited by John Bender and Simon Stern. Oxford University Press, 1996.

Fienus, Thomas. *De viribus imaginationis*. Louvain, 1608.

Filmer, Robert. *Patriarcha: or the Natural Power of Kings*. London, 1680.

The First Report of the Female Society for Birmingham, West-Bromwich, Wednesbury, Walsall, and Their Respective Neighbourhoods, for the Relief of British Negro Slaves. Birmingham, 1826.

Fisher, Joseph. *The Honour of Marriage*. London, 1695.

[Fitzgerald, Gerald]. *The Injured Islanders*. London, 1779.

Foote, Samuel. *The Nabob*. London, 1778.

Forbes, James. *Oriental Memoirs*. 2nd ed. 2 vols. London, 1834.

Fordyce, James. *Sermons to Young Women*. 4th ed. 2 vols. London, 1767.

[Formey, Jean Henri Samuel]. *Souvenirs d'un citoyen*. 2 vols. Berlin, 1789.

Forster, Georg. "Noch etwas über die Menschenraßen." *Der Teutsche Merkur* 56 (1786): 57–86, 150–66.

Forster, Georg. *A Voyage Round the World*. 2 vols. London, 1777.

Forster, Johann Reinhold. *Observations Made During a Voyage Round the World*. Edited by Nicholas Thomas, Harriet Guest, and Michael Dettelbach. University of Hawaii Press, 1996.

Forster, Johann Reinhold. *The Resolution Journal of Johann Reinhold Forster*. Edited by Michael Hoare. 4 vols. Hakluyt Society, 1982.

Fortrey, Samuel. *Englands Interest and Improvement*. Cambridge, 1663.

Foster, James. *Sermons on the Following Subjects*. 4 vols. London, 1732–44.

Franklin, Benjamin. *Autobiography and Other Writings*. Edited by Ormond Seavey. Oxford University Press, 1993.

Frisch, Johann Leonhard. *Beschreibung von allerley Insecten in Teutsch-Land*. 13 vols. Berlin, 1720–38.

Garrick, David. *The Letters of David Garrick*. Edited by David M. Little and George M. Karhl. 3 vols. Oxford University Press, 1963.

Gee, Joshua. *The Trade and Navigation of Great-Britain Considered*. London, 1729.

Gell, Robert. *Gell's Remaines, Or, Several Select Scriptures of the New Testament Opened and Explained*. 2 vols. London, 1676.

General State of the Marine Society. London, 1798.

[Gibbes, Philip]. *Instructions for the Treatment of Negroes*. London, 1797.

Gibbon, Edward. *The History of the Decline and Fall of the Roman Empire*. 6 vols. London, 1776–89.

Gill, John. *An Exposition Of the Book of Solomon's Song, Commonly Called Canticles*. London, 1776.

Godwin, William. *Of Population*. London, 1820.

Godwyn, Morgan. *The Negro's & Indians Advocate*. London, 1680.

Goethe, Johann Wolfgang von. *Elective Affinities*. Translated and edited by David Constantine. Oxford University Press, 1999.

Goldsmith, Oliver. *History of the Earth and Animated Nature*. 3rd ed. 8 vols. London, 1790.

Goldsmith, Oliver. *The Vicar of Wakefield*. Edited by Arthur Friedman and Robert L. Mack. Oxford University Press, 2006.

[Gott, Samuel]. *The Divine History of the Genesis of the World, Explicated & Illustrated*. London, 1670.

Graaf, Regnier de. *De mulierum organis generationi inservientibus*. Leiden, 1672.

Graham, James. *A Lecture on the Generation, Increase and Improvement of the Human Species*. London, [1780].

Grainger, James. *An Essay on the More Common West-India Diseases*. London, 1764.

Grant, John. *Notes of Cases Adjudged in Jamaica*. Edinburgh, 1794.

Grantham, Thomas. *A Marriage Sermon: Or, A Wife and no Wife*. London, 1681.

Graunt, John. *Natural and Political Observations*. London, 1662.

Gray, S. *The Happiness of States*. London, 1815.

Grégoire, Henri. *An Enquiry Concerning the Intellectual and Moral Faculties, and Literature of Negroes*. Translated by D. B. Warden. Brooklyn, NY, 1810.

Griffith, Elizabeth. *A Wife in the Right*. London, 1772.

Grose, John-Henry. *A Voyage to the East-Indies*. London, 1757.

The Guardian. Edited by John Calhoun Stephens. University Press of Kentucky, 1982.

Haighton, John. "An Experimental Inquiry concerning Animal Impregnation." *PTRS* 87 (1797): 159–96.

Hale, Matthew. *Discourse Touching Provision for the Poor*. London, 1683.

Hale, Matthew. *The Primitive Origination of Mankind*. London, 1677.

Haller, Albrecht von. "Vorrede." In *Allgemeine Historie der Natur*, by [Georges-Louis Leclerc, Comte de Buffon]. Pt. 2, vol. 1, sig. a2r-cv. Leipzig, 1752.

Handley, James. *Mechanical Essays on the Animal Oeconomy*. London, 1721.

Hanway, Jonas. *Candid Historical Account of the Hospital*. 2nd ed. London, 1760.

Hanway, Jonas. *Letter V to Robert Dingley*. London, 1758.

Hanway, Jonas. *Letters on the Importance of the Rising Generation*. 2 vols. London, 1767.

Hanway, Jonas. *Thoughts on the Duty of a Good Citizen*. London, 1756.

Harvey, William. *Anatomical Exercitations, Concerning the Generation of Living Creatures*. London, 1653.

Harvey, William. *Disputations Touching the Generation of Animals*. Edited by Gweneth Whitteridge. Oxford University Press, 1981.

Hawkesworth, John. *An Account of the Voyages Undertaken by the Order of His Present Majesty for Making Discoveries in the Southern Hemisphere*. 3 vols. London, 1773.

Heber, Reginald. *Narrative of a Journey Through the Upper Provinces of India*. 2 vols. London, 1844.

Henderson, Captain [George]. *An Account of the British Settlement of Honduras*. London, 1811.

Henderson, Captain [George]. *A Brief View of the Actual Condition and Treatment of the Negro Slaves, in the British Colonies*. London, 1816.

[Heyrick, Elizabeth]. *Immediate, Not Gradual Abolition*. 3rd ed. London, 1824.

Hickey, William. *Memoirs*. Edited by Alfred Spencer. 4 vols. Hurst & Blackett, 1913–25.

Highmore, Nathaniel. *The History of Generation*. London, 1651.

Hill, John [Ricgard Roe, pseud.]. *Concubitus sine Lucina*. London, 1750.

Hill, John [Richard Roe, pseud.]. *A Letter to Dr Abraham Johnson*. London, 1750.

Hill, John [Abraham Johnson, pseud.]. *Lucina sine concubitu*. London, 1750.

Hippocrates. *Hippocratic Writings*. Edited by G. E. R. Lloyd. Penguin, 1983.

Hoare, Prince. *Memoirs of Granville Sharp, Esq*. London, 1820.

Hobbes, Thomas. *De Cive: The English Version*. Edited by Howard Warrender. Oxford University Press, 1983.

Hobbes, Thomas. *Leviathan*. London, 1651.

Hogarth, William. *The Analysis of Beauty*. London, 1753.

Hole, Matthew. *Practical Discourses on the Offices of Baptism, Confirmation, and Matrimony*. 6 vols. Oxford, 1714–19.

Home, Everard. "Account of the Dissection of an Hermaphrodite Dog." *PTRS* 89 (1799): 157–78.

Hopkins, Ezekiel. *An Exposition on the Ten Commandments*. London, 1692.

Hume, David. *Essays, Moral, Political, and Literary*. Edited by Eugene F. Miller. Liberty Fund, 1987.

Hume, David. "Of the Populousness of Ancient Nations." In Hume, *Essays*, 377–464.

Humphreys, Thomas. *Marriage an Honourable Estate*. London, 1742.

Hunter, John (1728–1793). "Account of an Extraordinary Pheasant." *PTRS* 70 (1780): 527–35.

Hunter, John (1728–1793). *Observations on Certain Parts of the Animal Œconomy*. London, 1786.

Hunter, John (1728–1793). *Observations on Certain Parts of the Animal Oeconomy*. 2nd ed. London, 1792.

Hunter, John (1754–1809). *Disputatio Inauguralis Quaedam de Hominum Varietatibus.* Edinburgh, 1775.

Hunter, John (1754–1809). "Inaugural Disputation on the Varieties of Man (June 1775)." In Bendyshe, *Anthropological Treatises,* 357–94.

Hunter, William. "On the Uncertainty of the Signs of Murder, in the Case of Bastard Children." *Medical Observations and Inquiries* 6 (1784): 266–90.

Hutcheson, Francis. *An Inquiry into the Original of Our Ideas of Beauty and Virtue.* Liberty Fund, 2004.

Ingram, K. E. *Manuscript Sources for the History of the West Indies.* Kingston, 2000.

Jackson, William. *Thirty Letters on Various Subjects.* 2 vols. London, 1783.

[Jacob, Giles]. *Tractatus de Hermaphroditis.* London, 1718.

Jarrold, Thomas. *Anthropologia: or, Dissertation on the Form and Colour of Man.* London, 1808.

Jarrold, Thomas. "On National Character." *Memoirs of the Literary and Philosophical Society of Manchester* 2nd ser., 2 (1813): 328–53.

Jefferson, Thomas. *Notes on the State of Virginia.* Edited by William Preden. University of North Carolina Press, 1982.

Jerome. *Hebrew Questions on Genesis.* Edited by C. T. R. Hayward. Oxford University Press, 1995.

Jerome. "Liber Hebraicarum Quæstionum in Genesim." In *Patriologiæ cursus completus,* edited by J.-P. Migne. Vol. 23 of *S. Eusebii Hieronymi [. . .] Opera Omnia.* Paris, 1845.

Johnson, Samuel. *Dictionary of the English Language.* London, 1756.

Jones, John. *Medical, Philosophical, and Vulgar Errors.* London, 1792.

Journals of the Assembly of Jamaica. 14 vols. Jamaica, 1811–29.

K[eith], A[lexander]. *Observations on the Act for Preventing Clandestine Marriages.* London, 1753.

Kames, Henry Home, Lord. *Sketches of the History of Man.* Edited by James A. Harris. 3 vols. Liberty Fund, 2007 [1774].

Kant, Immanuel. *Anthropology, History, and Education.* Edited by Günter Zöller and Robert B. Louden. Cambridge University Press, 2007.

Kant, Immanuel. *Kant's gesammelte Schriften.* Edited by Royal Prussian (later German) Academy of Sciences. 29 vols. Berlin, 1900–.

King, Gregory. *A Scheme of the Rates and Duties Granted to His Majesty upon Marriages, Births and Burials, and upon Batchelors and Widowers.* London, 1695.

King, John. "Vitis Palatina [1613]." In *Conjugal Duty: Part II,* 282–302. London, 1736.

Knox, Vicesimus. *Essays, Moral and Literary.* 2 vols. London, 1778.

Kolben, Peter. *The Present State of the Cape of Good-Hope.* 2 vols. London, 1731.

La Beche, H. T. de. *Notes on the Present Condition of the Negroes in Jamaica.* London, 1825.

Lafitau, Joseph-François. *Customs of the American Indians Compared with the Customs of Primitive Times.* Edited by William N. Fenton and Elizabeth L. Moore. 2 vols. Champlain Society, 1974–77.

Larkin, James Francis, ed. *Stuart Royal Proclamations.* Vol. 2, *Proclamations of King Charles I, 1625–1646.* Oxford University Press, 1983.

Lascelles, Edwin, James Colleton, Edward Drax, Francis Ford, John Brathwaite, John Walter, William Thorpe Holder, James Holder, and Philip Gibbes. *Instructions for the Management of a Plantation in Barbadoes.* London, 1786.

Lavater, J. C. *Essays on Physiognomy.* Translated by Thomas Holcroft. 3 vols. London, 1789.

Lawrence, William. "Account of a Child Born Without a Brain." *Medico-Chirurgical Transactions* 5 (1814): 165–224.

[Lawrence, William]. "Generation." In *Cyclopedia; or, Universal Dictionary of Arts, Sciences, and Literature,* edited by Abraham Rees. 45 vols. London, 1819.

Lawrence, William. *Lectures on Physiology, Zoology and the Natural History of Man.* London, 1819.

Le Cat, Claude-Nicolas. *Traité de la couleur de la peau humaine.* Amsterdam, 1765.

Leibniz, Gottfried Wilhelm. *Esprit de Leibnitz.* 2 vols. Lyons, 1772.

Leibniz, Gottfried Wilhelm. *Opera Omnia.* 6 vols. Geneva, 1768.

Leman, Sir Tanfield. *Matrimony Analysed.* London, 1755.

Lemnius, Levinus. *A Discoruse [sic] Touching Generation.* London, 1667.

Lemnius, Levinus. *The Secret Miracles of Nature.* London, 1658.

Letsome, Sampson. *The Preacher's Assistant.* London, [1753].

A Letter from a By-Stander, Containing Remarks on and Objections to the Bill [. . .] for the better Preventing Clandestine Marriages. London, 1753.

Letter on Registering and Numbering the People of Great Britain. London, 1753.

A Letter to Sir John Phillips, Bart. London, 1747.

A Letter to the Public [. . .] upon the Subject of the Act of Parliament, For the better preventing of Clandestine Marriages. London, 1753.

Lewis, Matthew Gregory. *Journal of a Residence Among the Negroes in the West Indies.* London, 1845.

Linnaeus [Carl von Linné]. *The Animal Kingdom [. . .] Class 1. Mammalia [. . .] Being a Translation of That Part of the Systema Naturæ.* Translated by Robert Kerr. Edinburgh, 1792.

Linnaeus [Carl von Linné]. *Sponsalia plantarum.* Stockholm, 1749 [1746].

Lithgow, William. *The Totall Discourse, Of the Rare Aduentures, and painefull Peregrinations of long nineteene Yeares Trauayles.* London, 1632.

Lithgow, William. "W. Lithgow's Description of Ireland." In *Phœnix Britannicus,* edited by J. Morgan. London, 1731.

Locke, John. *Essay Concerning Human Understanding.* Edited by Peter H. Nidditch. Oxford University Press, 1975 [1690].

Locke, John. *Political Essays.* Edited by Mark Goldie. Cambridge University Press, 1997.

Locke, John. *Two Treatises of Government.* Edited by Peter Laslett. Cambridge University Press, 1988.

Lodwick, Francis. *On Language, Theology, and Utopia.* Edited by Felicity Henderson and William Poole. Oxford University Press, 2011.

[Long, Edward]. *The History of Jamaica.* 3 vols. London, 1774.

Lover of his Countrey. *The Grand Concern of England Explained.* London, 1673.

Luffman, John. *A Brief Account of the Island of Antigua.* 2nd ed. London, 1789.

Macaulay, Zachary. *Negro Slavery.* London, 1823.

MacCurdy, Edward. *The Notebooks of Leonardo da Vinci.* George Braziller, 1955.

Mackenzie, E. *Historical Account of Newcastle-Upon-Tyne.* 2 vols. Newcastle, 1827.

[MacLachlan, Daniel]. *Essay Upon Improving and Adding to the Strength of Great Britain and Ireland by Fornication.* London, 1735.

Maconochie, Allan. *Information for Joseph Knight [. . .] against John Wedderburn.* Edinburgh, 1775.

Madan, Martin. *Thelyphthora; or, a Treatise on Female Ruin.* 2 vols. London, 1780.

Madison, James. *The Debates in the Federal Convention of 1787 Which Framed the Constitution of the United States of America.* Edited by Gaillard Hunt and James Brown Scott. Oxford University Press, 1920.

Malpighi, Marcello. *De externo tactus organo anatomica observatio.* Naples, 1665.

Malthus, T. R. *An Essay on the Principle of Population.* 2nd ed. London, 1803.

Malthus, T. R. *An Essay on the Principle of Population.* 3rd ed. 2 vols. London, 1806.

Malthus, T. R. *An Essay on the Principle of Population [. . .]: The Version Published in 1803, with the Variora.* Edited by Patricia James. 2 vols. Cambridge University Press, 1989.

[Mandeville, Bernard]. *The Fable of the Bees.* Edited by F. B. Kaye. 2 vols. Oxford University Press, 1924.

[Mandeville, Bernard]. *A Modest Defence of Publick Stews.* London, 1724.

Manton, Thomas. *A Fourth Volume, Containing One hundred and fifty Sermons.* London, 1693.

Marly; or, the Life of a Planter in Jamaica. 2nd ed. Glasgow, 1828.

Marriage Promoted. London, 1690.

Marryat, Joseph. *More Thoughts, Occasioned by Two Publications.* 2nd ed. London, 1816.

Martin, R. Montgomery. *History of the West Indies.* 2 vols. London, 1836–37.

Marx, K. F. H. *Zum Andenken an Joh. Friedrich Blumenbach.* Göttingen, 1840.

Massey, Edmund. *A Sermon Against the Dangerous and Sinful Practice of Inoculation.* London, 1722.

Mathison, Gilbert. *Notices Respecting Jamaica, in 1809–1810–1811.* London, 1811.

Mauclerc, John Henry. *Dr. Blondel Confuted; or, the Ladies Vindicated.* London, 1747.

[Maupertuis, Pierre-Louis Moreau de]. *The Earthly Venus.* Translated by Simone Brangier Boas. Johnson Reprint, 1966.

[Maupertuis, Pierre-Louis Moreau de]. *Réflexions philosophiques sur l'origine des langues et la signification des mots.* Paris, 1748.

Maupertuis, Pierre-Louis Moreau de. *Vénus physique.* Paris, 1745.

McNeill, Hector. *Observations on the Treatment of the Negroes in the Island of Jamaica.* London, 1788.

Meiners, C. *Untersuchungen über die Verschiedenheiten der Menschennaturen.* 3 vols. Tübingen, 1811–15.

Mill, James. *The History of British India.* 3 vols. London, 1817.

Millar, John. *Observations Concerning the Distinction of Ranks in Society.* London, 1771.

Miller, Vincent [pseud.]. *The Man-Plant: or, Scheme for Increasing and Improving the British Breed.* London, 1752.

Mitchell, John. "An Essay upon the Causes of the different Colours of People in different Climates." *PTRS* 43 (1744–45): 102–50.

M'Mahon, Benjamin. *Jamaica Plantership.* London, 1839.

The Modern Part of an Universal History. 44 vols. London, 1759–66.

[Monboddo, James Burnett, Lord]. *Antient Metaphysics.* 6 vols. Edinburgh, 1779–99.

[Monboddo, James Burnett, Lord]. *The Origin and Progress of Language.* 6 vols. Edinburgh, 1773–92.

[Monboddo, James Burnett, Lord]. *The Origin and Progress of Language.* Vol. 1. 2nd ed. Edinburgh, 1774.

Montagu, Elizabeth. *Mrs. Montagu [. . .] Her Letters and Friendships from 1762 to 1800.* Edited by Reginald Blunt. 2 vols. Edinburgh, 1924.

Montaigne, Michel de. *The Complete Works of Montaigne.* Edited by Donald M. Frame. Hamish Hamilton, 1957.

[Montesquieu, Charles de Secondat, Baron de]. *Lettres persanes.* 2 vols. Amsterdam, 1721.

Montesquieu, Charles de Secondat, Baron de. *The Spirit of the Laws.* Translated and edited by Anne M. Cohler, Basia C. Miller, and Harold S. Stone. Cambridge University Press, 1989.

Moore, Francis. *Travels into the Inland Parts of Africa.* London, 1738.

Moreton, J. B. *West India Customs and Manners.* London, 1793.

[Morgann, Maurice]. *A Plan for the Abolition of Slavery in the West Indies.* London, 1772.

Morris, Ralph [pseud.]. *A Narrative of the Life and astonishing Adventures of John Daniel.* London, 1751.

Moscati, Pietro. *Von dem körperlichen wesentlichen Unterschiede zwischen der Structur der Thiere und der Menschen.* Translated by Johann Beckmann. Göttingen, 1771.

Munro, Innes. *Narrative of the Military Operations on the Coromandel Coast.* London, 1789.

[Neville, Henry]. *The Isle of Pines, or, A Late Discovery of a fourth Island near Terra Australis, Incognita.* London, 1668.

New Discoveries Concerning the World. London, 1778.

Newcome, Peter. *A Catechetical Course of Sermons.* 2 vols. London, 1702.

[Newton, Isaac]. *Opticks: or, A Treatise on the Reflexions, Refractions, Inflexions and Colours of Light.* London, 1704.

Newton, Thomas. *Dissertations on the Prophecies.* London, 1759.

Nichols, John. *Literary Anecdotes of the Eighteenth Century.* 6 vols. London, 1812.

Nott, J. C. "The Mulatto a Hybrid." *American Journal of the Medical Sciences* 6 (1843): 252–56.

Nugent, Maria. *Lady Nugent's Journal.* Edited by Philip Wright. University of the West Indies Press, 2002 [1966].

"Observations de phisique générale." *Histoire de l'Académie Royale des Sciences* (1734): 15–17.

Ogilby, John. *Africa.* London, 1670.

[O'Keeffe, John]. *Short Account of the New Pantomime Called Omai, or, A Trip round the World.* London, 1785.

Old Bailey Proceedings Online 1674–1913. Version 9.0. Fall 2023. www.oldbaileyonline.org.

Omiah's Farewell. London, 1776.

Onania: or, The Heinous Sin of Self-Pollution. 8th ed. London, 1723.

Original Papers Relative to the Establishment of a Society in Bengal. London, 1784.

Orme, Robert. *Historical Fragments of the Mogul Empire.* London, 1805.

Ovington, J. *A Voyage to Suratt.* London, 1696.

Paley, William. *The Principles of Moral and Political Philosophy.* London, 1785.

Pallas, Peter Simon. "Kurze Nachrichten." *Neue Nordische Beyträge* 1 (1781): 156–57.

Pallas, Peter Simon. "Memoire sur la variation des animaux." *Acta academiae scientiarum imperialis petropolitanae* (July–December 1780): 69–102.

Paré, Ambroise. *On Monsters and Marvels.* Edited by Janis L. Pallister. University of Chicago Press, 1983.

Parsons, James. "An Account of the White Negro Shewn before the Royal Society." *PTRS* 55 (1765): 45–53.

Parsons, James. "A Letter from James Parsons [. . .] Giving a Short Account of His Book Intituled, *A Mechanical Critical Inquiry into the Nature of Hermaphrodites*." *PTRS* 41 (1739–41): 650–52.

Parsons, James. "A Letter to the President, concerning the Hermaphrodite shewn in London." *PTRS* 47 (1751–52): 142–45.

Parsons, James. *Mechanical and Critical Enquiry into the Nature of Hermaphrodites*. London, 1741.

Parsons, James. *Philosophical Observations on the Analogy Between the Propagation of Animals and That of Vegetables*. London, 1752.

Parsons, James. *Prælecturi [. . .] Elenchus Gynaicopathologicus, et Obstetricarius*. London, 1741.

Parsons, James. *Remains of Japhet*. London, 1767.

Patot, Simon Tyssot de. *The Travels and Adventures of James Massey*. Translated by Stephen Whatley. London, 1733.

[Pauw, Cornelius de]. *Recherches philosophiques sur les Américains*. 2 vols. Berlin, 1768–69.

[Petty, William]. *Five Essays in Political Arithmetick*. London, 1698.

Petty, William. *The Petty Papers*. Edited by the Marquis of Lansdowne. 2 vols. Constable & Company, 1927.

Petty, William. *Political Arithmetick*. London, 1690.

Petty, William. *Treatise of Taxes & Contributions*. London, 1662.

Petyt, William. *Britannia Languens*. London, 1680.

[Peyrère, Isaac de la]. *Men before Adam*. London, 1656.

[Peyrère, Isaac de la]. *Prae-Adamitae*. Amsterdam, 1655.

Phillippo, James M. *Jamaica: Its Past and Present State*. Dawsons, 1969 [1843].

A Philosophical Essay on Fecundation. London, 1742.

Place, Francis. *Illustrations and Proofs of the Principle of Population*. London, 1822.

A Plan for the Abolition of Slavery. London, 1828.

[Pollexfen, John]. *England and East-India Inconsistent in Their Manufactures*. London, 1697.

Polwhele, Richard. *Discourses on Different Subjects*. 2 vols. London, 1791.

Pomponius Mela, *De situ orbis*. Edited by Isaac Vossius. London, 1658.

Poole, Matthew. *Annotations upon the Holy Bible*. 2 vols. London, 1683–85.

Populousness with Oeconomy. London, 1759.

Porta, John Baptista [Giambattista Della Porta]. *Natural Magick*. London, 1658.

Porteus, Beilby. *A Letter to the Clergy of the West-India Islands*. London, 1788.

Porteus, Beilby. *A Letter to the Governors, Legislatures, and Proprietors of Plantations, in the British West-India Islands*. London, 1808.

Postlethwayt, Malachy. *Universal Dictionary of Trade and Commerce*. 4th ed. 2 vols. London, 1774.

[Preston, William]. *Seventeen Hundred and Seventy-Seven*. London, 1777.

Preston, William Graham, Lord. *A Sermon, Preached at Mountfield in Sussex*. London, 1771.

[Price, Joseph]. *The Saddle Put on the Right Horse*. London, 1783.

Prichard, James Cowles. *Researches into the Physical History of Man*. Edited by George W. Stocking Jr. University of Chicago Press, 1973 [1813].

Prichard, James Cowles. *Researches into the Physical History of Mankind*. 2nd ed. 2 vols. London, 1826.

Prichard, James Cowles. *Researches into the Physical History of Mankind.* 3rd ed. 5 vols. London, 1836–47.

Prince, Mary. *The History of Mary Prince.* Edited by Sara Salih. Penguin, 2004.

Proisy, Chevalier de. *État des Finances de Saint-Domingue.* Port-au-Prince, 1790.

Purchas, Samuel. *Hakluytus Posthumus, or Purchas his Pilgrimes.* 4 vols. London, 1625.

Quilletus, Claudius. *Callipædia: or, The Art of Getting Pretty Children.* London, 1710.

Ramsay, James. *Essay on the Treatment and Conversion of African Slaves.* London, 1784.

Ray, John. *The Wisdom of God Manifested in the Works of the Creation.* London, 1691.

Raynal, Guillaume-Thomas. *A Philosophical and Political History of [. . .] the East and West Indies.* Translated by J. O. Justamond. 8 vols. London, 1783.

Reflections on the Repeal of the Marriage-Act. London, 1765.

Remarks on a Late Publication, Entitled, "An Essay on the Principle of Population." London, 1803.

Report of the State of the County-Hospital in Northampton. Northampton, 1766.

Richardson, Charles. *A New Dictionary of the English Language.* London, 1839.

[Richardson, Samuel]. *Pamela.* Edited by Thomas Keymer and Alice Wakely. Oxford University Press, 2001.

[Richardson, Samuel]. *Pamela; or, Virtue Rewarded.* 4th ed. 4 vols. London, 1742.

Ridley, Glocester. *Sermon Preached before the Honourable Trustees For establishing the Colony of Georgia.* London, 1746.

Ridley, Glocester. *A Sermon Preached [. . .] Before the Vice-Presidents, Treasurer, and Governours of the City of London Lying-In Hospital.* London, 1764.

Risley, Herbert. *The People of India.* 2nd ed. Thacker, 1915.

Robertson, William. *History of America.* 2 vols. London, 1777.

[La Roche, Charles-François Tiphaigne de]. *Amilec, or the Seeds of Mankind.* London, 1753.

Roughley, Thomas. *The Jamaica Planter's Guide.* London, 1823.

Rousseau, Jean-Jacques. *The* Discourses *and Other Early Political Writings.* Edited by Victor Gourevitch. Cambridge University Press, 1997.

Rousseau, Jean-Jacques. *Œuvres complètes.* Edited by Bernard Gagnebin and Marcel Raymond. 5 vols. Gallimard, 1959–95.

Rush, Benjamin. *An Account of the Bilious remitting Yellow Fever, as it appeared in the City of Philadelphia, in the Year 1793.* 2nd ed. Philadelphia, 1794.

Rush, Benjamin. "Observations Intended to Favour a Supposition That the Black Color [. . .] of the Negroes Is Derived from the Leprosy." *Transactions of the American Philosophical Society* 4 (1799): 289–97.

Rushton, Edward. *Collected Writings.* Edited by Paul Baines. Liverpool University Press, 2014.

Sadeur, Jacques [Gabriel de Foigny]. *A New Discovery of Terra Incognita Australis.* London, 1693.

Sadeur, Jacques [Gabriel de Foigny]. *The Southern Land, Known.* Edited by David Fausett. Syracuse University Press, 1993.

Saint-Méry, Moreau de. *Description [. . .] de la partie française de l'isle Saint-Domingue.* 2 vols. Philadelphia, 1797–98.

Saint-Vincent, Bory de. *L'homme.* 2nd ed. 2 vols. Paris, 1827.

[Salmon, Thomas]. *A Critical Essay Concerning Marriage.* London, 1724.

Sancho, Ignatius. *Letters of the Late Ignatius Sancho, an African.* Edited by Vincent Caretta. Penguin, 1998.

[Schaw, Janet]. *Journal of a Lady of Quality.* Edited by Evangeline Walker Andrews and Charles McLean Andrews. Yale University Press, 1923.

Schotte, J. P. *A Treatise on the Synochus Atrabiliosa.* London, 1782.

Scrafton, Luke. *Reflections on the Government of Indostan.* London, 1770.

Secker, Thomas. *Lectures on the Catechism of the Church of England.* 4th ed. 2 vols. London, 1771.

[Senior, Bernard Martin]. *Jamaica, as It Was, as It Is, and as It May Be.* London, 1835.

Shaftesbury, Anthony Ashley Cooper, Third Earl of. *Characteristics of Men, Manners, Opinions, Times.* Edited by Lawrence E. Klein. Cambridge University Press, 1999.

Sharp, Jane. *The Midwives Book.* London, 1671.

[Shebbeare, John]. *The Marriage Act. A Novel.* 2 vols. London, 1754.

[Sheridan, Thomas]. *A Discourse of the Rise & Power of Parliaments.* London, 1677.

Slave Law of Jamaica. London, 1828.

[Smalbroke], Richard, Lord Bishop of St David's. *Reformation necessary to prevent Our Ruine.* London, 1727.

Smith, Adam. *Wealth of Nations.* Edited by W. B. Todd. 2 vols. Oxford University Press, 1979.

Smith, Samuel Stanhope. *An Essay on the Causes of the Variety of Complexion and Figure in the Human Species.* Philadelphia, 1787.

Smith, Samuel Stanhope. *An Essay on the Causes of the Variety of Complexion and Figure in the Human Species.* Edited by Winthrop D. Jordan. Harvard University Press, 1965.

Smollett, Tobias. *The Adventures of Peregrine Pickle.* Edited by James L. Clifford and Paul Gabriel Boucé. Oxford University Press, 1983.

Smollett, Tobias. *Roderick Random.* Edited by Paul-Gabriel Boucé. Oxford University Press, 1979.

Some Reasons Humbly offer'd, why the Castration Of Persons found Guilty of Robbery and Theft, May be the best Method of Punishment for those Crimes. Dublin, 1725.

Sömmerring, Samuel Thomas. *Briefwechsel 1784–1792.* Edited by Franz Dumont. Gustav Fischer Verlag, 1997–98.

Sömmerring, Samuel Thomas. *Über die körperliche Verschiedenheit des Mohren vom Europäer.* Mainz, 1784.

Sömmerring, Samuel Thomas. *Ueber die körperliche Verschiedenheit des Negers vom Europäer.* Frankfurt and Mainz, 1785.

De Souligné. *A Comparison Between Old Rome In its Glory, As to the Extent and Populousness, and London, As it is at Present.* London, 1706.

The Speech of Sir John Knight of Bristol. London, [1694?].

Speech of the Right Hon. George Canning [. . .] on Wednesday, the 17th of March, 1824. London, 1824.

Speeches of Mr. Barrett; First Report of the Female Society for Birmingham. Birmingham, 1826.

A Spy on Mother Midnight. London, 1748.

Stanhope, Philip Dormer. *Genuine Memoirs of "Asiaticus."* London, 1784.

Stephen, James. *The Slavery of the West India Colonies Delineated.* 2 vols. London, 1824–30.

Sterne, Laurence. *The Life and Opinions of Tristram Shandy, Gentleman.* Edited by Ian Campbell Ross. Oxford University Press, 2000.

[Stewart, John]. *An Account of Jamaica, and Its Inhabitants.* London, 1808.

Stewart, John. *A View of the Past and Present States of the Island of Jamaica.* Edinburgh, 1823.

Stillingfleet, Edward. *Origines Sacrae.* London, 1662.

Supplément au Voyage de M. de Bougainville. Translated by M. de Fréville. 2nd ed. Neuchâtel, 1773.

A Supplement to the Morning-Exercise at Cripplegate. 2nd ed. London, 1676.

Supplement to the Onania. London, [1725?].

Swammerdam, Johann. *Miraculum Naturae.* Leiden, 1672.

[Swift, Jonathan]. *The Examiner* 45 (June 7, 1711).

Swift, Jonathan. *Gulliver's Travels.* Edited by Claude Rawson and Ian Higgins. Oxford University Press, 2005.

Tavernier, Jean Baptiste. *The Six Voyages of John Baptista Tavernier.* London, 1677.

Taylor, Jeremy. *ENIAYTOΣ: A Course of Sermons for All the Sundays of the year.* 3rd ed. London, 1668.

[Temple, William]. *A Vindication of Commerce and the Arts.* London, 1758.

Tench, Watkin. *A Complete Account of the Settlement at Port Jackson.* London, 1793.

The Tendencies of the Foundling Hospital in Its Present Extent Considered. London, 1760.

Thackeray, William Makepeace. *Vanity Fair.* Edited by John Carey. Penguin, 2001.

Thomson, James. *A Treatise of the Diseases of Negroes.* [Kingston], 1820.

Tissot, S. A. D. *Onanism: or, A Treatise upon the Disorders produced by Masturbation.* London, 1766.

[Tobin, James]. *Cursory Remarks upon the Reverend Mr. Ramsay's Essay.* London, 1785.

[Toland, John]. *Two Essays Sent in a Letter from Oxford to a Nobleman in London.* London, 1695.

Tornielli, Agostino. *Annales Sacri.* 2 vols. [Milan], 1610.

[Towers, Joseph]. *Dialogues Concerning the Ladies.* London, 1785.

The Travels of Hildebrand Bowman, Esquire. Edited by Lance Bertelsen. Broadview, 2017 [1778].

Treytorens, M de. "Observations de phisique générale." *Histoire de l'Académie royale des sciences* (1734): 15–17.

Tucker, Henry Carre. *Tucker's Note Book of Rules and Regulations.* 3rd ed. Calcutta, 1857.

Tucker, Josiah. *The Elements of Commerce.* [Bristol?], 1755.

Turner, Daniel. *The Force of the Mother's Imagination upon Her Foetus in Utero.* London, 1730.

Turner, Daniel. *De Morbis Cutaneis. A Treatise of Diseases Incident to the Skin.* London, 1714.

Two Reports from the Committee of the Honourable House of Assembly of Jamaica on the Subject of the Slave Trade. Kingston, 1789.

Tyson, Edward. *Orang-Outang, sive Homo Silvestris.* London, 1699.

D'Urville, J. "Sur les îles du grand océan." *Bulletin de la société de géographie* 105 (January 1832): 1–21.

Ussher, James. *The Annals of the World.* London, 1658.

Van Leeuwenhoek, Antoni. "Part of a Letter." *PTRS* 21 (1699): 270–72.

Vandermonde, Charles-Augustin. *Essai sur la manière de perfectionner l'espèce humaine.* 2 vols. Paris, 1756.

Varaisse, Denis. *The History of the Sevarites or Sevarambi.* London, 1675.

[Venette, Nicolas]. *Conjugal Love Reveal'd*. 7th ed. London, [1720?].

Venette, Nicolas. *La Generation de l'Homme, ou Tableau de l'Amour Conjugal*. 2 vols. London, 1751.

Venette, Nicolas. *Tableau de l'amour Considéré dans l'Éstat du mariage*. Parma, 1687.

Vico, Giambattista. *The First New Science*. Edited by Leon Pompa. Cambridge University Press, 2002.

Vico, Giambattista. *New Science: Unabridged Translation of the Third Edition (1744)*. Translated by Thomas Goddard Bergin and Max Harold Fisch. 3rd ed. Cornell University Press, 1984.

Virey, Julien-Joseph. *Histoire naturelle du genre humain*. 2nd ed. 3 vols. Paris, 1824.

Virey, Julien-Joseph. "Du Lait." *Journal de la société des pharmaciens de Paris* 1 (An VIII [1799–1800]): 469–71.

Virey, Julien-Joseph. *Natural History of the Negro Race*. Translated by J. H. Guenebault. Charleston, 1837.

Vogt, Carl. *Lectures on Man*. London, 1864.

Vogt, Carl. *Vorlesungen über den Menschen*. Gießen, 1863.

Voltaire. *Letters Concerning the English Nation*. London, 1733.

Voltaire. *Lord Chesterfield's Ears*. London, 1826.

Voltaire. "Les Oreilles du comte de Chesterfield." Edited by Gerhard Stenger. In *Œuvres de 1774–1775*, 125–207. Vol. 76 of *Œuvres complètes de Voltaire*. Voltaire Foundation, 2013.

Voltaire. "Relation touchant un maure blanc, amené d'Afrique à Paris en 1744." Edited by Jean Mayer. In *Œuvres de 1742–1745 (II)*, 187–200. Vol. 28B of *Œuvres complètes de Voltaire*. Voltaire Foundation, 2008.

Vossii, Isaaci. *Variarum Observationum Liber*. London, 1685.

The Voyages and Travels of Sir John Mandevile, Knight. London, 1705.

Wafer, Lionel. *A New Voyage and Description of the Isthmus of America*. London, 1699.

[Wallace, Robert]. *A Dissertation on the Numbers of Mankind in antient and modern Times*. Edinburgh, 1753.

[Wallace, Robert]. *Various Prospects of Mankind, Nature, and Providence*. London, 1761.

Waller, John Augustine. *A Voyage in the West Indies*. London, 1820.

Ward, Rev. W. *A View of the History, Literature, and Religion of the Hindoos*. 4 vols. London, 1817–20.

[Webb, Daniel]. *Selections from Les recherches philosophiques [. . .] of M. Pauw*. Bath, 1789.

Webster, Noah. *A Dictionary for Primary Schools*. New York, 1833.

Wells, William Charles. *Two Essays [. . .] and an Account of a Female of the White Race of Mankind*. London, 1818.

Wesley, John. *An Extract of the Rev. Mr. John Wesley's Journal, From Sept. 13, 1773, to Jan. 2, 1776*. London, 1791.

West Indian. *Notes in Defence of the Colonies*. Jamaica, 1826.

White, Charles. *An Account of the Regular Gradation in Man*. London, 1799.

Wilberforce, William. *An Appeal to the Religion, Justice, and Humanity of the Inhabitants of the British Empire*. London, 1823.

Williamson, John. *Medical and Miscellaneous Observations*. 2 vols. Edinburgh, 1817.

Williamson, Thomas. *The East India Vade-Mecum*. 2 vols. London, 1810.

Wilson, Alexander. *Some Observations Relative to the Influence of Climate on Vegetable and Animal Bodies.* London, 1780.

Wimpffen, Francis Alexander Stanislaus, Baron de. *A Voyage to Santo Domingo, in the Years 1788, 1789, and 1790.* Translated by J. Wright. London, 1797.

Winckelmann, Johann Joachim. *History of the Art of Antiquity.* Translated by Harry Francis Mallgrave. Getty Research Institute, 2006.

Wollstonecraft, Mary. *Posthumous Works.* 4 vols. London, 1798.

Zimmermann, Eberhard August Wilhelm von. *Geographische Geschichte des Menschen.* 3 vols. Leipzig, 1778–83.

3. Secondary Literature

Aarsleff, Hans. *From Locke to Saussure: Essays on the Study of Language and Intellectual History.* University of Minnesota Press, 1982.

Aarsleff, Hans. *The Study of Language in England, 1780–1860.* Athlone, 1983.

Adelmann, Howard. *Marcello Malpighi and the Evolution of Embryology.* 5 vols. Cornell University Press, 1966.

Adorno, Theodor, and Max Horkheimer. *Dialectic of Enlightenment.* Verso, 1997 [1944].

Aldridge, Alfred Owen. "Population and Polygamy in Eighteenth-Century Thought." *Journal of the History of Medicine and Allied Science* 4 (1949): 129–48.

Alexander, Michael. *Omai, Noble Savage.* Collins & Harvill, 1977.

Altink, Henrice. *Representations of Slave Women in Discourses on Slavery and Abolition, 1780–1838.* Routledge, 2007.

Amussen, Susan. *An Ordered Society: Gender and Class in Early Modern England.* Basil Blackwell, 1988.

Andrew, Donna T. "'Adultery à-la-mode': Privilege, the Law, and Attitudes to Adultery, 1770–1809." *History* 82 (1997): 5–23.

Andrew, Donna T., ed. *London Debating Societies, 1776–1799.* London Record Society, 1994.

Andrew, Donna T. *Philanthropy and Police: London Charity in the Eighteenth Century.* Princeton University Press, 1989.

Anstey, Roger. *The Atlantic Slave Trade and British Abolition.* Macmillan, 1975.

Appiah, Kwame Anthony. "Racisms." In *Anatomy of Racism*, edited by David Theo Goldberg. University of Minnesota Press, 1990.

Armstrong, Meg. "'The Effects of Blackness': Gender, Race, and the Sublime in Aesthetic Theories of Burke and Kant." *Journal of Aesthetics and Art Criticism* 54 (1996): 213–36.

Arnold, David. *Colonizing the Body: State Medicine and Epidemic Disease.* University of California Press, 1993.

Arnold, David. "Race, Place and Bodily Difference in Early Nineteenth-Century India." *Historical Research* 77 (2004): 254–73.

Arthur, Paul Longley. *Virtual Voyages: Travel Writing and the Antipodes, 1605–1837.* Anthem, 2010.

Ashcraft, Richard. "Leviathan Triumphant: Thomas Hobbes and the Politics of Wild Men." In *The Wild Man Within*, edited by Edward Dudley and Maximilian E. Novak. University of Pittsburgh Press, 1972.

Aucante, Vincent. "Descartes's Experimental Method and the Generation of Animals." In Smith, *Problem of Animal Generation*, 70–75.

Augstein, Hannah Franziska. *James Cowles Prichard's Anthropology*. Brill, 1999.

Austen-Leigh, R. A. "Philip Dormer Stanhope: Author of the 'Genuine Memoirs of Asiaticus.'" *Notes and Queries* 11 (1922): 165–67.

Bailyn, Bernard. *Voyagers to the West*. Knopf, 1986.

Banerjee, Pompa. *Burning Women*. Palgrave, 2003.

Bannet, Eve Tavor. *The Domestic Revolution*. Johns Hopkins University Press, 2000.

Banton, Michael. *The Idea of Race*. Tavistock, 1977.

Banton, Michael. *Racial Theories*. 2nd ed. Cambridge University Press, 1998.

Barker, Anthony J. *The African Link: British Attitudes to the Negro in the Era of the Atlantic Slave Trade, 1550–1807*. Frank Cass, 1978.

Barnard, Alan. "*Orang Outang* and the Definition of *Man*: The Legacy of Lord Monboddo." In *Fieldwork and Footnotes: Studies in the History of European Anthropology*, edited by Han F. Vermeulen and Arturo Alvarez Roldán. Routledge, 1995.

Bashford, Alison, and Joyce E. Chaplin. *The New Worlds of Thomas Robert Malthus: Rereading the* Principle of Population. Princeton University Press, 2016.

Bayly, Susan. *Caste, Society and Politics in India from the Eighteenth Century to the Modern Age*. Cambridge University Press, 1999.

Beckles, Hilary McD. *Natural Rebels: A Social History of Enslaved Black Women in Barbados*. Zed Books, 1989.

Beckles, Hilary McD. "Property Rights in Pleasure: The Marketing of Enslaved Women's Sexuality." In *Caribbean Slavery in the Atlantic World*, edited by Verene A. Shepherd and Hilary McD. Beckles. James Currey, 2000.

Beckles, Hilary McD. "White Women and Slavery in the Caribbean." *History Workshop Journal* 36 (1993): 66–82.

Bederman, Gail. "Sex, Scandal, Satire, and Population in 1798: Revisiting Malthus's First Essay." *JBS* 47 (2008): 768–95.

Beeson, David. *Maupertuis: An Intellectual Biography*. Voltaire Foundation, 1992.

Begiato, Joanne. *Manliness in Britain, 1760–1900*. Manchester University Press, 2020.

Bennett, Michael. "Inoculation of the Poor Against Smallpox in Eighteenth-Century England." In *Experiences of Poverty in Late Medieval and Early Modern England and France*, edited by Anne M. Scott. Ashgate, 2012.

Berg, Maxine, and Elizabeth Eger. "The Rise and Fall of the Luxury Debates." In *Luxury in the Eighteenth Century*, edited by Maxine Berg and Elizabeth Eger. Palgrave, 2007.

Berlin, Ira. *Many Thousands Gone: The First Two Centuries of Slavery in North America*. Harvard University Press, 1998.

Bernal, Martin. *Black Athena: The Afroasiatic Roots of Classical Civilization*. 3 vols. Rutgers University Press, 1987.

Bernasconi, Robert. "Kant and Blumenbach's Polyps: A Neglected Chapter in the History of the Concept of Race." In Eigen and Larrimore, *German Invention*, 73–90.

Bernasconi, Robert, ed. *Race*. Blackwell, 2001.

Berridge, Virginia. "Health and Medicine." In *The Cambridge Social History of Britain, 1750–1950*, edited by F. M. L. Thompson, 3:171–242. 3 vols. Cambridge University Press, 1990.

Berry, Christopher. *The Idea of Luxury*. Cambridge University Press, 1994.

Bertoletti, Stefano Fabbri. "The Anthropological Theory of Johann Friedrich Blumenbach." In *Romanticism in Science: Science in Europe, 1790–1840,* edited by Stefano Poggi and Maurizio Bossi. Kluwer Academic, 1994.

Bethencourt, Francisco. *Racisms: From the Crusades to the Twentieth Century.* Princeton University Press, 2013.

Beusterien, John L. "Jewish Male Menstruation in Seventeenth-Century Spain." *Bulletin of the History of Medicine* 73 (1999): 447–56.

Bewell, Alan. "'On the Banks of the South Sea': Botany and Sexual Controversy in the Late Eighteenth Century." In *Visions of Empire: Voyages, Botany, and Representations of Nature,* edited by David Philip Miller and Peter Hanns Reill. Cambridge University Press, 1996.

Bewell, Alan. *Romanticism and Colonial Disease.* Johns Hopkins University Press, 1999.

Biller, Peter. *The Measure of Multitude.* Oxford University Press, 2000.

Bindman, David. *Ape to Apollo: Aesthetics and the Idea of Race in the 18th Century.* Cornell University Press, 2002.

Bindman, David. *"Race Is Everything": Art and Human Difference.* Reaktion, 2023.

Bindman, David, and Henry Louis Gates Jr., eds. *The Image of the Black in Western Art, Volume III: From the "Age of Discovery" to the Age of Abolition, Part 3: The Eighteenth Century.* Harvard University Press, 2011.

Biskup, Thomas. "The University of Göttingen and the Personal Union, 1737–1837." In *The Hanoverian Dimension in British History, 1714–1837,* edited by Brendan Simms and Torsten Riotte. Cambridge University Press, 2007.

Bitbol-Hespériès, Annie. "Monsters, Nature, and Generation from the Renaissance to the Early Modern Period: The Emergence of Medical Thought." In Smith, *Problem of Animal Generation,* 47–62.

Blake, Randolph, and Robert Sekuler. *Perception.* 5th ed. McGraw-Hill, 2006.

Blakley, Christopher Michael. *Empire of Brutality: Enslaved People and Animals in the British Atlantic World.* Louisiana State University Press, 2023.

Bloch, Iwan. *Sexual Life in England.* Translated by William H. Forstern. F. Aldor, 1938.

Bloch, Marc. *The Historian's Craft.* Translated by Peter Putman. Manchester University Press, 1992.

Bohls, Elizabeth A. *Women Travel Writers and the Language of Aesthetics, 1716–1818.* Cambridge University Press, 1995.

Bonar, James. *Theories of Population from Raleigh to Arthur Young.* George Allen & Unwin, 1931.

Borsay, Peter. *The English Urban Renaissance.* Oxford University Press, 1989.

Boucé, Paul-Gabriel. "The Rape of Gulliver Reconsidered." *Swift Studies* 11 (1996): 98–114.

Boucé, Paul-Gabriel, ed. *Sexuality in Eighteenth-Century Britain.* Manchester University Press, 1982.

Boucé, Paul-Gabriel. "Some Sexual Beliefs and Myths in Eighteenth-Century England." In Boucé, *Sexuality,* 37–41.

Boulukos, George. *The Grateful Slave: The Emergence of Race in Eighteenth-Century British and American Culture.* Cambridge University Press, 2008.

Bowler, Peter J. "Bonnet and Buffon: Theories of Generation and the Problem of Species." *Journal of the History of Biology* 6 (1973): 259–81.

Bowler, Peter J. "Preformation and Pre-Existence in the Seventeenth Century: A Brief Analysis." *Journal of the History of Biology* 4 (1971): 221–44.

Braddick, Michael J. *State Formation in Early Modern England, c. 1550–1700.* Cambridge University Press, 2000.

Braude, Benjamin. "The Sons of Noah and the Construction of Ethnic and Geographical Identities in the Medieval and Early Modern Periods." *W&MQ*, 3rd ser., 54 (1997): 103–42.

Braudy, Leo. *The Frenzy of Renown: Fame and Its History.* Oxford University Press, 1986.

Brewer, John. *The Sinews of Power: War, Money, and the English State, 1688–1783.* Routledge, 1989.

Brooks, Colin. "Projecting, Political Arithmetic and the Act of 1695." *EHR* 97 (1982): 31–53.

Brown, Christopher Leslie. *Moral Capital: Foundations of British Abolitionism.* University of North Carolina Press, 2006.

Brown, Kathleen. "'Changed . . . into the Fashion of Man': The Politics of Sexual Difference in a Seventeenth-Century Anglo-American Settlement." *Journal of the History of Sexuality* 6 (1995): 171–93.

Brown, Kathleen. *Good Wives, Nasty Wenches, and Anxious Patriarchs.* University of North Carolina Press, 1996.

Brown, Kathleen. "Native Americans and Early Modern Concepts of Race." In *Empire and Others: British Encounters with Indigenous Peoples, 1600–1850*, edited by Martin Daunton and Rick Halpern. University of Pennsylvania Press, 1999.

Brown, Michael. *A Political Biography of John Toland.* Pickering & Chatto, 2012.

Brubaker, Rogers, and Frederick Cooper. "Beyond 'Identity.'" *Theory and Society* 29 (2000): 1–47.

Buck, Peter. "People Who Counted: Political Arithmetic in the Eighteenth Century." *Isis* 73 (1982): 28–45.

Bullough, Vern L. "Prostitution and Reform in Eighteenth-Century England." In *'Tis Nature's Fault: Unauthorized Sexuality During the Enlightenment*, edited by Robert Purks Maccubbin. Cambridge University Press, 1987.

Burkhardt, Richard W., Jr. "Closing the Door on Lord Morton's Mare: The Rise and Fall of Telegony." *Studies in the History of Biology* 3 (1979): 1–21.

Burnard, Trevor. *Jamaica in the Age of Revolution.* University of Pennsylvania Press, 2020.

Burnard, Trevor. *Mastery, Tyranny and Desire: Thomas Thistlewood and His Slaves in the Anglo-Jamaican World.* University of North Carolina Press, 2004.

Burnard, Trevor. "Slave Naming Patterns: Onomastics and the Taxonomy of Race in Eighteenth-Century Jamaica." *Journal of Interdisciplinary History* 31 (2001): 325–46.

Burnard, Trevor. "Tropical Hospitality, British Masculinity, and Drink in Late Eighteenth-Century Jamaica." *Historical Journal* 65 (2022): 202–23.

Bush, Barbara. "'Sable Venus,' 'She Devil,' or 'Drudge'? Black Slavery and the 'Fabulous Fiction' of Black Women's Identities, c. 1650–1838." *Women's History Review* 9 (2000): 761–89.

Bush, Barbara. *Slave Women in Caribbean Society, 1650–1838.* James Currey, 1990.

Butler, Judith. *Bodies That Matter: On the Discursive Limits of "Sex."* Routledge, 1993.

Butler, Judith. *Gender Trouble.* Routledge, 1990.

Buzon, Frédéric de. "Leibniz: étymologie et origine des nations." *Révue Française d'Histoire des Idées Politiques* 36 (2012): 383–400.

Cadden, Joan. *Meanings of Sex Difference in the Middle Ages.* Cambridge University Press, 1993.

Caplan, Jane. Introduction to *Written on the Body: The Tattoo in European and American History*, xi–xxiii. Edited by Jane Caplan. Reaktion, 2000.

Caplan, Jane. "'This or That Particular Person': Protocols of Identification in Nineteenth-Century Europe." In *Documenting Individual Identity: The Development of State Practices in the Modern World*, edited by Jane Caplan and John Torpey. Princeton University Press, 2001.

Capp, Bernard. "The Double Standard Revisited: Plebeian Women and Male Sexual Reputation in Early Modern England." *P&P* 162 (1999): 70–100.

Carter, Harold B. *Sir Joseph Banks, 1743–1820*. British Museum, 1988.

Carter, Philip. *Men and the Emergence of Polite Society*. Longman, 2001.

Chaplin, Joyce E. *Subject Matter: Technology, the Body, and Science on the Anglo-American Frontier, 1500–1676*. Harvard University Press, 2001.

Chapman, Richard Allen. "*Leviathan* Writ Small: Thomas Hobbes on the Family." *American Political Science Review* 69 (1975): 76–90.

Charbit, Yves. *The Classical Foundations of Population Thought*. Springer, 2011.

Charters, Erica. "Making the Body Modern: Race, Medicine, and the Colonial Soldier in the Mid Eighteenth Century." *Patterns of Prejudice* 46 (2012): 214–31.

Chater, Kathleen. *Untold Histories: Black People in England and Wales During the Period of the British Slave Trade, c. 1660–1807*. Manchester University Press, 2009.

Chatterjee, Indrani. "Colouring Subalternity: Slaves, Concubines and Social Orphans in Early Colonial India." In *Subaltern Studies X: Writings on South Asian History and Society*, edited by Gautam Bhadra, Gyan Prakash, and Susie Tharu. Oxford University Press, 1999.

Cheek, Pamela. *Sexual Antipodes: Enlightenment Globalization and the Placing of Sex*. Stanford University Press, 2003.

Cherry, S. "The Voluntary General Hospitals, Mortality and Local Populations in the English Provinces in the Eighteenth and Nineteenth Centuries." *Population Studies* 34 (1980): 59–75, 251–65.

Chilton, Lisa. *Agents of Empire: British Female Emigration to Canada and Australia, 1860s–1930*. University of Toronto Press, 2007.

Claeys, Gregory, ed. *Utopias of the British Enlightenment*. Cambridge University Press, 2003.

Clark, Alice. *Working Life of Women in the Seventeenth Century*. Routledge, 1919.

Clark, Thomas Blake. *Omai, First Polynesian Ambassador to England*. University of Hawaii Press, 1940.

Cloyd, E. L. *James Burnett: Lord Monboddo*. Oxford University Press, 1972.

Cody, Lisa Forman. *Birthing the Nation: Sex, Science, and the Conception of Eighteenth-Century Britons*. Oxford University Press, 2005.

Cody, Lisa Forman. "The Doctor's in Labour; or, a New Whim Wham from Guildford." *Gender and History* 4 (1992): 175–96.

Cohen, Michèle. "The Grand Tour: Constructing the English Gentleman in Eighteenth-Century France." *History of Education* 21 (1992): 241–57.

Cole, F. J. *Early Theories of Sexual Generation*. Oxford University Press, 1930.

Coleman, Deirdre. "Janet Schaw and the Complexions of Empire." *ECS* 36 (2003): 169–93.

Colley, Linda. *Britons: Forging the Nation*. Pimlico, 1992.

Collingham, E. J. *Imperial Bodies: The Physical Experience of the Raj, c. 1800–1947.* Cambridge University Press, 2001.

Combe, Thomas, and Bruce Buchan. "Among 'Savage and Brutal Nations': Instructing Identity and Science in the Pacific." *Journal for Eighteenth-Century Studies* 45 (2022): 29–41.

Cookson, J. E. "Political Arithmetic and War in Britain, 1793–1815." *War & Society* 1 (1983): 37–60.

Cooley, Mackenzie. *The Perfection of Nature: Animals, Breeding, and Race in the Renaissance.* University of Chicago Press, 2022.

Corfield, P. J. *The Impact of English Towns, 1700–1800.* Oxford University Press, 1982.

Courcy, Anne de. *The Fishing Fleet: Husband-Hunting in the Raj.* Weidenfeld & Nicolson, 2012.

Crais, Clifton, and Pamela Scully. *Sara Baartman and the Hottentot Venus.* Princeton University Press, 2009.

Craton, Michael. *Testing the Chains: Resistance to Slavery in the British West Indies.* Cornell University Press, 1982.

Crawford, Julie. *Marvelous Protestantism: Monstrous Births in Post-Reformation England.* Johns Hopkins University Press, 2005.

Crawford, Patricia. "Attitudes to Menstruation in Seventeenth-Century England." *P&P* 91 (1981): 47–73.

Crawford, Patricia. *Blood, Bodies and Families in Early Modern England.* Longman, 2004.

Cressy, David. *Coming Over: Migration and Communication Between England and New England in the Seventeenth Century.* Cambridge University Press, 1987.

Crome, Andrew. "The 1753 'Jew Bill' Controversy: Jewish Restoration to Palestine, Biblical Prophecy, and English National Identity." *EHR* 130 (2015): 1449–78.

Crump, Margaret. M. *James Cowles Prichard of the Red Lodge.* University of Nebraska Press, 2025.

Curran, Andrew. *The Anatomy of Blackness.* Johns Hopkins University Press, 2011.

Curran, Andrew. "Logics of the Human in the *Supplément au Voyage de Bougainville.*" In *New Essays on Diderot,* edited by James Fowler. Cambridge University Press, 2011.

Curran, Andrew. "Rethinking Race History: The Role of the Albino in the French Enlightenment Life Sciences." *History and Theory* 48 (2009): 151–79.

Curtin, Philip D. *Two Jamaicas: The Role of Ideas in a Tropical Colony, 1830–1865.* Harvard University Press, 1955.

Dabhoiwala, Faramerz. "A Man of Parts and Learning." *London Review of Books,* November 21, 2024. https://www.lrb.co.uk/the-paper/v46/n22/fara-dabhoiwala/a-man -of-parts-and-learning.

Dabhoiwala, Faramerz. *The Origins of Sex: A History of the First Sexual Revolution.* Allen Lane, 2012.

Dabydeen, David. *Hogarth's Blacks: Images of Blacks in Eighteenth-Century English Art.* Manchester University Press, 1987.

Dantas, Mariana L. R. "Humble Slaves and Loyal Vassals: Free Africans and Their Descendants in Eighteenth-Century Minas Gerais, Brazil." In *Imperial Subjects: Race and Identity in Colonial Latin America,* edited by Andrew B. Fisher and Matthew D. O'Hara. Duke University Press, 2009.

Daston, Lorraine, and Peter Galison. *Objectivity.* Zone, 2007.

Daston, Lorraine, and Katharine Park. "The Hermaphrodite and the Orders of Nature: Sexual Ambiguity in Early Modern France." In *Premodern Sexualities*, edited by Louise Fradenburg and Carla Freccero. Routledge, 1995.

Daston, Lorraine, and Katharine Park. *Wonders and the Order of Nature*. Zone, 1998.

Davenport, Romola, Leonard Schwartz, and Jeremy Boulton. "The Decline of Adult Smallpox in Eighteenth-Century London." *Economic History Review* 64 (2011): 1289–1314.

Davidoff, Leonore, and Catherine Hall. *Family Fortunes: Men and Women of the English Middle Class, 1780–1850*. Routledge, 1987.

Davidson, Jenny. *Breeding: A Partial History of the Eighteenth Century*. Columbia University Press, 2009.

Davies, Owen. *Witchcraft, Magic and Culture, 1736–1951*. Manchester University Press, 1999.

Davis, David Brion. *The Problem of Slavery in the Age of Revolution, 1770–1823*. Cornell University Press, 1975.

Dawson, Mark S. *Bodies Complexioned: Human Variation and Racism in Early Modern English Culture, c. 1600–1750*. Manchester University Press, 2019.

Dawson, Virginia P. *Nature's Enigma: The Problem of the Polyp in the Letters of Bonnet, Trembley and Réaumur*. American Philosophical Society, 1987.

Dayan, Joan. *Haiti, History, and the Gods*. University of California Press, 1995.

Dekkers, Midas. *Dearest Pet: On Bestiality*. Translated by Paul Vincent. Verso, 2000.

Delbourgo, James. "The Newtonian Slave Body." *Atlantic Studies: Global Currents* 9 (2012): 185–207.

D'Elia, Donald J. "Dr. Benjamin Rush and the Negro." *JHI* 30 (1969): 413–22.

Dening, Greg. *Mr Bligh's Bad Language*. Cambridge University Press, 1994.

DeVun, Leah. "Heavenly Hermaphrodites: Sexual Difference at the Beginning and End of Time." *Postmedieval* 9 (2018): 132–46.

Dickinson, H. T. "The Poor Palatines and the Parties." *EHR* 82 (1967): 464–85.

Dikötter, Frank. *The Discourse of Race in Modern China*. Hurst, 1992.

Dorlin, Elsa. *La matrice de la race: Généalogie sexuelle et coloniale de la Nation française*. La Découverte, 2006.

Doron, Claude-Olivier. *L'homme altéré: Races et dégénérescence (XVIIᵉ-XIXᵉ siècles)*. Champ Vallon, 2016.

Doron, Claude-Olivier. "Race and Genealogy: Buffon and the Formation of the Concept of 'Race.'" *HUMANA.MENTE Journal of Philosophical Studies* 22 (2012): 75–109.

Douglas, Bronwen. "Art as Ethno-Historical Text: Science, Representation, and Indigenous Presence in Eighteenth- and Nineteenth-Century Oceanic Voyage Literature." In *Double Vision: Art Histories and Colonial Histories in the Pacific*, edited by Nicholas Thomas and Diane Losche. Cambridge University Press, 1999.

Douglas, Bronwen. "Climate to Crania: Science and the Racialization of Human Difference." In Douglas and Ballard, *Foreign Bodies*, 33–98.

Douglas, Bronwen. "'Novus Orbis Australis': Oceania in the Science of Race, 1750–1850." In Douglas and Ballard, *Foreign Bodies*, 99–155.

Douglas, Bronwen. *Science, Voyages, and Encounters in Oceania, 1511–1850*. Palgrave, 2014.

Douglas, Bronwen. "Slippery Word, Ambiguous Praxis: 'Race' and Late-18th-Century Voyagers in Oceania." *Journal of Pacific History* 41 (2006): 1–29.

Douglas, Bronwen, and Chris Ballard, eds. *Foreign Bodies: Oceania and the Science of Race, 1750–1940*. Australian National University E Press, 2008.

Douthwaite, Julia V. *The Wild Girl, Natural Man, and the Monster: Dangerous Experiments in the Age of Enlightenment.* University of Chicago Press, 2002.

Drescher, Seymour. *Econocide: British Slavery in the Era of Abolition.* 2nd ed. University of North Carolina Press, 2010.

Drescher, Seymour. *The Mighty Experiment: Free Labor versus Slavery in British Emancipation.* Oxford University Press, 2002.

Droixhe, Daniel. *La Linguistique et l'appel de l'histoire (1600–1800).* Droz, 1978.

Duden, Barbara. *The Woman Beneath the Skin: A Doctor's Patients in Eighteenth-Century Germany.* Translated by Thomas Dunlap. Harvard University Press, 1991.

Dunmore, John. *Monsieur Baret: First Woman Around the World, 1766–68.* Heritage, 2002.

Earle, Rebecca. *The Body of the Conquistador: Food, Race and the Colonial Experience in Spanish America, 1492–1700.* Cambridge University Press, 2012.

Edmond, Rod. *Representing the South Pacific: Colonial Discourse from Cook to Gaugin.* Cambridge University Press, 1997.

Edwards, Philip. *The Story of the Voyage.* Cambridge University Press, 1994.

Eigen, Sara, and Mark Larrimore, eds. *The German Invention of Race.* State University of New York Press, 2006.

Eliav-Feldon, Miriam, Benjamin Isaac, and Joseph Ziegler, eds. *The Origins of Racism in the West.* Cambridge University Press, 2009.

Ellingson, Ter. *The Myth of the Noble Savage.* University of California Press, 2001.

Elliott, J. H. *Empires of the Atlantic World: Britain and Spain in America, 1492–1830.* Yale University Press, 2006.

Ellis, Markman, Catherine Hall, Miles Ogborn, and Silvia Sebastiani. "Edward Long and Other Animals: The Orangutan and Race-Making in the Late Eighteenth Century." *Modern Intellectual History* 21 (2024): 753–82.

Eltis, David. *Economic Growth and the Ending of the Transatlantic Slave Trade.* Oxford University Press, 1987.

Eltis, David. "The Traffic in Slaves Between the British West Indian Colonies, 1807–1833." *Economic History Review* 25 (1972): 55–64.

Eltis, David, and Stanley L. Engerman, eds. *The Cambridge World History of Slavery*, vol. 3, *AD 1420–AD 1804.* Cambridge University Press, 2011.

Engelman, Todd M. *The Jews of Georgian England, 1714–1830.* Jewish Publication Society, 1979.

Engerman, Stanley J., and B. W. Higman. "The Demographic Structure of the Caribbean Slave Societies in the Eighteenth and Nineteenth Centuries." In *General History of the Caribbean: 3. The Slave Societies of the Caribbean*, edited by Franklin W. Knight. UNESCO, 2003.

Essner, Cornelia. *Die "Nürnberger Gesetze" oder Die Verwaltung des Rassenwahns 1933–1945.* Ferdinand Schöningh, 1992.

Everill, Bronwen. *Not Made by Slaves: Ethical Capitalism in the Age of Abolition.* Harvard University Press, 2020.

Eze, Emmanuel Chukwudi. "Hume, Race, and Human Nature." *JHI* 61 (2000): 691–98.

Eze, Emmanuel Chukwudi. *Race and the Enlightenment: A Reader.* Blackwell, 1997.

Fabian, Bernhard. "The Importation of English Books into Germany in the Eighteenth Century." *Studies on Voltaire and the Eighteenth Century* 193 (1980): 1780–82.

Fairchild, Hoxie Neale. *The Noble Savage: A Study in Romantic Naturalism*. Columbia University Press, 1928.

Fara, Patricia. *Sex, Botany, and Empire*. Cambridge University Press, 2003.

Fausett, David. *Images of the Antipodes in the Eighteenth Century*. Rodopi, 1994.

Fausto-Sterling, Anne. "Gender, Race, and Nation: The Comparative Anatomy of 'Hottentot' Women in Europe, 1815–1817." In *Deviant Bodies: Critical Perspectives on Difference in Science and Popular Culture*, edited by Jennifer Terry and Jacqueline Urla. Indiana University Press, 1995.

Fausto-Sterling, Anne. *Sexing the Body: Gender Politics and the Construction of Sexuality*. Basic, 2000.

Figal, Sara Eigen. *Heredity, Race, and the Birth of the Modern*. Routledge, 2008.

Finn, Margot. "Slaves Out of Context: Domestic Slavery and the Anglo-Indian Family, *c*. 1780–1830." *Transactions of the Royal Historical Society* 19 (2009): 181–203.

Finucci, Valerie. "Maternal Imagination and Monstrous Birth: Tasso's *Gerusalemme liberata*." In Finucci and Brownlee, *Generation and Degeneration*, 41–77.

Finucci, Valerie, and Kevin Brownlee, eds. *Generation and Degeneration: Tropes of Reproduction in Literature and History from Antiquity to Early Modern Europe*. Duke University Press, 2001.

Fisch, Jörg. *Burning Women: A Global History of Widow-Sacrifice from Ancient Times to the Present*. Translated by Rekha Kamath Rajan. Seagull, 2006.

Fischer, Kirsten. *Suspect Relations: Sex, Race, and Resistance in Colonial North Carolina*. Cornell University Press, 2002.

Fissell, Mary E. *Vernacular Bodies: The Politics of Reproduction in Early Modern England*. Oxford University Press, 2004.

Fletcher, Anthony. *Gender, Sex and Subordination in England, 1500–1800*. Yale University Press, 1995.

Floyd-Wilson, Mary. *English Ethnicity and Race in Early Modern Drama*. Cambridge University Press, 2003.

Foucault, Michel. *The History of Sexuality: The Will to Knowledge*. Translated by Robert Hurley. Penguin, 1998.

Fox-Genovese, Elizabeth. *Within the Plantation Household: Black and White Women of the Old South*. University of North Carolina Press, 1988.

Francus, Marilyn. "Monstrous Mothers, Monstrous Societies: Infanticide and the Rule of Law in Restoration and Eighteenth-Century England." *ECL* 21 (1997): 133–56.

Fredrickson, George M. *Racism: A Short History*. Princeton University Press, 2002.

Friedman, John Block. *The Monstrous Races in Medieval Art and Thought*. Syracuse University Press, 2000.

Fryer, Peter. *Staying Power: The History of Black People in Britain*. Pluto, 1984.

Fuchs, Barbara. "A Mirror Across the Water: Mimetic Racism, Hybridity, and Cultural Survival." In *Writing Race Across the Atlantic World: Medieval to Modern*, edited by Philip D. Beidler and Gary Taylor. Palgrave, 2005.

Fulford, Tim, Debbie Lee, and Peter J. Kitson. *Literature, Science and Exploration in the Romantic Era*. Cambridge University Press, 2004.

Fullagar, Kate. *The Savage Visit: New World People and Popular Imperial Culture in Britain, 1710–1795*. University of California Press, 2012.

Gagnon, John H. "Others Have Sex with Others: Captain Cook and the Penetration of the Pacific." In *Sexual Cultures and Migration in the Era of AIDS*, edited by Gilbert Herdt. Oxford University Press, 1997.

Garraway, Doris. "Race, Reproduction and Family Romance in Moreau de Saint-Méry's *Description . . . de la partie française de l'isle Saint-Domingue*." *ECS* 38 (2005): 227–46.

Garrigus, John D. *Before Haiti: Race and Citizenship in French Saint-Domingue*. Palgrave, 2006.

Garrigus, John D. "Redrawing the Colour Line: Gender and the Social Construction of Race in Pre-Revolutionary Haiti." *Journal of Caribbean History* 30 (1996): 28–50.

Gascoigne, John. "The German Enlightenment and the Pacific." In *The Anthropology of the Enlightenment*, edited by Larry Wolff and Marco Cipolloni. Stanford University Press, 2007.

Gascoigne, John. *Joseph Banks and the English Enlightenment*. Cambridge University Press, 1994.

Gates, Henry Louis, Jr., and Andrew S. Curran. *Who's Black and Why?* Harvard University Press, 2022.

Gattrell, Vic. *City of Laughter: Sex and Satire in Eighteenth-Century London*. Atlantic Books, 2006.

Gélis, Jacques. "Refaire le corps: Les déformations volontaires du corps de l'enfant à la naissance." *Ethnologie française* 14 (1984): 7–28.

George, M. Dorothy. *London Life in the Eighteenth Century*. Peregrine Books, 1966.

Gera, Deborah Levine. *Ancient Greek Ideas on Speech, Language, and Civilization*. Oxford University Press, 2003.

Gerzina, Gretchen. *Black England*. John Murray, 1995.

Ghosh, Durba. *Sex and the Family in Colonial India*. Cambridge University Press, 2006.

Ghosh, Suresh Chandra. *The Social Condition of the British Community in Bengal*. Brill, 1970.

Gibson, William, and Joanne Begiato. *Sex and the Church in the Long Eighteenth Century*. I. B. Tauris, 2017.

Gilbert, Geoffrey. Introduction to *T. R. Malthus: Critical Responses*, 1:1–22. Edited by Geoffrey Gilbert. 4 vols. Routledge, 1998.

Gilbert, Ruth. *Early Modern Hermaphrodites: Sex and Other Stories*. Palgrave, 2002.

Gilbert, Ruth. "Seeing and Knowing: Science, Pornography and Early Modern Hermaphrodites." In *At the Borders of the Human: Beasts, Bodies and Natural Philosophy in the Early Modern Period*, edited by Erica Fudge, Ruth Gilbert, and Susan Wiseman. Macmillan, 2002.

Glacken, Clarence J. *Traces on the Rhodian Shore: Nature and Culture in Western Thought from Ancient Times to the End of the Eighteenth Century*. University of California Press, 1967.

Glass, D. V. *Numbering the People*. Saxon House, 1973.

Goetz, Rebecca Anne. *The Baptism of Early Virginia*. Johns Hopkins University Press, 2012.

Goldberg, David Theo. *Racist Culture*. Blackwell, 1993.

Goldberg, Jonathan. *Sodometries*. Stanford University Press, 1992.

Goldenberg, David M. *The Curse of Ham: Race and Slavery in Early Christianity, Judaism and Islam*. Princeton University Press, 2003.

Gould, Stephen Jay. *The Mismeasure of Man*. Rev. ed. Norton, 1996.

Gould, Stephen Jay. "On Mental and Visual Geometry." *Isis* 89 (1998): 502–4.

Goveia, Elsa V. *Slave Society in the British Leeward Islands at the End of the Eighteenth Century*. Yale University Press, 1965.

Gowing, Laura. *Common Bodies: Women, Touch, and Power in the Seventeenth Century*. Yale University Press, 2003.

Graham, John. "Lavater's Physiognomy in England." *JHI* 22 (1961): 561–72.

Gray, B. Kirkman. *History of English Philanthropy*. P. S. King & Son, 1905.

Griffin, Emma. "A Conundrum Resolved? Rethinking Courtship, Marriage and Population Growth in Eighteenth-Century England." *P&P* 215 (2012): 125–64.

Gross, Ariela J. *What Blood Won't Tell: A History of Race on Trial in America*. Harvard University Press, 2008.

Groves, J. Percy. *The 66th Berkshire Regiment*. Reading, 1887.

Guasco, Michael. *Slaves and Englishmen*. University of Pennsylvania Press, 2014.

Guerrini, Anita. "Advertising Monstrosity: Broadsides and Human Exhibition in Early Eighteenth-Century London." In *Ballads and Broadsides in Britain, 1500–1800*, edited by Patricia Fumerton and Anita Guerrini. Ashgate, 2010.

Guerrini, Anita. "The Hermaphrodite of Charing Cross." In *The Uses of Humans in Experiment*, edited by Erika Dyck and Larry Stewart. Brill, 2016.

Guest, Harriet. *Empire, Barbarism, and Civilisation: Captain Cook, William Hodges and the Return to the Pacific*. Cambridge University Press, 2007.

Guest, Harriet. "The Great Distinction: Figures of the Exotic in the Work of William Hodges." *Oxford Art Journal* 12 (1989): 36–58.

Guest, Harriet. "Ornament and Use: Mai and Cook in London." In Wilson, *New Imperial History*, 317–44.

Guicciardi, Jean-Pierre. "Hermaphrodite et le prolétaire." *Dix-huitième siècle* 12 (1980): 49–77.

Gwilliam, Tassie. "Female Fraud: Counterfeit Maidenheads in the Eighteenth Century." *Journal of the History of Sexuality* 6 (1996): 518–48.

Hackforth-Jones, Jocelyn. "Mai/Omai in London and the South Pacific: Performativity, Cultural Entanglement, and Indigenous Appropriation." In *Material Identities*, edited by Joanna Sofaer. Blackwell, 2007.

Hall, Catherine. *Civilising Subjects: Metropole and Colony in the English Imagination*. Cambridge University Press, 2002.

Hall, Catherine. *Lucky Valley: Edward Long and the History of Racial Capitalism*. Cambridge University Press, 2024.

Hall, Catherine. "Whose Memories? Edward Long and the Work of Re-Remembering." In *Britain's History and Memory of Transatlantic Slavery*, edited by Katie Donington, Ryan Hanley, and Jessica Moody. Liverpool University Press, 2016.

Hall, Catherine, and Sonya O. Rose, eds. *At Home with the Empire: Metropolitan Culture and the Imperial World*. Cambridge University Press, 2007.

Hall, Douglas. *In Miserable Slavery: Thomas Thistlewood in Jamaica, 1750–86*. Macmillan, 1992.

Hall, Max. *Benjamin Franklin and Polly Baker*. University of North Carolina Press, 1960.

Hall, Stuart. *Representation: Cultural Representations and Signifying Practices*. Sage, 1997.

Hall, Stuart. "Who Needs 'Identity'?" In *Identity: A Reader*, edited by Paul Du Gay, Jessica Evans, and Peter Redman. Sage, 2000.

Hamlin, Christopher. "Public Sphere to Public Health: The Transformation of 'Nuisance.'" In *Medicine, Health, and the Public Sphere in Britain, 1600–2000*, edited by S. Sturdy. Routledge, 2002.

Hammett, Ian Maxwell. "Lord Monboddo and the Lockian Science of Man." *Bulletin of Hiroshima Jogakuin College* 39 (1989): 63–80.

Hanley, Ryan. *Beyond Slavery and Abolition: Black British Writing, c. 1770–1830*. Cambridge University Press, 2019.

Hannaford, Ivan. *Race: The History of an Idea in the West*. Johns Hopkins University Press, 1996.

Haraway, Donna. *Primate Visions*. Verso, 1992.

Hardgrave, Robert L., Jr. *A Portrait of the Hindus: Balthazar Solvyns & the European Image of India, 1760–1824*. Oxford University Press, 2004.

Harol, Corinne. *Enlightened Virginity in Eighteenth-Century Literature*. Palgrave Macmillan, 2006.

Harris, Bob. "The *London Evening Post* and Mid-Eighteenth-Century British Politics." *EHR* 110 (1995): 1132–56.

Harrison, Mark. *Climates and Constitutions: Health, Race, Environment and British Imperialism in India, 1600–1850*. Oxford University Press, 1999.

Harrison, Mark. *Medicine in an Age of Commerce and Empire*. Oxford University Press, 2010.

Harrison, Mark. "'The Tender Frame of Man': Disease, Climate, and Racial Difference in India and the West Indies, 1760–1860." *Bulletin of the History of Medicine* 70 (1996): 68–93.

Harvey, Karen. "A Century of Sex? Gender, Bodies, and Sexuality in the Long Eighteenth Century." *Historical Journal* 45 (2002): 899–916.

Harvey, Karen. *The Impostress Rabbit Breeder: Mary Toft and Eighteenth-Century England*. Oxford University Press, 2021.

Harvey, Karen. *The Little Republic: Masculinity and Domestic Authority in Eighteenth-Century Britain*. Oxford University Press, 2012.

Harvey, Karen. *Reading Sex in the Eighteenth Century: Bodies and Gender in English Erotic Culture*. Cambridge University Press, 2004.

Harvey, Karen. "Sexuality and the Body." In *Women's History, 1700–1850*, edited by Hannah Barker and Elaine Chalus. Routledge, 2005.

Harvey, Karen. "The Substance of Sexual Difference: Change and Persistence in Representations of the Body in Eighteenth-Century England." *Gender and History* 14 (2002): 202–23.

Hawes, C. J. *Poor Relations: The Making of a Eurasian Community in British India, 1773–1833*. Curzon, 1996.

Heckscher, Eli F. *Mercantilism*. Edited by E. F. Söderlund. 2nd ed. Allen & Unwin, 1955.

Hegarty, Neil. "Unruly Subjects: Sexuality, Science, and Discipline in Eighteenth-Century Pacific Exploration." In Lincoln, *Science and Exploration*, 183–97.

Henderson, Tony. *Disorderly Women in Eighteenth-Century London: Prostitution and Control in the Metropolis, 1730–1830*. Longman, 1999.

Heng, Geraldine. *The Invention of Race in the European Middle Ages*. Cambridge University Press, 2018.

Heuman, Gad. *Between Black and White: Race, Politics, and the Free Coloreds in Jamaica, 1792–1865*. Clio, 1981.

Higman, B. W. "Demography and Family Structures." In Eltis and Engerman, *Cambridge World History of Slavery*, 479–511.

Higman, B. W. "Lady Nugent's Social History." *Journal of Caribbean History* 48 (2014): 65–85.

Higman, B. W. *Plantation Jamaica, 1750–1850*. University of the West Indies Press, 2005.

Higman, B. W. *Proslavery Priest: The Atlantic World of John Lindsay, 1729–1788*. University of the West Indies Press, 2011.

Higman, B. W. *Slave Population and Economy in Jamaica, 1807–1834*. Cambridge University Press, 1976.

Higman, B. W. *Slave Populations of the British Caribbean, 1807–1834*. Johns Hopkins University Press, 1984.

Higman, B. W. "Slavery and the Development of Demographic Theory in the Age of the American Revolution." In *Slavery and British Society, 1776–1846*, edited by James Walvin. Macmillan, 1982.

Higman, B. W. "The West India 'Interest' in Parliament, 1807–1833." *Historical Studies* 13 (1967): 1–19.

Hill, Christopher. *The World Turned Upside Down: Radical Ideas During the English Revolution*. Penguin, 1975.

Hill, Thomas E., Jr., and Bernard Boxill. "Kant and Race." In *Race and Racism*, edited by Bernard Boxill. Oxford University Press, 2001.

Hilton, Boyd. *A Mad, Bad, & Dangerous People? England 1783–1846*. Oxford University Press, 2008.

Hitchcock, Tim. *English Sexualities, 1700–1800*. Macmillan, 1997.

Hitchcock, Tim, and Michèle Cohen, eds. *English Masculinities, 1660–1800*. Longman, 1999.

Hoare, Michael E. *The Tactless Philosopher: Johann Reinhold Forster (1729–98)*. Hawthorn, 1976.

Hodgen, Margaret T. *Early Anthropology in the Sixteenth and Seventeenth Centuries*. University of Pennsylvania Press, 1964.

Hodson, Chrisopher. "Weird Science: Identity in the Atlantic World." *W&MQ*, 3rd ser., 68 (2011): 227–32.

Hoffmeier, Michael H. "Maupertuis and the Eighteenth-Century Critique of Preexistence." *Journal of the History of Biology* 15 (1982): 119–44.

Hogarth, Rana A. *Medicalizing Blackness: Making Racial Difference in the Atlantic World, 1780–1840*. University of North Carolina Press, 2017.

Hollander, Samuel. "On Malthus's Population Principle and Social Reform." *History of Political Economy* 18 (1986): 187–235.

Holzman, James M. *The Nabobs in England*. New York, 1926.

Hoppit, Julian. "Political Arithmetic in Eighteenth-Century England." *Economic History Review* 49 (1996): 516–40.

Horowitz, Maryanne Cline, ed. *Race, Gender and Rank: Early Modern Ideas of Humanity*. University of Rochester Press, 1992.

Horsman, Reginald. "Origins of Racial Anglo-Saxonism in Great Britain before 1850." *JHI* 37 (1976): 387–410.

Hudson, Nicholas. "From 'Nation' to 'Race': The Origin of Racial Classification in Eighteenth-Century Thought." *ECS* 29 (1996): 247–64.

Hudson, Nicholas. "'Hottentots' and the Evolution of European Racism." *Journal of European Studies* 34 (2004): 308–32.

Hudson, Nicholas. "The 'Hottentot Venus,' Sexuality, and the Changing Aesthetics of Race, 1650–1850." *Mosaic* 41 (2008): 19–41.

Hudson, Nicholas. *Writing and European Thought 1600–1830*. Cambridge University Press, 1994.

Huet, Marie-Hélène. *Monstrous Imagination*. Harvard University Press, 1993.

Hunt, Lynn, Margaret Jacob, and Wijnand Mijnhardt, eds. *The Book That Changed Europe: Picart & Bernard's* Religious Ceremonies of the World. Harvard University Press, 2010.

Hunt, Lynn, and Margaret Jacob. Introduction to *Bernard Picart and the First Global Vision of Religion*, 1–13. Edited by Lynn Hunt, Margaret Jacob, and Wijnand Mijnhardt. Getty Research Institute, 2010.

Huse, William. "A Noble Savage on the Stage." *Modern Philology* 33 (1936): 303–16.

Hutchinson, E. P. *The Population Debate*. Houghton Mifflin, 1967.

Hyam, Ronald. *Empire and Sexuality*. Manchester University Press, 1990.

Innes, Joanna. *Inferior Politics: Social Problems and Social Policies in Eighteenth-Century Britain*. Oxford University Press, 2009.

Isaac, Benjamin. *The Invention of Racism in Classical Antiquity*. Princeton University Press, 2003.

Jacob, François. *The Logic of Life: A History of Heredity*. Pantheon, 1973.

Jacoby, Karl. "Slaves by Nature? Domestic Animals and Human Slaves." *Slavery and Abolition* 15 (1994): 89–99.

James, Patricia. *Population Malthus: His Life and Times*. Routledge, 2006.

Jenkins, Bill. "Race Before Darwin: Variation, Adaptation and the Natural History of Man in Post-Enlightenment Edinburgh, 1790–1835." *British Journal for the History of Science* 53 (2020): 333–50.

Jimack, Peter. *Diderot: Supplément au Voyage de Bougainville*. Grant & Cutler, 1988.

Johnson, Alan L. "Reproduction in the Female." In *Sturkie's Avian Physiology*, edited by Colin G. Scanes. 6th ed. Elsevier, 2015.

Jolly, Margaret. "From Point Venus to Bali Ha'i: Eroticism and Exoticism in Representations of the Pacific." In *Sites of Desire, Economies of Pleasure: Sexualities in Asia and the Pacific*, edited by Lenore Manderson and Margaret Jolly. University of Chicago Press, 1997.

Jolly, Margaret. "'Ill-Natured Comparisons': Racism and Relativism in European Representation of ni-Vanuata from Cook's Second Voyage." *Anthropology and History* 5 (1993): 331–64.

Jones, Cecily. "Contesting the Boundaries of Gender, Race and Sexuality in Barbadian Plantation Society." *Women's History Review* 12, no. 2 (2003): 195–232.

Jooken, Lieve. "Lord Monboddo and Adam Smith on the Origin and Development of Language." In *Florilegium Historiographiae Linguisticae: Études d'historographie de la linguistique et de grammaire comparée à la mémoire de Maurice Leroy*, edited by Jan de Clercq and Piet Desmet. Peeters, 1994.

Jooken, Lieve. "Lord Monboddo's 'Of the Origin and Progress of Language' (1773–1792) and Condillac's 'Essai sur l'origine des connoisances humaines' (1746): Two Distinct Views on Language and Thought." *Leuvense Bijdragen (Leuven Contributions in Linguistics and Philology* 80 (1991): 307–27.

Jooken, Lieve. "'A Sound Like the Gaggling of Geese': Lord Monboddo's Discussion of Languages of the New World." In *Orbis Supplementa: Perspectives on English*, edited by Keith Carlon, Kristin Davidse, and Brygida Rudzka-Ostyn. Peeters, 1994.

Joppien, Rüdiger. "Philippe Jacques de Loutherbourg's Pantomime 'Omai, or a Trip Round the World' and the Artists of Captain Cook's Voyages." In *Captain Cook and the South Pacific*, edited by T. C. Mitchell. British Museum, 1979.

Jordan, Winthrop D. "American Chiaroscuro: The Status and Definition of Mulattoes in the British Colonies." *W&MQ*, 3rd ser., 19 (1962): 183–200.

Jordan, Winthrop D. *White over Black: American Attitudes Toward the Negro, 1550–1812*. University of North Carolina Press, 1968.

Jordanova, Ludmilla. "The Popularization of Medicine: Tissot on Onanism." *Textual Practice* 1 (1987): 68–79.

Junker, Thomas. "Blumenbach's Racial Geometry." *Isis* 89 (1998): 498–501.

Kahn, Fritz. "Das Versehen der Schwangeren in Volksglaube und Dichtung." *Sexual-Probleme* 8 (1912): 300–328, 398–435.

Kalof, Linda, and William Bynum, eds. *A Cultural History of the Human Body*. 6 vols. Berg, 2010.

Kates, Gary. *Monsieur d'Eon Is a Woman*. Johns Hopkins University Press, 2001.

Katz, David S. "Shylock's Gender: Jewish Male Menstruation in Early Modern England." *Review of English Studies* 50 (1999): 440–62.

Katzew, Ilona. *Casta Painting: Images of Race in Eighteenth-Century Mexico*. Yale University Press, 2004.

Keller, Eve. "Embryonic Individuals: The Rhetoric of Seventeenth-Century Embryology and the Construction of Early-Modern Identity." *ECS* 33 (2000): 321–48.

Keller, Eve. *Generating Bodies and Gendered Selves: The Rhetoric of Reproduction in Early Modern England*. University of Washington Press, 2007.

Kettler, Andrew. *The Smell of Slavery*. Cambridge University Press, 2020.

Kidd, Colin. *British Identities Before Nationalism*. Cambridge University Press, 1999.

Kidd, Colin. *The Forging of Races: Race and Scripture in the Protestant Atlantic World, 1600–2000*. Cambridge University Press, 2006.

Kiernan, V. G. "Noble and Ignoble Savages." In Rousseau and Porter, *Exoticism*, 86–116.

Kilday, Anne-Marie. *A History of Infanticide in Britain, c. 1600 to the Present*. Palgrave, 2013.

King, Helen. *The One-Sex Body on Trial*. Ashgate, 2013.

Kitson, Peter J. *Romantic Literature, Race, and Colonial Encounter*. Palgrave, 2007.

Klein, Lawrence E. *Shaftesbury and the Culture of Politeness*. Cambridge University Press, 1994.

Kleingeld, Pauline. "Kant's Second Thoughts on Race." *Philosophical Quarterly* 57 (2007): 573–92.

Knights, Mark. *The Devil in Disguise: Deception, Delusion, and Fanaticism in the Early English Enlightenment*. Oxford University Press, 2011.

Knott, Sarah, and Barbara Taylor, eds. *Women, Gender, and Enlightenment*. Palgrave, 2005.

Kriz, Kay Dian. "Marketing Mulatresses in the Paintings and Prints of Agostino Brunias." In *The Global Eighteenth Century*, edited by Felicity A. Nussbaum. Johns Hopkins University Press, 2003.

Kudlien, Fridolf. "The Seven Cells of the Uterus: The Doctrine and Its Roots." *Bulletin of the History of Medicine* 39 (1965): 415–23.

Lamb, Jonathan. "Circumstances Surrounding the Death of John Hawkesworth." *ECL* 18, no. 3 (1994): 97–113.

Lamb, Jonathan. *Preserving the Self in the South Seas 1680–1840.* University of Chicago Press, 2001.

Lamb, Jonathan, Vanessa Smith, and Nicholas Thomas, eds. *Exploration and Exchange: A South Seas Anthology.* University of Chicago Press, 2000.

Langford, Paul. *A Polite and Commercial People: England, 1727–1783.* Oxford University Press, 1989.

Laqueur, Thomas. *Making Sex: Body and Gender from the Greeks to Freud.* Harvard University Press, 1990.

Laqueur, Thomas. "The Rise of Sex in the Eighteenth Century: Historical Context and Historiographical Implications." *Signs* 37 (2012): 802–12.

Laqueur, Thomas. "Sex in the Flesh." *Isis* 94 (2003): 300–306.

Laqueur, Thomas. *Solitary Sex: A Cultural History of Masturbation.* Zone, 2003.

Larson, James L. *Interpreting Nature: The Science of Living Form from Linnaeus to Kant.* Johns Hopkins University Press, 1994.

Lawson, Philip, and Jim Phillips. "'Our Execrable Banditti': Perceptions of Nabobs in Mid-Eighteenth Century Britain." *Albion* 16 (1984): 225–41.

Lennox, James G. "The Comparative Study of Animal Development: William Harvey and Aristotelianism." In Smith, *Problem of Animal Generation,* 21–46.

Lenoir, Timothy. "Kant, Blumenbach, and Vital Materialism in German Biology." *Isis* 71 (1980): 77–108.

Lettow, Susanne, ed. *Reproduction, Race, and Gender in Philosophy and the Early Life Sciences.* State University of New York Press, 2014.

Levene, Alysa, Jonathan Reinarz, and Andrew Williams. "Child Patients, Hospitals and the Home in Eighteenth-Century England." *Family & Community History* 15 (2012): 15–33.

Levine, Joseph M. *The Battle of the Books: History and Literature in the Augustan Age.* Cornell University Press, 1991.

Levine, Philippa. *Gender and Empire.* Oxford University Press, 2004.

Levine, Philippa. "Sexuality, Gender, and Empire." In *Gender and Empire,* edited by Philippa Levine. Oxford University Press, 2007.

Lewis, Gordon K. *Main Currents in Caribbean Thought: The Historical Evolution of Caribbean Society in Its Ideological Aspects, 1492–1900.* Johns Hopkins University Press, 1983.

Lewis, Rhodri. *William Petty on the Order of Nature: An Unpublished Manuscript Treatise.* Arizona Center for Medieval and Renaissance Studies, 2012.

Lewis, Rhodri. "William Petty's Anthropology: Religion, Colonialism, and the Problem of Human Diversity." *Huntington Library Quarterly* 74 (2011): 261–88.

Lifschitz, Avi. *Language and Enlightenment: The Berlin Debates of the Eighteenth Century.* Oxford University Press, 2012.

Lincoln, Margarette, ed. *Science and Exploration in the Pacific: European Voyages to the Southern Oceans in the 18th Century.* Boydell, 2001.

Lipski, Jakub. "Three Mid-Eighteenth-Century Mash-Ups: Hybridity and Conflicted Discourse in Robert Paltock's *Peter Wilkins* and Its Early Imitations." *1650–1850: Ideas, Aesthetics, and Inquiries in the Early Modern Era* 28 (2023): 119–39.

Livesay, Daniel. *Children of Uncertain Fortune: Mixed-Race Jamaicans in Britain and the Atlantic Family, 1733–1833*. University of North Carolina Press, 2018.

Loomba, Ania. *Shakespeare, Race, and Colonialism*. Oxford University Press, 2002.

Loomba, Ania, and Jonathan Burton. *Race in Early Modern England*. Palgrave Macmillan, 2007.

López-Beltrán, Carlos. "Forging Heredity: From Metaphor to Cause, a Reification Story." *Studies in the History and Philosophy of Science* 25 (1994): 211–35.

Lorenz, Maren. *Menschenzucht*. Wallstein, 2018.

Love, Henry Davidson. *Vestiges of Old Madras, 1640–1800*. 4 vols. John Murray, 1913.

Lowe, Richard G., and Randolph B. Campbell. "The Slave-Breeding Hypothesis: A Demographic Comment on the 'Buying' and 'Selling' States." *Journal of Southern History* 42 (1976): 401–12.

Luke, Hugh J., Jr. "Sir William Lawrence: Physician to Shelley and Mary." *Papers on English Language and Literature* 1 (1965): 141–52.

Maccubbin, Robert Purks, ed. *'Tis Nature's Fault: Unauthorised Sexuality During the Enlightenment*. Cambridge University Press, 1985.

MacDougall, Hugh A. *Racial Myth in English History*. University Press of New England, 1982.

MacLeod, Duncan J. *Slavery, Race, and the American Revolution*. Cambridge University Press, 1974.

MacMillan, Margaret. *Women of the Raj: The Mothers, Wives, and Daughters of the British Empire in India*. Thames & Hudson, 1988.

Magnusson, Lars. *Mercantilism: The Shaping of an Economic Language*. Routledge, 1994.

Major, Andrea. *Pious Flames: European Encounters with Sati in India, 1500–1830*. Oxford University Press, 2006.

Major, Andrea. *Slavery, Abolitionism and Empire in India, 1772–1843*. Liverpool University Press, 2012.

Major, Andrea. *Sovereignty and Social Reform in India: British Colonialism and the Campaign Against Sati, 1830–1860*. Routledge, 2011.

Malchow, H. L. *Gothic Images of Race in Nineteenth-Century Britain*. Stanford University Press, 1996.

Malcolmson, Cristina. *Studies of Skin Color in the Early Royal Society*. Ashgate, 2013.

Mander, Jenny. "No Woman Is an Island: The Female Figure in French Enlightenment Anthropology." In Knott and Taylor, *Women, Gender and Enlightenment*, 97–116.

Manyoni, Joseph R. "Extra-Marital Mating Patterns in Caribbean Family Studies: A Methodological Excursus." *Anthropologica* 22 (1980): 104–11.

Marshall, P. J. "British Society in India Under the East India Company." *Modern Asian Studies* 31 (1997): 89–108.

Marshall, P. J. *Edmund Burke and the British Empire in the West Indies*. Oxford University Press, 2019.

Martensen, Robert. "The Transformation of Eve: Women's Bodies, Medicine and Culture in Early Modern England." In Porter and Teich, *Sexual Knowledge*, 107–33.

Martin-Allanic, Jean-Etienne. *Bougainville navigateur et les découvertes de son temps*. 2 vols. Presses Universitaires de France, 1964.

Martínez, María Elena. *Genealogical Fictions: Limpieza de Sangre, Religion, and Gender in Colonial Mexico*. Stanford University Press, 2008.

Mason, Michael. *The Making of Victorian Sexuality.* Oxford University Press, 1994.

Mayr, Ernst. *The Growth of Biological Thought.* Harvard University Press, 1982.

Mazzolini, Renato. "Albinos, Leucoæthiopes, Dondos, Kakerlakken: Sulla storia dell'albinismo dal 1609 al 1812." In *La natura e il corpo,* edited by Giuseppe Olmi and Giuseppe Papagno. L. S. Olschki, 2006.

Mazzolini, Renato. "A Defense of Human Rights: Blumenbach on Albinism." In *Johann Friedrich Blumenbach: Race and Natural History,* edited by Nicolaas Rupke and Gerhard Lauer. Routledge, 2019.

Mazzolini, Renato. "*Las Castas*: Interracial Crossing and Social Structure, 1770–1835." In *Heredity Produced: At the Crossroads of Biology, Politics, and Culture, 1500–1870,* edited by Staffan Müller-Wille and Hans-Jörg Rheinberger. MIT Press, 2007.

Mazzoni, Cristina. *Maternal Impressions: Pregnancy and Childbirth in Literature and Theory.* Cornell University Press, 2002.

Mc Inerney, Tim. *Nobility and the Making of Race in Eighteenth-Century Britain.* Bloomsbury, 2023.

McCalman, Iain. *Radical Underworld: Prophets, Revolutionaries and Pornographers in London, 1795–1840.* Oxford University Press, 1993.

McClure, Ruth K. *Coram's Children: The London Foundling Hospital in the Eighteenth Century.* Yale University Press, 1981.

McCormack, Matthew, ed. *Public Men: Masculinity and Politics in Modern Britain.* Palgrave, 2007.

McCormick, E. H. *Omai: Pacific Envoy.* Auckland University Press, 1977.

McCormick, Ted. "Political Arithmetic's 18th Century Histories: Quantification in Politics, Religion, and the Public Sphere." *History Compass* 12 (2014): 239–51.

McCormick, Ted. *William Petty and the Ambitions of Political Arithmetic.* Oxford University Press, 2009.

McLaren, Angus. *Reproductive Rituals: The Perception of Fertility in England from the Sixteenth Century.* Methuen, 1984.

McNeil, Peter. *Pretty Gentlemen: Macaroni Men and the Eighteenth-Century Fashion World.* Yale University Press, 2018.

Meek, Ronald L. *Social Science and the Ignoble Savage.* Cambridge University Press, 1976.

Meijer, Miriam Claude. "Petrus Camper on the Origin and Color of Blacks." *History of Anthropology Newsletter* 24, no. 2 (1997): 3–9.

Meijer, Miriam Claude. *Race and Aesthetics in the Anthropology of Petrus Camper (1722–1789).* Rodopi, 1999.

Mendelson, Sara, and Patricia Crawford. *Women in Early Modern England, 1550–1720.* Oxford University Press, 1998.

Mensch, Jennifer. "Kant and the Skull Collectors: German Anthropology from Blumenbach to Kant." In *Kant and His German Contemporaries. Volume 1: Logic, Mind, Epistemology, Science and Ethics,* edited by Corey W. Dyck and Falk Wunderlich. Cambridge University Press, 2017.

Merians, Linda. *Envisioning the Worst: Representations of "Hottentots" in Early-Modern England.* University of Delaware Press, 2001.

Meusel, Johann Georg. *Lexikon der vom Jahr 1750 bis 1800 verstorbenen teutschen Schriftsteller.* 15 vols. Leipzig, 1802–16.

Midgley, Clare. *Women Against Slavery: The British Campaigns, 1780–1870.* Routledge, 1992.

Mikkelsen, Jon M., ed. *Kant and the Concept of Race.* State University of New York Press, 2013.

Miller, Genevieve. *The Adoption of Inoculation for Smallpox in England and France.* University of Pennsylvania Press, 1957.

Miller, Genevieve. "Putting Lady Mary in Her Place: A Discussion of Historical Causation." *Bulletin of the History of Medicine* 55 (1981): 2–16.

Miller, Leo. *John Milton Among the Polygamophiles.* Loewenthal, 1974.

Mills, Charles W. "Kant's *Untermenschen.*" In *Race and Racism in Modern Philosophy,* edited by Andrew Valls. Cornell University Press, 2005.

Mills, Charles W. *The Racial Contract.* Cornell University Press, 1997.

Minardi, Margot. "The Boston Inoculation Controversy of 1721–1722: An Incident in the History of Race." *W&MQ,* 3rd ser., 61 (2004): 47–76.

Minchinton, Walter E., ed. *Mercantilism: System or Expediency?* D. C. Heath, 1969.

Miramon, Charles de. "Noble Dogs, Noble Blood: The Invention of the Concept of Race in the Late Middle Ages." In Eliav-Feldon, Isaac, and Ziegler, *Origins of Racism,* 200–216.

Moes, Robert J., and C. D. O'Malley. "Realdo Colombo: 'On Those Things Rarely Found in Anatomy.'" *Bulletin of the History of Medicine* 34 (1960): 508–28.

Mohammed, Patricia. "'But Most of All Mi Love Me Browning': The Emergence in Eighteenth and Nineteenth-Century Jamaica of the Mulatto Woman as the Desired." *Feminist Review* 65 (2000): 22–48.

Mole, Tom. *Byron's Romantic Celebrity: Industrial Culture and the Hermeneutic of Intimacy.* Palgrave, 2007.

Mole, Tom, ed. *Romanticism and Celebrity Culture, 1750–1850.* Cambridge University Press, 2009.

Molineux, Catherine. *Faces of Perfect Ebony: Encountering Atlantic Slavery in Imperial Britain.* Harvard University Press, 2012.

Moorehead, Alan. *The Fatal Impact.* Penguin, 2000 [1969].

Morgan, Jennifer L. *Laboring Women: Reproduction and Gender in New World Slavery.* University of Pennsylvania Press, 2004.

Morgan, Jennifer L. "'Some Could Suckle over Their Shoulder': Male Travelers, Female Bodies, and the Gendering of Racial Ideology, 1500–1770." *W&MQ,* 3rd ser., 54 (1997): 167–92.

Morgan, Kenneth. "Slave Women and Reproduction in Jamaica, c. 1776–1834." *History* 91 (1996): 231–53.

Morgan, Philip D. "Slaves and Livestock in Eighteenth-Century Jamaica: Vineyard Pen, 1750–1751." *W&MQ,* 3rd ser., 52 (1995): 47–76.

Morieux, Renaud. "Diplomacy from Below and Belonging: Fishermen and Cross-Channel Relations in the Eighteenth Century." *P&P* 202 (2009): 83–125.

Morris, Robert J. "*Aikāne:* Accounts of Hawaiian Same-Sex Relationships in the Journals of Captain Cook's Third Voyage (1776–80)." *Journal of Homosexuality* 19, no. 4 (1990): 21–54.

Morris, Robert J. "Same-Sex Friendships in Hawaiian Lore: Constructing the Canon." In *Oceanic Homosexualities,* edited by Stephen O. Murray. Garland, 1992.

Mosse, George L. *Nationalism and Sexuality*. Howard Fertig, 1985.

Mosse, George L. *Toward the Final Solution: A History of European Racism*. J. M. Dent, 1978.

Müller-Wille, Staffan. "Cabbage, Tulips, Ethiopians—'Experiments' in Early Modern Heredity." In *A Cultural History of Heredity I: 17th and 18th Centuries*, edited by Hans-Jörg Rheinberger, Peter McLaughlin, and Staffan Müller-Wille. Max-Planck-Institut für Wissenschaftsgeschichte, 2002.

Nanda, Serena. *Neither Man Nor Woman: The Hijras of India*. 2nd ed. Wadsworth, 1999.

Nechtman, Tillman W. *Nabobs: Empire and Identity in Eighteenth-Century Britain*. Cambridge University Press, 2010.

Needham, Joseph. *History of Embryology*. 2nd ed. Cambridge University Press, 1959.

Nelson, William Max. *Enlightenment Biopolitics: A History of Race, Eugenics, and the Making of Citizens*. University of Chicago Press, 2024.

Nelson, William Max. "Making Men: Enlightenment Ideas of Racial Engineering." *American Historical Review* 115 (2010): 1364–94.

Newman, Brooke N. *A Dark Inheritance: Blood, Race, and Sex in Colonial Jamaica*. Yale University Press, 2018.

Newman, Gerald. *Rise of English Nationalism*. St. Martin's, 1997.

Newton, Melanie J. *The Children of Africa in the Colonies: Free People of Color in Barbados in the Age of Emancipation*. Louisiana State University Press, 2008.

Nichols, R. H., and F. A. Wray. *The History of the Foundling Hospital*. Oxford University Press, 1935.

Nicholson, Linda. "Interpreting Gender." *Signs* 20 (1994): 79–105.

Nussbaum, Felicity A. *The Limits of the Human: Fictions of Anomaly, Race, and Gender in the Long Eighteenth Century*. Cambridge University Press, 2003.

Nussbaum, Felicity A. *Torrid Zones: Maternity, Sexuality, and Empire in Eighteenth-Century English Narratives*. Johns Hopkins University Press, 1995.

O'Brien, Karen. *Narratives of Enlightenment: Cosmopolitan History from Voltaire to Gibbon*. Cambridge University Press, 1997.

O'Connell, Lisa. "'Matrimonial Ceremonies Displayed': Popular Ethnography and Enlightened Imperialism." *ECL* 26, no. 3 (2002): 98–116.

Obeyesekere, Gananath. *Cannibal Talk: The Man-Eating Myth and Human Sacrifice in the South Seas*. University of California Press, 2005.

Odumosu, Temi. *Africans in English Caricature, 1769–1819*. Harvey Miller, 2017.

Oliver, Douglas L. *Ancient Tahitian Society*. 2 vols. University Press of Hawaii, 1974.

Orr, Bridget. "'Southern Passions Mix with Northern Art': Miscegenation and the *Endeavour* Voyage." *Eighteenth-Century Life* 18, no. 3 (1994): 212–31.

Otterness, Philip. *Becoming German: The 1709 Palatine Migration to New York*. Cornell University Press, 2004.

Outhwaite, R. B. *Clandestine Marriage in England, 1500–1850*. Hambledon, 1995.

Outram, Dorinda. *The Enlightenment*. 3rd ed. Cambridge University Press, 2013.

Owen, David. *English Philanthropy, 1660–1960*. Harvard University Press, 1964.

Pagden, Anthony. *The Fall of Natural Man: The American Indian and the Origins of Comparative Ethnology*. Cambridge University Press, 1990.

Pagel, Walter. *New Light on William Harvey*. Karger, 1976.

Park, Katharine. "Cadden, Laqueur, and the 'One-Sex Body.'" *Medieval Feminist Forum* 46, no. 1 (2010): 96–100.

Park, Katharine, and Robert A. Nye. "Destiny Is Anatomy." *New Republic*, February 18, 1991.

Passmann, Dirk F. "Mud and Slime: Some Implications of the Yahoos' Genealogy and the History of an Idea." *British Journal for Eighteenth-Century Studies* 11 (1988): 1–17.

Pateman, Carole. *The Sexual Contract*. Polity, 1988.

Patterson, Orlando. *The Sociology of Slavery*. Macgibbon & Kee, 1967.

Paugh, Katherine. "The Politics of Childbearing in the British Caribbean and the Atlantic World During the Age of Abolition, 1776–1838." *P&P* 221 (2013): 119–160.

Paugh, Katherine. *The Politics of Reproduction: Race, Medicine and Fertility in the Age of Abolition*. Oxford University Press, 2017.

Peabody, Sue. "Slavery, Freedom, and the Law in the Atlantic World, 1420–1807." In Eltis and Engerman, *Cambridge World History of Slavery*, 594–630.

Peakman, Julie. *Lascivious Bodies: A Sexual History of the Eighteenth Century*. Atlantic Books, 2004.

Pearce, S. J. "The Inquisitor and the Moseret: *The Invention of Race in the European Middle Ages* and the New English Colonialism in Jewish Historiography." *Medieval Encounters* 26 (2020): 145–90.

Pearsall, Sarah M. S. *Polygamy: An Early American History*. Yale University Press, 2019.

Pearson, W. H. "Hawkesworth's Alterations." *Journal of Pacific History* 7 (1972): 45–72.

Perry, Ruth. "Colonizing the Breast: Sexuality and Maternity in Eighteenth-Century England." *ECS* 2 (1996): 204–34.

Perry, Thomas W. *Public Opinion, Propaganda, and Politics in Eighteenth-Century England*. Harvard University Press, 1962.

Petley, Christer. "Gluttony, Excess, and the Fall of the Planter Class in the British Caribbean." *Atlantic Studies* 9 (2012): 85–106.

Petley, Christer. "Slavery, Emancipation and the Creole World View of Jamaican Colonists, 1800–1834." *Slavery and Abolition* 26 (2005): 93–114.

Pettit, Alexander, and Patrick Spedding, eds. *Eighteenth-Century British Erotica*. 5 vols. Pickering & Chatto, 2002.

Pinches, Sylvia. "Women as Agents and Objects of Charity in Eighteenth-Century Birmingham." In *Women and Urban Life in Eighteenth-Century England*, edited by Rosemary Sweet and Penelope Lane. Ashgate, 2003.

Pincus, Steven. *1688: The First Modern Revolution*. Yale University Press, 2009.

Pincus, Steven. "Rethinking Mercantilism." *W&MQ*, 3rd ser., 69 (2012): 3–34.

Pinto-Correia, Clara. *The Ovary of Eve: Egg and Sperm and Preformation*. University of Chicago Press, 1997.

Pinto-Correia, Clara, and João Lourenço Monteiro. "Science in Support of Racial Mixture: Charles-Augustin Vandermonde's Enlightenment Program for Improving the Health and Beauty of the Human Species." *Endeavour* 38 (2014): 19–26.

Poliakov, Léon. *The Aryan Myth*. Translated by Edmund Howard. Chatto & Heinemann, 1974.

Pomata, Gianna. "Menstruating Men: Similarity and Difference of the Sexes in Early Modern Medicine." In Finucci and Brownlee, *Generation and Degeneration*, 109–52.

Poole, William. "Francis Lodwick's Creation: Theology and Natural Philosophy in the Early Royal Society." *JHI* 66 (2005): 245–63.

Popkin, Richard H. "Hume's Racism." *Philosophical Forum* 9 (1977–78): 211–26.

Popkin, Richard H. *Isaac la Peyrère (1596–1676): His Life, Work and Influence.* Brill, 1987.

Popkin, Richard H. "Medicine, Racism, Anti-Semitism: A Dimension of Enlightenment Culture." In *The Languages of Psyche: Mind and Body in Enlightenment Thought*, edited by G. S. Rousseau. University of California Press, 1990.

Popkin, Richard H. "The Philosophical Basis of Eighteenth-Century Racism." In *Racism in the Eighteenth Century*, edited by Harold E. Pagliaro. Case Western Reserve University Press, 1973.

Porter, Martin. *Windows of the Soul: The Art of Physiognomy in European Culture, 1470–1780.* Oxford University Press, 2005.

Porter, Roy. "Bodies of Thought: Thoughts About the Body in Eighteenth-Century England." In *Interpretation and Cultural History*, edited by Joan H. Pittock and Andrew Wear. Macmillan, 1991.

Porter, Roy. "Cleaning Up the Great Wen: Public Health in Eighteenth-Century London." *Medical History* Suppl. 11 (1991): 61–75.

Porter, Roy. "The Exotic as Erotic: Captain Cook at Tahiti." In Rousseau and Porter, *Exoticism*, 117–44.

Porter, Roy. "History of the Body Reconsidered." In *New Perspectives on Historical Writing*, edited by Peter Burke. 2nd ed. Cambridge University Press, 2001.

Porter, Roy. "The Literature of Sexual Advice before 1800." In Porter and Teich, *Sexual Knowledge*, 134–57.

Porter, Roy. "Mixed Feelings: The Enlightenment and Sexuality in Eighteenth-Century Britain." In Boucé, *Sexuality*, 1–27.

Porter, Roy. "The Sexual Politics of James Graham." *British Journal for Eighteenth-Century Studies* 5 (1984): 199–206.

Porter, Roy. "Spreading Carnal Knowledge or Selling Dirt Cheap? Nicolas Venette's *Tableau de l'Amour Conjugal* in Eighteenth-Century England." *Journal of European Studies* 14 (1984): 233–55.

Porter, Roy. "A Touch of Danger: The Man-Midwife as Sexual Predator." In Porter and Rousseau, *Sexual Underworlds*, 206–32.

Porter, Roy, and Lesley Hall. *The Facts of Life: The Creation of Sexual Knowledge in England, 1650–1950.* Yale University Press, 1995.

Porter, Roy, and G. S. Rousseau, eds. *Sexual Underworlds of the Enlightenment.* Manchester University Press, 1987.

Porter, Roy, and Mikuláš Teich, eds. *Sexual Knowledge, Sexual Science: The History of Attitudes to Sexuality.* Cambridge University Press, 1994.

Potter, George Reuben. "Unpublished Marginalia in Coleridge's Copy of Malthus's *Essay on Population.*" *Proceedings of the Modern Language Association of America* 51 (1936): 1061–68.

Pratt, Mary Louise. *Imperial Eyes: Travel Writing and Transculturation.* Routledge, 1992.

Probert, Rebecca. "The Judicial Interpretation of Lord Hardwicke's Act 1753." *Legal History* 23 (2002): 129–51.

Probert, Rebecca. *Marriage Law and Practice in the Long Eighteenth Century.* Cambridge University Press, 2009.

Prochaska, Frank. *Women and Philanthropy in Nineteenth-Century England.* Oxford University Press, 1990.

Rabin, Dana. "Beyond 'Lewd Women' and 'Wanton Wenches': Infanticide and Child-Murder in the Long Eighteenth Century." In *Writing British Infanticide: Child Murder, Gender, and Print, 1722–1859*, edited by Jennifer Thorn. Associated University Presses, 2003.

Ramey, Lynn T. *Black Legacies: Race and the European Middle Ages*. University Press of Florida, 2014.

Rawson, Claude. *God, Gulliver, and Genocide*. Oxford University Press, 2001.

Razzell, Peter. *Essays in English Population History*. Caliban, 1994.

Read, Sara. *Menstruation and the Female Body in Early Modern England*. Palgrave, 2013.

Reddy, Gayatri. *With Respect to Sex: Negotiating Hijra Identity in South India*. University of Chicago Press, 2005.

Reill, Peter Hanns. *Vitalizing Nature in the Enlightenment*. University of California Press, 2005.

Reis, Elizabeth. *Bodies in Doubt: An American History of Intersex*. 2nd ed. Johns Hopkins University Press, 2021.

Rennie, Neil. *Far-Fetched Facts: The Literature of Travel and the Idea of the South Seas*. Oxford University Press, 1995.

Rensch, Karl H. "Appendix II: Forster's Polynesian Linguistics." In Forster, *Observations*, 383–400.

Rensch, Karl H. *The Language of the Noble Savage*. Archipelago, 2000.

Rensch, Karl H. "The Language of the Noble Savage: Early European Perceptions of Tahitian." In *Currents in Pacific Linguistics*, edited by Robert Blust. Australian National University, 1991.

Richards, Robert J. "Kant and Blumenbach on the *Bildungstrieb*: A Historical Misunderstanding." *SHPBBMS* 31 (2000): 11–32.

Richards, Robert J. *The Romantic Conception of Life*. University of Chicago Press, 2002.

Riddle, John M. *Contraception and Abortion from the Ancient World to the Renaissance*. Harvard University Press, 1992.

Ridley, Glynis. *The Discovery of Jeanne Baret*. Crown, 2011.

Riley, James C. *The Eighteenth-Century Campaign to Avoid Disease*. Macmillan, 1987.

Riley, James C. *Population Thought in the Age of the Demographic Revolution*. Carolina Academic Press, 1985.

Ritchie, Robert C. *The Duke's Province*. University of North Carolina Press, 1977.

Ritvo, Harriet. "Barring the Cross: Miscegenation and Purity in Eighteenth- and Nineteenth-Century Britain." In *Human, All Too Human*, edited by Diana Fuss. Routledge, 1996.

Ritvo, Harriet. "Possessing Mother Nature: Genetic Capital in Eighteenth-Century Britain." In *Early Modern Conceptions of Property*, edited by John Brewer and Susan Staves. Routledge, 1995.

Robb, Peter, ed. *The Concept of Race in South Asia*. Oxford University Press, 1995.

Robbins, Caroline. "A Note on General Naturalization Under the Later Stuarts." *Journal of Modern History* 34 (1962): 168–77.

Roberts, Justin. *Slavery and the Enlightenment in the British Atlantic, 1750–1807*. Cambridge University Press, 2013.

Roderick, Colin. "Sir Joseph Banks, Queen Oberea and the Satirists." In *Captain James Cook: Image and Impact*, edited by Walter Veit. Hawthorn, 1972.

Roe, Shirley A. *Matter, Life, and Generation: Eighteenth-Century Embryology and the Haller-Wolff Debate.* Cambridge University Press, 1981.

Roger, Jacques. *Buffon.* Translated by Sarah Lucille Bonnefoi. Cornell University Press, 1997.

Roger, Jacques. *The Life Sciences in Eighteenth-Century French Thought.* Translated by Robert Ellrich. Stanford University Press, 1997.

Rogers, Nicholas. *Mayhem: Post-War Crime and Violence in Britain, 1748–1753.* Yale University Press, 2012.

Roper, Lyndal. *Oedipus and the Devil: Witchcraft, Sexuality and Religion in Early Modern Europe.* Routledge, 1994.

Rosen, George. *From Medical Police to Social Medicine: Essays on the History of Health Care.* Science History, 1974.

Rosenberg, Aubrey. *Simon Tyssot de Patot and His Work, 1655–1738.* Springer, 1972.

Rousseau, G. S. *The Notorious Sir John Hill.* Lehigh University Press, 2012.

Rousseau, G. S. "Pineapples, Pregnancy, Pica, and *Peregrine Pickle.*" In *Tobias Smollett: Bicentennial Essays,* edited by G. S. Rousseau and Paul-Gabriel Boucé. Oxford University Press, 1971.

Rousseau, G. S, and Roy Porter, eds. *Exoticism and the Enlightenment.* Manchester University Press, 1990.

Rousseau, G. S, and Roy Porter. Introduction to *Exoticism and the Enlightenment,* 1–22. Edited by Rousseau and Porter.

Rusnock, Andrea A. *Vital Accounts: Quantifying Health and Population in Eighteenth-Century England and France.* Cambridge University Press, 2002.

Russell, Gillian. "An 'Entertainment of Oddities': Fashionable Sociability and the Pacific in the 1770s." In Wilson, *New Imperial History,* 48–70.

Russell, Nicholas. *Like Engend'ring Like: Heredity and Animal Breeding in Early Modern England.* Cambridge University Press, 1986.

Ryden, David Beck. *West Indian Slavery and British Abolition, 1783–1807.* Cambridge University Press, 2009.

Sagar, Paul. "Of Mushrooms and Method: History and the Family in Hobbes's Science of Politics." *European Journal of Political Theory* 14 (2015): 98–117.

Said, Edward W. *Orientalism.* Penguin, 2003.

Salih, Sara. *Representing Mixed-Race.* Routledge, 2011.

Salmond, Anne. *Trial of the Cannibal Dog: Captain Cook in the South Seas.* Allen Lane, 2003.

Sayre, Gordon M. *Les Sauvages Américains: Representations of Native Americans in French and English Colonial Literature.* University of North Carolina Press, 1997.

Sbiroli, Lynn Salkin. *Libertine o madri illibate.* Marsilio Editori, 1989.

Schaffer, Simon. "'On Seeing Me Write': Inscription Devices in the South Seas." *Representations* 97 (2007): 90–122.

Schapera, I., ed. *The Early Cape Hottentots.* Cape Town: Van Riebeeck Society, 1933.

Schaub, Jean-Frédéric, and Silvia Sebastiani. *Race et histoire dans les sociétés occidentales (XVe-XVIIIe siècle).* Albin Michel, 2021.

Schiebinger, Londa. "The Anatomy of Difference: Race and Sex in Eighteenth-Century Science." *ECS* 23 (1990): 387–405.

Schiebinger, Londa. "Mammals, Primatology and Sexology." In Porter and Teich, *Sexual Knowledge,* 184–209.

Schiebinger, Londa. *The Mind Has No Sex? Women in the Origins of Modern Science.* Harvard University Press, 1989.

Schiebinger, Londa. *Nature's Body: Sexual Politics and the Making of Modern Science.* Pandora, 1993.

Schiebinger, Londa. *Plants and Empire: Colonial Bioprospecting in the Atlantic World.* Harvard University Press, 2004.

Schiebinger, Londa. "Skelettestreit." *Isis* 94 (2003): 307–13.

Schwartz, Stuart B., and Frank Salomon. "New Peoples and New Kinds of People: Adaptation, Readjustment, and Ethnogenesis in South American Indigenous Societies. Colonial Era)." In *The Cambridge History of the Native Peoples of the Americas, Vol. 3: South America, Part 2,* edited by Stuart B. Schwartz and Frank Salomon. Cambridge University Press, 1999.

Scobie, Ruth. *Celebrity Culture and the Myth of Oceania in Britain, 1770–1823.* Boydell, 2019.

Scott, Hew, ed. *Fasti Ecclesiæ Scoticanæ.* 12 vols. Oliver & Boyd, 1915–.

Scott, Joan Wallach. "Gender: A Useful Category of Historical Analysis." *American Historical Review* 91 (1986): 1053–75.

Scott, Joan Wallach. *Gender and the Politics of History.* Rev. ed. Columbia University Press, 1999.

"Scraps of English Folklore, XII." *Folklore* 37 (1926): 79.

Scutt, Ronald, and Christopher Gotch. *Skin Deep: The Mystery of Tattooing.* P. Davies, 1974.

Sebastiani, Silvia. "Challenging Boundaries: Apes and Savages in Enlightenment." In *Simianization: Apes, Gender, Class, and Race,* edited by Wulf D. Hund, Charles W. Mills, and Silvia Sebastiani. Lit, 2015.

Sebastiani, Silvia. "Monboddo's 'Ugly Tail': The Question of Evidence in Enlightenment Sciences of Man." *History of European Ideas* 48 (2022): 45–65.

Sebastiani, Silvia. "A 'Monster with Human Visage': The Orangutan, Savagery, and the Borders of Humanity in the Global Enlightenment." *History of the Human Sciences* 32 (2019): 80–99.

Sebastiani, Silvia. "'Race,' Women and Progress in the Late Scottish Enlightenment." In Knott and Taylor, *Women, Gender and Enlightenment,* 75–96.

Sebastiani, Silvia. *The Scottish Enlightenment: Race, Gender, and the Limits of Progress.* Palgrave, 2013.

Sedgwick, Eve Kosofsky. *Epistemology of the Closet.* Rev. ed. University of California Press, 2008.

Seigel, Jerrold. *The Idea of the Self: Thought and Experience in Western Europe Since the Seventeenth Century.* Cambridge University Press, 2005.

Sen, Sudipta. "Colonial Aversions and Domestic Desires: Blood, Race, Sex and the Decline of Intimacy in Early British India." In *Sexual Sites, Seminal Attitudes: Sexualities, Masculinities and Culture in South Asia,* edited by Sanjay Srivastava. Sage, 2004.

Sergeant, Mark J. T. "Female Perception of Male Body Odor." *Vitamins & Hormones* 83 (2010): 25–45.

Seth, Suman. *Difference and Disease: Medicine, Race, and the Eighteenth-Century British Empire.* Cambridge University Press, 2018.

Seth, Suman. "Materialism, Slavery, and *The History of Jamaica.*" *Isis* 105 (2014): 764–72.

Seth, Vanita. *Europe's Indians: Producing Racial Difference.* Duke University Press, 2010.

Seth, Vanita. "The Origins of Racism: A Critique of the History of Ideas." *History and Theory* 59 (2020): 343–68.

Shail, Andrew, and Gilliam Howie, eds. *Menstruation: A Cultural History.* Palgrave Macmillan, 2005.

Sheridan, Richard B. *Doctors and Slaves: A Medical and Demographic History of Slavery in the British West Indies, 1680–1834.* Cambridge University Press, 1985.

Shoemaker, Robert B. *Gender in English Society, 1650–1850.* Longman, 1998.

Shuttleton, David E. *Smallpox and the Literary Imagination.* Cambridge University Press, 2007.

Shyllon, Folarin. *Black People in Britain, 1555–1833.* Oxford University Press, 1977.

Sidbury, James, and Jorge Cañizares-Esguerra. "On the Genesis of Destruction, and Other Missing Subjects." *W&MQ*, 3rd ser., 68 (2011): 240–46.

Siraisi, Nancy G. *Medieval and Early Renaissance Medicine: An Introduction to Knowledge and Practice.* University of Chicago Press, 1990.

Slack, Paul. "Government and Information in Seventeenth-Century England." *P&P* 184 (2004): 33–68.

Slack, Paul. *The Invention of Improvement: Information and Material Progress in Seventeenth-Century England.* Oxford University Press, 2015.

Sloan, Phillip R. "From Logical Universals to Historical Individuals: Buffon's Idea of Biological Species." In *Histoire du concept d'espèce de la vie,* edited by Scott Atran. Fondation Singer-Polignac, 1987.

Sloan, Phillip R. "Preforming the Categories: Eighteenth-Century Generation Theory and the Biological Roots of Kant's A Priori." *Journal of the History of Philosophy* 40 (2002): 229–53.

Smith, Bernard. *European Vision and the South Pacific.* 2nd ed. Yale University Press, 1985.

Smith, Bruce R. *Homosexual Desire in Shakespeare's England.* University of Chicago Press, 1994.

Smith, Justin E. H. "Imagination and the Problem of Heredity in Mechanist Embryology." In Smith, *Problem of Animal Generation,* 80–99.

Smith, Justin E. H., ed. *The Problem of Animal Generation in Early Modern Philosophy.* Cambridge University Press, 2006.

Smith, Lisa Wynne. "The Body Embarrassed? Rethinking the Leaky Male Body in Eighteenth-Century England and France." *Gender & History* 23 (2010): 26–46.

Smithers, Gregory D. *Slave Breeding: Sex, Violence, and Memory in African American History.* University Press of Florida, 2012.

Sollors, Werner. *Neither Black Nor White Yet Both: Thematic Explorations of Interracial Literature.* Oxford University Press, 1997.

Somerville, Margaret R. *Sex and Subjection: Attitudes to Women in Early Modern Society.* Arnold, 1995.

Sowerby, Scott. *Making Toleration: The Repealers and the Glorious Revolution.* Harvard University Press, 2013.

Spear, Jennifer M. *Race, Sex, and Social Order in Early New Orleans.* Johns Hopkins University Press, 2009.

Spear, Percival. *The Nabobs: A Study of the Social Life of the English in 18th Century India.* Oxford University Press, 1963.

Sramek, Joseph. *Gender, Morality, and Race in Company India, 1765–1858.* Palgrave, 2011.

Stangeland, Charles Emil. *Pre-Malthusian Doctrines of Population.* Columbia University Press, 1904.

Statt, Daniel. *Foreigners and Englishmen: The Controversy over Immigration and Population, 1660–1760.* University of Delaware Press, 1995.

Steel, M. J. "A Philosophy of Fear: The World View of the Jamaican Plantocracy in a Comparative Perspective." *Journal of Caribbean History* 27 (1993): 1–20.

Stein, Claudia. "The Birth of Biopower in Eighteenth-Century Germany." *Medical History* 55 (2011): 331–37.

Stepan, Nancy. "Biological Degeneration: Races and Proper Places." In *Degeneration: The Dark Side of Progress,* edited by J. Edward Chamberlin and Sander L. Gilman. Columbia University Press, 1985.

Stepan, Nancy. *The Idea of Race in Science: Great Britain, 1800–1960.* Macmillan, 1982.

Stepan, Nancy. "Race and Gender: The Role of Analogy in Science." *Isis* 77 (1986): 261–77.

Stepan, Nancy. "Race, Gender, Science and Citizenship." *Gender and History* 10 (1998): 26–52.

Stephanson, Raymond. "*Tristram Shandy* and the Art of Conception." In *Vital Matters: Eighteenth-Century Views of Conception, Life, and Death,* edited by Helen Deutsch and Mary Terrall. University of Toronto Press, 2012.

Stephanson, Raymond. *The Yard of Wit.* University of Pennsylvania Press, 2004.

Stephens, Frederic George, and M. Dorothy George. *Catalogue of Personal and Political Satires.* 12 vols. British Museum, 1870–1954.

Stern, Philip, and Carl Vennerlind, eds. *Mercantilism Reimagined: Political Economy in Early Modern Britain and Its Empire.* Oxford University Press, 2014.

Stewart, Ian B. "James Cowles Prichard and the Linguistic Foundations of Ethnology." *Beiträge zur Wissenschaftsgeschichte* 46 (2023): 76–91.

Stewart, Ian B. "The Mother Tongue: Historical Study of the Celts and Their Language(s) in Eighteenth-Century Britain and Ireland." *P&P* 243 (2019): 71–107.

Stewart, Ian B. "William Frédéric Edwards and the Study of Human Races in France, from the Restoration to the July Monarchy." *History of Science* 58 (2020): 273–300.

Stocking, George W., Jr. "From Chronology to Ethnology: James Cowles Prichard and British Anthropology, 1800–1850." In Prichard, *PHM,* ix–cx.

Stolberg, Michael. "A Woman Down to Her Bones: The Anatomy of Sexual Difference in the Sixteenth and Early Seventeenth Centuries." *Isis* 94 (2003): 274–99.

Stoler, Ann Laura. *Race and the Education of Desire.* Duke University Press, 1995.

Stoler, Ann Laura. "Racial Histories and Their Regimes of Truth." *Political Power and Social Theory* 11 (1997): 183–206.

Stollberg, Jochen, and Wolfgang Böcker, eds. "*. . . die Kunst zu Seyn*": *Arthur Schopenhauers Mitschriften der Vorlesungen Johann Friedrich Blumenbachs (1809–1811).* Universitätsverlag, 2013.

Stone, Lawrence. *The Family, Sex and Marriage in England, 1500–1800.* Penguin, 1990.

Stone, Lawrence. *The Road to Divorce: England, 1530–1987.* Oxford University Press, 1990.

Sturma, Michael. *South Sea Maidens: Western Fantasy and Sexual Politics in the South Pacific.* Greenwood, 2002.

Stuurman, Siep. "François Bernier and the Invention of Racial Classification." *History Workshop Journal* 50 (2000): 1–21.

Stuurman, Siep. *The Invention of Humanity: Equality and Cultural Difference in World History*. Harvard University Press, 2017.

Sudai, Maayan. "Sex Ambiguity in Early Modern Common Law." *Law & Social Inquiry* 47 (2022): 478–513.

Tadman, Michael. "The Demographic Cost of Sugar: Debates on Slaves Societies and Natural Increase in the Americas." *American Historical Review* 105 (2000): 1534–75.

Tadman, Michael. *Speculators and Slaves: Masters, Traders, and Slaves in the Old South*. University of Wisconsin Press, 1996.

Taylor, Charles. *Sources of the Self: The Making of the Modern Identity*. Harvard University Press, 1989.

Teltscher, Kate. *India Inscribed: European and British Writing on India, 1600–1800*. Oxford University Press, 1995.

Terrall, Mary. *The Man Who Flattened the Earth: Maupertuis and the Sciences in the Enlightenment*. University of Chicago Press, 2002.

Thomas, Keith. "The Double Standard." *JHI* 20 (1959): 195–216.

Thomas, Nicholas. *Colonialism's Culture: Anthropology, Travel and Government*. Cambridge University Press, 1994.

Thomas, Nicholas. *Discoveries: The Voyages of Captain Cook*. Penguin, 2004.

Thomas, Nicholas. Introduction to *Tattoo: Bodies, Art and Exchange in the Pacific and the West*, 7–29. Edited by Nicholas Thomas, Anna Cole, and Bronwen Douglas. Reaktion, 2005.

Thompson, E. P. *Customs in Common*. Penguin, 1993.

Thompson, Peter. "Henry Drax's Instructions on the Management of a Seventeenth-Century Barbadian Sugar Plantation." *W&MQ*, 3rd ser., 66 (2009): 565–604.

Tinland, Franck. *L'Homme Sauvage: Homo Ferus et Homo Sylvestris—De l'animal à l'homme*. Payot, 1968.

Tise, Larry E. *Proslavery: A History of the Defense of Slavery in America, 1701–1840*. University of Georgia Press, 1987.

Tomaselli, Sylvana. "The Enlightenment Debate on Women." *History Workshop Journal* 20 (1985): 101–24.

Tomaselli, Sylvana. "Moral Philosophy and Population Questions in Eighteenth-Century Europe." *Population and Development Review* 14, Suppl. (1988): S7–S29.

Toulalan, Sarah. *Imagining Sex: Pornography and Bodies in Seventeenth-Century England*. Oxford University Press, 2007.

Toulalan, Sarah, and Kate Fisher, eds. *The Routledge History of Sex and the Body, 1500 to the Present*. Routledge, 2013.

Treggiari, Susan. "Social Status and Social Legislation." In *The Cambridge Ancient History, Vol. 10: The Augustan Empire, 43 BC–AD 69*, edited by Alan K. Bowman, Edward Champlin and Andrew Lintott. Cambridge University Press, 1996.

Trumbach, Randolph. *Sex and the Gender Revolution*. Vol. 1, *Heterosexuality and the Third Gender in Enlightenment London*. University of Chicago Press, 1998.

Tullett, William. "Grease and Sweat: Race and Smell in Eighteenth-Century English Culture." *Cultural and Social History* 13 (2016): 307–22.

Turley, David. *The Culture of English Antislavery, 1780–1860*. Routledge, 1991.

Turnbull, Paul. "British Anatomists, Phrenologists and the Construction of the Aboriginal Race, *c.* 1790–1830." *History Compass* 5 (2007): 26–50.

Turner, David M. "Birth Anomaly and Childhood Disability." In *The Secrets of Generation: Reproduction in the Long Eighteenth Century*, edited by Raymond Stephanson and Darren N. Wagner. University of Toronto Press, 2015.

Turner, Sasha. *Contested Bodies: Pregnancy, Childrearing, and Slavery in Jamaica.* University of Pennsylvania Press, 2017.

Tuttle, Leslie. *Conceiving the Old Regime: Pronatalism and the Politics of Reproduction in Early Modern France.* Oxford University Press, 2010.

Tzoref-Ashkenazi, Chen. "German Voices from India: Officers of the Hanoverian Regiments in East India Company Service." *South Asia* 32 (2009): 189–211.

Venturino, Diego. "Race et histoire: Le paradigme nobiliare de la distinction sociale au début du XVIIIe siècle." In *L'idée de 'race' dans les sciences humaines et la littérature (XVIIIe et XIXe siècles)*, edited by Sarga Moussa. L'Harmattan, 2003.

Vickery, Amanda. "Golden Age to Separate Spheres? A Review of the Categories and Chronology of English Women's History." *Historical Journal* 36 (1993): 383–414.

Visaria, Leela, and Pravin Visaria. "Population (1757–1947)." In *Cambridge Economic History of India: Volume 2, c. 1757–c. 1970*, edited by Dharma Kumar and Meghnad Desai. Cambridge University Press, 1983.

Voegelin, Eric. *The History of the Race Idea.* Translated by Ruth Hein. Edited by Klaus Vondung. Louisiana State University Press, 1998 [1933].

Wagner, Peter. *Eros Revived: Erotica of the Enlightenment in England and America.* Paladin, 1990.

Wahrman, Dror. "Change and the Corporeal in Seventeenth- and Eighteenth-Century Gender History: Or, Can Cultural History Be Rigorous?" *Gender and History* 20 (2008): 584–602.

Wahrman, Dror. *The Making of the Modern Self: Identity and Culture in Eighteenth-Century England.* Yale University Press, 2004.

Walde, Alois. *Lateinisches etymologisches Wörterbuch.* 2nd ed. Carl Winter, 1910.

Wallace, Lee. *Sexual Encounters: Pacific Texts, Modern Sexualities.* Cornell University Press, 2003.

Wallace, Lee. "Too Darn Hot: Sexual Contact in the Sandwich Islands on Cook's Third Voyage." *ECL* 18, no. 3 (1994): 232–42.

Walvin, James. *Black and White: The Negro and English Society.* Penguin, 1973.

Walvin, James. *Fruits of Empire.* Macmillan, 1997.

Ward, J. R. *British West Indian Slavery, 1750–1834.* Oxford University Press, 1988.

Weil, Rachel. *Political Passions: Gender, the Family, and Political Argument in England, 1680–1714.* Manchester University Press, 1999.

Wells, Andrew. "Blurred Lines: Bestiality and the Human Ape in Enlightenment Scotland." In *Interspecies Interactions: Animals and Humans Between the Middle Ages and Modernity*, edited by Sarah Cockram and Andrew Wells. Routledge, 2018.

Wells, Andrew. "Masculinity and Its Other in Eighteenth-Century Racial Thought." In *Masculinity and the Other: Historical Perspectives*, edited by Heather Ellis and Jessica Meyer. Cambridge Scholars, 2009.

Wells, Andrew. "Misplaced Politeness and Anglo-German Racial Thought: Charles White, Edward Holme, and the *Account of the Regular Gradation in Man* (1799)."

In *Crossing the Channel—British-German Historical and Cultural Dialogues*, edited by Rudolf Boch, Marian Nebelin, and Cecile Sandten. Duncker & Humblot, 2022.

Wells, Andrew. "Race Fixing: Improvement and Race in Eighteenth-Century Britain." *History of European Ideas* 36 (2010): 134–38.

Wells, Kentwood D. "Sir William Lawrence (1783–1867): A Study of Pre-Darwinian Ideas on Heredity and Variation." *Journal of the History of Biology* 4 (1971): 319–61.

West, Hugh. "The Limits of Enlightenment Anthropology: Georg Forster and the Tahitians." *History of European Ideas* 10 (1989): 147–60.

Wheeler, Roxann. *The Complexion of Race: Categories of Difference in Eighteenth-Century British Culture.* University of Pennsylvania Press, 2000.

Whelan, Frederick G. "Population and Ideology in the Enlightenment." *History of Political Thought* 12 (1991): 49–57.

White, Hayden. *Tropics of Discourse: Essays in Cultural Criticism.* Johns Hopkins University Press, 1978.

Whyte, Iain. *Zachary Macaulay 1768–1838: The Steadfast Scot in the British Anti-Slavery Movement.* Liverpool University Press, 2011.

Wiegman, Robyn. *American Anatomies: Theorizing Race and Gender.* Duke University Press, 1995.

Wilkins, John S. *Species: A History of the Idea.* University of California Press, 2009.

Wilkinson, Lise. "'The Other' John Hunter, M.D., F.R.S. (1754–1809): His Contributions to the Medical Literature, and to the Introduction of Animal Experiments into Infectious Disease Research." *Notes and Records of the Royal Society of London* 36 (1982): 227–41.

Williams, Eric. "The British West Indian Slave Trade After Its Abolition in 1807." *Journal of Negro History* 27 (1942): 175–91.

Williams, Eric. *Capitalism and Slavery.* University of North Carolina Press, 1944.

Williams, Raymond. *Keywords.* Rev. ed. Oxford University Press, 1983.

Williamson, Karina. Introduction to *Marly: or, The Life of Planter in Jamaica*, xi–xxxii. Edited by Karina Williamson. Macmillan Caribbean, 2005.

Wilson, Adrian. *The Making of Man-Midwifery: Childbirth in England, 1660–1760.* Harvard University Press, 1995.

Wilson, Dudley. *Signs and Portents: Monstrous Births from the Middle Ages to the Enlightenment.* Routledge, 1993.

Wilson, Kathleen. *The Island Race: Englishness, Empire, and Gender in the Eighteenth Century.* Routledge, 2003.

Wilson, Kathleen, ed. *The New Imperial History.* Cambridge University Press, 2004.

Wilson, Kathleen. *The Sense of the People: Politics, Culture and Imperialism in England, 1715–1785.* Cambridge University Press, 1995.

Wilson, Kathleen. "Thinking Back: Gender Misrecognition and Polynesian Subversions Aboard the Cook Voyages." In Wilson, *New Imperial History*, 345–62.

Wilson, Philip K. "Eighteenth-Century 'Monsters' and Nineteenth-Century 'Freaks': Reading the Maternally Marked Child." *Literature and Medicine* 21 (2002): 1–25.

Wilson, Philip K. "'Out of Sight, Out of Mind?': The Daniel Turner-James Blondel Dispute over the Power of the Maternal Imagination." *Annals of Science* 49 (1992): 63–85.

Winch, Donald. *Riches and Poverty: An Intellectual History of Political Economy in Britain, 1750–1834.* Cambridge University Press, 1996.

Wokler, Robert. "Apes and Races in the Scottish Enlightenment: Monboddo and Kames on the Nature of Man." In *Philosophy and Science in the Scottish Enlightenment*, edited by Peter Jones. John Donald, 1988.

Wokler, Robert. "From *l'homme physique* to *l'homme morale* and Back: Towards a History of Enlightenment Anthropology." *Journal of the Human Sciences* 6 (1993): 121–38.

Wokler, Robert. "Perfectible Apes in Decadent Cultures: Rousseau's Anthropology Revisited." *Dædalus* 107, no. 3 (1978): 107–34.

Wokler, Robert. "Tyson and Buffon on the Orang-utan." *Studies on Voltaire and the Eighteenth Century* 155 (1976): 2301–19.

Wood, Betty, and Roy Clayton. "Jamaica's Struggle for a Self-Perpetuating Slave Population: Demographic, Social, and Religious Changes on Golden Grove Plantation, 1812–1832." *Journal of Caribbean Studies* 6 (1988): 287–308.

Wood, Marcus. *Blind Memory: Visual Representations of Slavery in England and America, 1780–1865.* Manchester University Press, 2000.

Wood, P. B. "Buffon's Reception in Scotland: The Aberdeen Connection." *Annals of Science* 44 (1987): 169–90.

Wood, P. B. "The Natural History of Man in the Scottish Enlightenment." *History of Science* 28 (1990): 89–123.

Worrall, David. *Harlequin Empire: Race, Ethnicity and the Drama of the Popular Enlightenment.* Pickering & Chatto, 2007.

Wright, Thomas. *Caricature History of the Georges.* London, 1868.

Wrightson, Keith. *English Society 1580–1680.* Unwin Hyman, 1982.

Wrigley, E. A. "British Population During the 'Long' Eighteenth Century, 1680–1840." In *The Cambridge Economic History of Modern Britain*, edited by Roderick Floud and Paul Johnson, 1:57–95. 3 vols. Cambridge University Press, 2004.

Wrigley, E. A., and R. S. Schofield. *The Population History of England, 1541–1871.* Cambridge University Press, 1989.

Yule, Henry, and A. C. Burnell. *Hobson-Jobson: A Glossary of Colloquial Anglo-Indian Words and Phrases.* John Murray, 1903.

Yuval-Davis, Nira. "Theorizing Identity: Beyond the 'Us' and 'Them' Dichotomy." *Patterns of Prejudice* 44 (2010): 261–80.

Zammito, John H. *The Gestation of German Biology.* University of Chicago Press, 2018.

Zammito, John H. *Kant, Herder, and the Birth of Anthropology.* University of Chicago Press, 2002.

Zammito, John H. "The Lenoir Thesis Revisited: Blumenbach and Kant." *SHPBBMS* 43 (2012): 120–32.

Zammito, John H. "Policing Polygeneticism in Germany, 1775: Kames, Kant, and Blumenbach." In Eigen and Larrimore, *German Invention*, 35–54.

Zirkle, Conway. *The Beginnings of Plant Hybridization.* University of Pennsylvania Press, 1935.

Zirkle, Conway. "The Early History of the Idea of the Inheritance of Acquired Characters and of Pangenesis." *Transactions of the American Philosophical Society* 35, no. 2 (1946): 91–151.

4. Unpublished Dissertations and Theses

Augstein, Franziska A. "James C. Prichard's Views of Mankind." PhD diss., University College London, 1996.

Brunton, Deborah. "Pox Britannica: Smallpox Inoculation in Britain, 1721–1730." PhD diss., University of Pennsylvania, 1990.

Hammett, Ian Maxwell. "Lord Monboddo's *The Origin and Progress of Language*: Its Sources, Genesis and Background, with Special Attention to the Advocates' Library." PhD diss., University of Edinburgh, 1985.

Jooken, Lieve. "The Linguistic Conceptions of Lord Monboddo (1714–1799)." PhD diss., Katholieke Universiteit Leuven, 1996.

López-Beltrán, Carlos. "Human Heredity, 1750–1870: The Construction of a Scientific Domain." PhD diss., King's College London, 1992.

Van Lieshout, Carry. "London's Changing Waterscapes: The Management of Water in Eighteenth-Century London." PhD diss., King's College London, 2013.

Wells, Andrew. "Sex and Racial Theory in Britain, 1690–1833." DPhil diss., University of Oxford, 2009.